DATE DUE

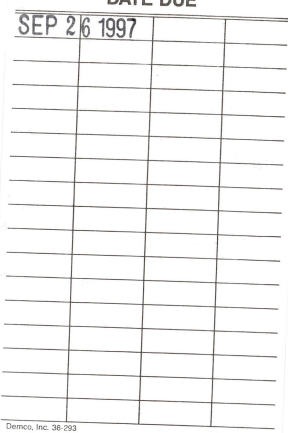

SEP 2 6 1997			

Demco, Inc. 38-293

Irradiation of Polymers

ACS SYMPOSIUM SERIES **620**

Irradiation of Polymers

Fundamentals and Technological Applications

Roger L. Clough, EDITOR
Sandia National Laboratories

Shalaby W. Shalaby, EDITOR
Poly-Med, Inc.

Developed from a symposium sponsored
by the Division of Polymer Chemistry, Inc.,
at the 208th National Meeting
of the American Chemical Society,
Washington, DC,
August 21–26, 1994

American Chemical Society, Washington, DC 1996

Library of Congress Cataloging-in-Publication Data

Irradiation of polymers: fundamentals and technological applications / Roger L. Clough, editor, Shalaby W. Shalaby, editor.

    p.   cm.—(ACS symposium series, ISSN 0097–6156; 620)

"Developed from a symposium sponsored by the Division of Polymer Chemistry, Inc., at the 208th National Meeting of the American Chemical Society, Washington, D.C., August 21–25, 1994."

Includes bibliographical references and indexes.

ISBN 0–8412–3377–2

1. Polymers—Effect of radiation on—Congresses.

I. Clough, Roger L. (Roger Lee), 1949–   . II. Shalaby, Shalaby W. III. American Chemical Society. Division of Polymer Chemistry, Inc. IV. American Chemical Society. Meeting (208th: 1994: Washington, D.C.) V. Series.

QD381.9.R3I784   1995
547.7'04582—dc20   95–47865
                                                         CIP

This book is printed on acid-free, recycled paper.

PRINTED IN THE UNITED STATES OF AMERICA

Foreword

THE ACS SYMPOSIUM SERIES was first published in 1974 to provide a mechanism for publishing symposia quickly in book form. The purpose of this series is to publish comprehensive books developed from symposia, which are usually "snapshots in time" of the current research being done on a topic, plus some review material on the topic. For this reason, it is necessary that the papers be published as quickly as possible.

Before a symposium-based book is put under contract, the proposed table of contents is reviewed for appropriateness to the topic and for comprehensiveness of the collection. Some papers are excluded at this point, and others are added to round out the scope of the volume. In addition, a draft of each paper is peer-reviewed prior to final acceptance or rejection. This anonymous review process is supervised by the organizer(s) of the symposium, who become the editor(s) of the book. The authors then revise their papers according to the recommendations of both the reviewers and the editors, prepare camera-ready copy, and submit the final papers to the editors, who check that all necessary revisions have been made.

As a rule, only original research papers and original review papers are included in the volumes. Verbatim reproductions of previously published papers are not accepted.

Contents

STRUCTURAL AND PHYSICOCHEMICAL EFFECTS OF IRRADIATION

MODELING AND MONITORING OF RADIATION EVENTS

RADIATION PROCESSING: POLYMERIZATION, CURING,
AND STERILIZATION

STABILITY AND STABILIZATION OF POLYMERS
TO IONIZING RADIATION

Preface

CONSISTENT INTEREST IN UNDERSTANDING the fundamental aspects of radiation effects in polymers and the growing importance of high-energy radiations as a critical means in the production of electronic components and biomedical devices justified an update of this area of strategic technology since our 1991 book, *Radiation Effects on Polymers*, ACS Symposium Series 475. About two-thirds of the chapters in this book are based on an ACS symposium and are written by participants whose diverse expertise covers a vast area of traditional and less traditional radiation-related technologies. The remaining book chapters were solicited from distinguished experts in the field to further enhance the comprehensiveness of this volume and to underscore the important role of radiation in electronics and sterilization. Many of the chapters were designed to provide the reader with a brief review of major progress in particular segments of the field, along with the most current research results. Thus, scientists and engineers with interest in polymers and radiation technology are given a wide range of valuable information in a single volume.

The first section of the book presents a brief review of current understanding of radiation effects in polymers, mechanisms of pertinent events (such as charge transport and photoionization), and typical and atypical responses and changes in properties of homochain [e.g., ultrahigh-molecular-weight polyethylene (UHMW-PE) and polybutene-1] and heterochain (e.g., polyhydroxybutyrate) polymers. Sophisticated computational and analytical methods have been used successfully in modeling and monitoring radiation events as evidenced in the second section of the book. Discussions in this section pertain to network formation, chain scission, and end-group determination.

The significance of radiation processing and the growing interest in radiation as an effective means of sterilization in the biomedical, pharmaceutical, and food industries are reflected in the third section of the book. Four of the chapters are dedicated to sterilization, including a chapter on the first report of a novel radiochemical sterilization process for medical devices. Other chapters cover the use of radiation in solid-state and thin-film polymerization; preparation of cross-linked pharmaceutically useful gels; and development of interpenetrating polymeric systems, fiber-reinforced composites, and improved synthetic rubbers.

Although most of the fourth section is dedicated to traditional topics pertinent to stability and stabilization, contemporary topics of growing significance are also addressed. These entail the effect of radiation on UHMW-PE for its significant biomedical role in prosthetic implants and on organic insulators for their critical use in fusion reactors. To underscore the importance of radiation technology in the electronics industry, advances in the use of radiation in lithography are given in the final section of the book. These chapters deal with photosensitive metathesis polymers, chemical amplification, and dry-develop resist technology.

Acknowledgments

To all authors who contributed to this book and to those who participated in the symposium, we express our sincere appreciation. Thanks are also due to those who assisted us in organizing or chairing the symposium, particularly K. Gillen, D. Hill, V. Markovic, A. Pla, J. Silverman, and G. Taylor. For his initial efforts in co-organizing the symposium and for being an outstanding polymer scientist, good friend, and valuable colleague, we dedicate this book to J. O'Donnell, who passed away recently.

We also express our appreciation to the ACS Division of Polymer Chemistry, Inc., for sponsoring the symposium. We thank A. Wilson and M. Althuis of ACS Books and K. Burg, a bioengineering graduate student at Clemson University, for their conscientious efforts to ensure a timely review and editing of the book.

ROGER L. CLOUGH
Organic Materials Division
Sandia National Laboratories
Albuquerque, NM 87185

SHALABY W. SHALABY
Poly-Med, Inc.
6309 Highway 187
Anderson, SC 29625

November 13, 1995

Dedication

THIS BOOK IS DEDICATED to James O'Donnell, former professor at the University of Queensland, Australia, for his valuable contributions to the field of polymers and radiation technology and for guidance to his worldwide colleagues during the past three decades. The global polymer community will certainly miss his leadership in cementing an international bond among scientists and engineers of many professional interests.

STRUCTURAL AND PHYSICOCHEMICAL EFFECTS OF IRRADIATION

Chapter 1

High-Energy Ion Irradiation Effects on Polymer Materials

H. Kudoh, T. Sasuga, and T. Seguchi

Takasaki Radiation Chemistry Research Establishment, Japan Atomic Energy Research Institute, Takasaki, Gunma 370–12, Japan

The changes in mechanical properties of several polymers induced by high energy protons (10, 30 and 45 MeV protons) were studied and compared with those induced by 2 MeV electrons and Co-60 gamma rays. Changes in elongation at break showed the same trends for 10 MeV proton and electron irradiations in the case of both polyethylene (PE) and polytetrafluoroethylene (PTFE). The flexural strength at break of polymethylmethacrylate (PMMA) and glass fiber reinforced plastic (GFRP) also showed no difference between 30 or 45 MeV proton and Co-60 gamma ray irradiations.

The evaluation of radiation resistance of polymer materials to high energy ion irradiation is necessary for the selection of materials applied to space environments. The radiation resistance of polymer materials has been studied extensively with gamma rays and electron beams. Radiation effects may be imagined to be different among radiation sources such as gamma rays, electrons and ions, because the linear energy transfer (LET), which is the energy deposited along the track of ionizing particles, is different for ions, electrons and gamma rays. In this case, the evaluation of radiation resistance to ions would be necessary. However, reports on this subject are very limited because there are few facilities for uniform dose irradiation of sufficiently wide areas to allow reasonable measurements of mechanical property changes such as tensile elongation and flexural strength. Based on this, the Japan Atomic Energy Research Institute, Takasaki Radiation Chemistry Research Establishment, constructed ion accelerators (Takasaki Ion Accelerators for Advanced Radiation Application; TIARA) and irradiation chambers for the irradiation of polymer films or sheets uniformly over wide areas. In this paper, the changes in mechanical properties induced by proton irradiation were studied for several polymers such as polyethylene, polytetrafluoroethylene, polymethylmethacrylate and glass fiber reinforced plastic and compared with those by 2 MeV electron or Co-60 gamma rays irradiations.

EXPERIMENTAL

Materials. Polymers used in this work are polyethylene (PE, medium density), polytetrafluoroeythylene (PTFE), polymethylmethacrylate (PMMA), and bis phenol A type epoxy resin matrix glass fiber reinforced plastic (GFRP). PE and PTFE in the form of 0.5 and 0.1 mm thick film were cut into JIS No. 4 type dumbbell specimens for tensile tests. PMMA and GFRP were 3 mm and 2 mm thick specimens for flexural tests, respectively.

Irradiation. Proton beams of 10 mm diameter spot with 10, 30 and 45 MeV from a cyclotron accelerator were used for uniform irradiation over a 100 mm x 100 mm wide area by beam scanning. The frequency of beam scanning was 50 Hz in the horizontal and 0.5 Hz in the vertical direction. Specimens were placed on a graphite plate cooled by water in the chamber. Irradiations were carried out under high vacuum of around 10^{-7} torr at room temperature. 10 MeV protons were used for irradiation of PE and PTFE, and 30 MeV protons were used in the case of PMMA and GFRP. 45 MeV protons were also applied to the PMMA. The proton energy for irradiation was selected as shown in Table I so that ions can pass through the materials entirely for all irradiation conditions. The mass stopping power and the range of proton penetration in the polymers were calculated by Bethe's formula and Seltzer and Berger's method as shown also in Table I. The mass stopping power depends on the proton accelerating energy and is higher than that of 2 MeV electrons. The stopping power defines the LET. The theoretical absorbed dose (D) is calculated as the product of mass stopping power (S) and the fluence (Q);

$$D(kGy)=S(MeVcm^2g^{-1})xQ(\mu Ccm^{-2}).$$

Proton current was 500 nA ($3.1x10^{12}$ p/s) for PE, 50 nA for PTFE, and 200-300 nA for PMMA and GFRP. The current is equivalent to the average dose rate of 0.2 kGy/s for PE, 0.02 kGy/s for PTFE and 0.03 kGy/s for PMMA and GFRP. A cellulose tri-acetate (CTA) film dosimeter, which was developed for electron beam dosimetry, was applied to dosimetry. The observed dose determined from optical density at 280 nm showed good agreement with the calculated dose by the above formula. For comparison with proton irradiation effects, 2 MeV electron beam irradiation was carried out for PE and PTFE under vacuum at room temperature, and for PMMA and GFRP Co-60 gamma ray irradiation was conducted in nitrogen gas atmosphere at RT.

Measurements. Tensile tests were carried out for PE and PTFE, and three-point flexural tests for PMMA and GFRP were carried out at room temperature. By the analysis of differential scanning calorimeter (DSC) measurements, the molecular weight changes in PTFE were determined. Also the glass transition temperature (Tg) of GFRP was measured. For PMMA, the molecular weight and its distribution were measured by gel permeation chromatography (GPC).

Table I Stopping Powers and Ranges of Ions in Polymer (CTA[*])

Ion(Energy, MeV)	S(MeVcm^2/g)	R(mm)	Irradiated Polymer
e$^-$(2)	1.77	-	
H$^+$(10)	40.8	1.1	PE, PTFE
H$^+$(30)	16.1	7.8	GFRP, PMMA
H$^+$(45)	11.3	16.6	PMMA

*CTA; Cellulose tri-acetate

RESULTS and DISCUSSION

PE and PTFE. Figures 1 and 2 show the change in elongation at break for PE and PTFE as a function of dose. Tensile tests for PE and PTFE showed the same degradation behavior with 10 MeV proton and 2 MeV electron irradiations at equivalent absorbed dose. Figure 3 shows the change in molecular weight of PTFE determined by DSC and Suwa's method (*1*). The number average molecular weight (Mn) decreases with dose, suggesting that PTFE undergoes chain scission. Changes in Mn were similar for both irradiations, suggesting that the probability of chain scission for PTFE is not sensitive to LET in spite of the large differences in LET (2 MeV electron; 1.77 MeVcm^2/g, 10 MeV proton; 40.8 MeVcm^2/g for CTA).

PMMA. Figure 4 shows the change in flexural strength of PMMA measured at RT as a function of absorbed dose using protons (30 and 45 MeV) and Co-60 gamma rays (H. Kudoh et al., *Polymer*, submitted, 1995). The flexural strength of PMMA decreases with dose over 0.1 MGy, and the difference between Co-60 gamma rays and protons was very small. The molecular weight of PMMA decreased similarly for both protons and Co-60 gamma rays. Figure 5 shows the change in number and weight average molecular weight as a function of dose. The reciprocal of Mn increased linearly with dose, showing only chain scission. The G value of chain scission is 1.7 from the slope of the line in Fig. 5, which agrees well with values in literature (*2*). Figure 6 shows the change in

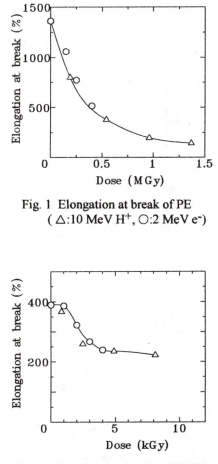

Fig. 1 Elongation at break of PE
(△:10 MeV H⁺, ○:2 MeV e⁻)

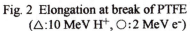

Fig. 2 Elongation at break of PTFE
(△:10 MeV H⁺, ○:2 MeV e⁻)

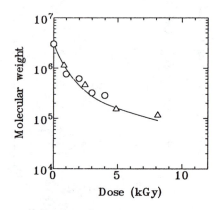

Fig. 3 Molecular weight of PTFE
($\triangle$:10 MeV H$^+$, $\bigcirc$:2 MeV e$^-$)

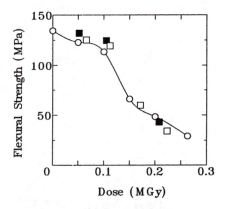

Fig. 4 Flexural strength of PMMA
($\square$:30 MeV H$^+$, $\blacksquare$:45 MeV H$^+$, $\bigcirc$:gamma ray)

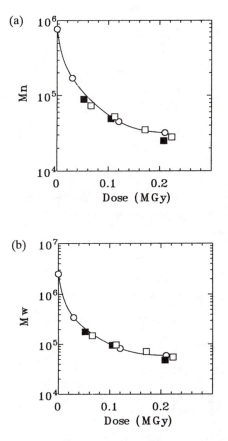

Fig. 5 Molecular weight of PMMA,
(a)number average, (b)weight average
($\square$:30 MeV H^+, $\blacksquare$:45 MeV H^+, $\bigcirc$:gamma ray)

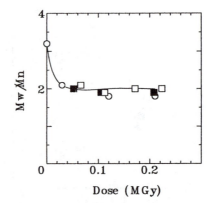

Fig. 6 Molecular weight distribution, the ratio of
        weight average to number average molecular weight
        of PMMA
($\square$:30 MeV H$^+$, $\blacksquare$:45 MeV H$^+$, $\bigcirc$:gamma ray)

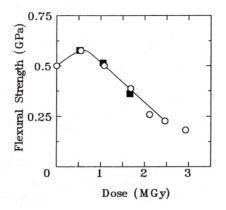

Fig. 7 Flexural strength of GFRP
        ($\blacksquare$:30 MeV H$^+$, $\bigcirc$:gamma ray)

the ratio of weight average to number average molecular weight. Mw/Mn is about 3 before irradiation, and becomes around 2 after irradiation even with a small dose of around 50 kGy. This means that the molecular weight distribution is random; that is, the chain scission occurs randomly. The decrease of flexural strength is related to the decrease of the molecular weight. The probability of main chain scission occurs in the same way in number and spatial distribution for Co-60 gamma rays, and for both 45 MeV and 30 MeV protons. Thus, LET effects on PMMA degradation are not observed in the LET range of this work: that is, 1.8-16 $MeVg^{-1}cm^2$ (ca. 0.02 - 0.15 eV/A). On the other hand, Yates et al.,(2) reported that the G value of scission for PMMA decreases with large LET. In Yates' work also, G (scission) seems to be almost constant below 0.2 eV/A.

GFRP. Figure 7 shows the change in flexural strength of GFRP measured at RT as a function of dose (H. Kudoh et al., *Polymer*, submitted, 1995.). The strength increases by about 20% at around 0.5 MGy (probably because dangling molecular chains allow larger displacement after undergoing chain scission by radiation), then decreases with dose above 1 MGy. This behavior is the same for 30 MeV proton and gamma-ray irradiation. The radiation degradation of GFRP is caused mainly by the degradation of matrix resin. Figure 8 shows the change in Tg of epoxy resin determined by DSC. The decrease of Tg indicates the destruction of network structure of epoxy resin with main chain scission. The scission probability deduced by DSC is the same for gamma rays and 30 MeV proton irradiation, and therefore the degradation of mechanical properties induced by chain scission is the same for the two radiation types.

LET effects. It has been assumed that a high LET radiation forms active species in high density in organic materials, that can result in different chemical reactions when compared to low LET radiation. For example (3), dimer formation in benzene increases with LET, whereas this is not observed in cyclohexane. This phenomenon is interpreted as being due to the recombination of excited molecules formed in the aromatic compound. Studies (4,5) of changes in mechanical properties induced by 8 MeV proton and 2 MeV electron irradiations for several polymers indicated that LET effects were scarcely observed in aliphatic polymers such as PE and polypropylene, whereas the radiation deterioration of the aromatic polymers such as polyethersulphone and bisphenol A type polysulphone was reduced in 8 MeV proton irradiation. These results indicate that the LET effect appears in aromatic compounds. In our experiments, there seem to be little or no LET effects in four different polymers in terms of mechanical properties and chemical reactions such as crosslinking and chain scission. However, studies using much higher LET radiations and higher ion masses are necessary to more fully evaluate LET effects.

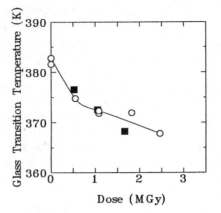

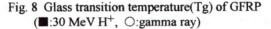

Fig. 8 Glass transition temperature(Tg) of GFRP
(■:30 MeV H⁺, ○:gamma ray)

CONCLUSION

Cyclotron and ion-irradiation-chamber   facilities were prepared at JAERI,
Takasaki for evaluation of the radiation resistance of polymers to ion
irradiation. By using them, high energy ion irradiation effects on polymer
materials such as PE, PTFE, PMMA and GFRP were studied by changes in
mechanical properties, molecular weight, and glass transition temperature.
The elongation at break for both PE and PTFE showed the same degradation
behavior with 10 MeV proton and 2 MeV electron irradiation. The molecular
weight of PTFE also decreased in a similar way upon irradiation with either
protons or electrons. The degradation of flexural strength at break of PMMA
and GFRP showed no difference between Co-60 gamma rays and 30 MeV
protons. The scission probability was the same with both radiation types. We
concluded that in the degradation of these polymers, LET effects are very small
for 10-45 MeV proton, electron and gamma ray irradiation.

Literature Cited

1. Suwa T., Takehisa M. and Machi S., *J. of Appl. Polym. Sci.*, **1973**, *17*, 3253
2. Yates B. W. and Shinozaki D. M., *J. Polym. Sci. B*, **1993**, *31*, 179
3. Burns W. G., *Trans. Faraday Soc.*, **1962**, *58*, 961
4. Sasuga T., Kawanishi S., Nishii M., Seguchi T. and Kohno I., *Polymer*, **1989**,
30, 2054
5. Sasuga T., Kawanishi S., Nishii M., Seguchi T. and Kohno I., *Radiat. Phys.
Chem.*, **1991**, *37*, 135

RECEIVED August 26, 1995

Chapter 2

Mechanistic Studies on the Radiation Chemistry of Poly(hydroxybutyrate)

Trudy Carswell-Pomerantz, Limin Dong, David J. T. Hill[1], James H. O'Donnell[†], and Peter J. Pomery

Polymer Materials and Radiation Group, Department of Chemistry, University of Queensland, Brisbane, Queensland 4072, Australia

The radiation chemical yields for formation of volatile products, for scission and cross-linking and for the formation of new chemical structures in poly(hydroxybutyrate) and two poly(hydroxybutyrate-co-valerate)s with mole fractions valerate of 0.184 and 0.263 have been investigated. The major volatile products formed on radiolysis of the polymers were carbon monoxide, carbon dioxide and hydrogen. For poly(hydroxybutyrate) at 303 K the yields were carbon monoxide (G = 0.74), carbon dioxide (G = 0.96) and hydrogen (G = 0.42). Scission dominated over crosslinking for the polymers, with $G(S) = 1.3$ and $G(X) = 0.0$ for gamma radiolysis of poly(hydroxybutyrate) at 303 K at low dose rate. The value of $G(S)$ was somewhat higher for electron beam irradiation of poly(hydroxybutyrate) at 303 K, with $G(S) = 2.2$ and $G(X) = 0.0$, indicative of a dose rate effect for this polymer. For the electron beam irradiation of the copolymers with valerate mole fractions of 0.184 and 0.263, $G(S)$ and $G(X)$ were found to be 2.7 and 0.3 and 3.4 and 0.5, respectively. Solution NMR studies revealed the formation of new saturated chain ends resulting from the loss of ester linkages. Based on the yields of new end-group structures the radiation chemical yield for loss of ester was estimated to be 1.4 for poly(hydroxybutyrate).

In a previous paper (*1*) we reported on the radical intermediates which are formed at 77K and 303K on the gamma radiolysis of poly(hydroxybutyrate) (PHB) and its copolymers with hydroxyvalerate (HV) which were observed by Electron Spin Resonance spectroscopy (ESR). In this study we found that at 77K both radical anions and neutral radicals were formed on radiolysis. The observed neutral radicals indicated that scission of the polyester chain occurred adjacent to the ester group, and that abstraction radicals were generated on the carbon atom adjacent to the carbonyl group.

[1]Corresponding author
[†]Deceased

0097–6156/96/0620–0011$12.00/0

In this paper the radiation chemical yields for the volatile products, scission and cross-linking and new structures are reported and a mechanism is proposed to account for the major degradation pathways leading to the observed products.

Experimental

The poly(hydroxybutyrate) and two of its copolymers with hydroxyvalerate were obtained from ICI (Australia). The copolymers contained hydroxyvalerate mole fractions of 0.184 and 0.263. The polymers were purified according to the procedure given in our previous paper (1). The crystallinities of the polymers were measured by X-ray diffraction and found to be $62\pm0.5\%$ for poly(hydroxybutyrate) and $47\pm0.4\%$ for the two copolymers.

For radiolysis, powdered samples of the polymers were placed in small glass ampoules and evacuated at a pressure of less than 0.1 mPa for at least 24 hours at ambient temperature (1). They were then sealed under vacuum prior to irradiation. The polymer samples for volatile product and molecular weight analysis were irradiated at 303K in a ^{60}Co Gammacell at a dose rate of 3.9 kGy hr^{-1} to various doses up to 200 kGy. The polymer samples for NMR analysis were irradiated in the irradiation pond at the Australian Nuclear Science and Technology Organization. Irradiations were carried out using ^{60}Co at ambient temperature to various doses up to 2 MGy.

Molecular weight studies on polymer films were undertaken using electron beam irradiation. The films were cast from chloroform solvent and were approximately 15 μm in thickness. After casting was complete, the films were allowed to stand for several weeks at ambient temperature to ensure the removal of all residual solvent. The samples were irradiated using a 3MeV Van de Graaf electron beam accelerator at a dose rate of 1.03 MGy hr^{-1}. The samples were irradiated under a flow of nitrogen (to exclude oxygen) using conditions which minimized any increase in temperature during the radiolysis.

The volatile products were analysed using a Hewlett Packard Model 5730A gas chromatograph. A modified injection system (2) allowed the irradiated ampoules to be broken in the carrier gas flow, thus releasing the evolved gases into the chromatograph for analysis. The column temperature was programmed, with the initial temperature being set at 80°C and held at that temperature for 8 minutes. The column was then heated at a rate of 8°C per minute to a temperature of 150 C, where the temperature was held for a further 16 minutes. Nitrogen was used as the carrier gas for analysis of hydrogen and helium was used for analysis of the other gases. Both the thermal conductivity (TC) and the flame ionization (FID) detectors were used in series to analyse the products. The gas chromatograph was interfaced to an IBM compatible computer to acquire and manipulate the data and to calculate the yield of the product gases. The number average and weight average molecular weights of the polymers after gamma irradiation were measured using a Waters model 6000 A liquid chromatograph with dichloromethane as the solvent. Five ultrastyragel columns ($10^2, 10^3, 10^4, 10^5$, and 10^6) were used for separation. The column set was calibrated with narrow molecular weight polystyrene standards obtained from the Pressure

Chemical Company. The Q factor method was used to correct for the difference between the copolyesters and the polystyrene standards.

The number and weight average molecular weights were determined after electron beam irradiation using a Waters model 712 wisp liquid chromatograph with dichloromethane as the solvent.

The changes in the chemical structure of the polymers were studied using ^1H and ^{13}C NMR. The ^1H and ^{13}C NMR spectra were recorded on either a Jeol GX400 or a Bruker AMX 500 NMR spectrometer at ambient temperature. ^1H homodecoupling, DEPT and 2D COSY analyses were used to confirm the assignments of the new resonances appearing after irradiation. Spectra were obtained in $CDCl_3$ solutions at 2% (w/v) for ^1H and 5% (w/v) for ^{13}C NMR.

Results and Discussion

Volatile Product Analysis. The major products observed on irradiation of these polyesters at ambient temperature were carbon dioxide, carbon monoxide, methane and hydrogen. Small amounts of ethane, propane and butane were also observed. Blank runs showed that no degradation products arose as a result of thermal degradation during sample preparation on the vacuum line. The G-values for the observed gases were obtained from the slope of a plot of gas yield versus the absorbed dose. The linearity of the plots for the production of carbon dioxide and hydrogen after irradiation of the homopolymer at 303K are shown in Figure 1. These plots are typical of those obtained for all the evolved gases for all of the polymers. The G-values for gas formation for the homopolymer and the copolymer containing 0.263 mole fraction hydroxyvalerate are given in Table I.

Table I. G-Values of volatile products following gamma radiolysis of poly(hydroxybutyrate-co-valerate) at 303K

PRODUCT GAS	MOLE FRACTION HYDROXYVALERATE	
	0.0	0.263
CO	0.74 ± 0.07	1.00 ± 0.11
CO_2	0.96 ± 0.07	0.64 ± 0.04
H_2	0.42 ± 0.02	0.53 ± 0.08
CH_4	0.056 ± 0.003	0.051 ± 0.001
C_2H_6	0.0079 ± 0.0009	0.040 ± 0.001
C_3H_8	0.0005 ± 0.0001	0.0051 ± 0.0004
C_4H_{10}	-	0.0013 ± 0.0001

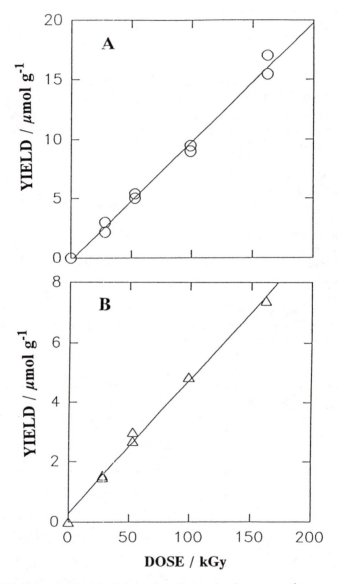

Figure 1: Yield versus absorbed dose plots for evolved gases on the gamma radiolysis of poly(hydroxybutyrate) at 303K. (A) Carbon dioxide (B) Hydrogen.

The G-values for carbon monoxide and carbon dioxide were significantly larger than those for the other product gases. The replacement of the methyl side chain with an ethyl group in the copolymer has the effect of increasing the carbon monoxide yield and decreasing the carbon dioxide yield. This observation might be related to the donation of electrons from the methyl side chain of the homopolymer, as suggested by D'Alelio (*3*). This would lead to a larger carbon-oxygen bond strength for the hydroxybutyrate units in the polymer chain, thus increasing the probability for formation of carbon dioxide relative to carbon monoxide from these units.

The sum of the G-values for formation of carbon dioxide and carbon monoxide was approximately the same for the homopolymer and the copolymer, 1.6 and 1.7 respectively. The formation of carbon dioxide and carbon monoxide can only arise as a result of degradation of the backbone ester group, so G(-COO) gives a minimum estimate for G-scission.

The susceptibility of an ester group to high energy radiation damage, as measured by the yield of carbon monoxide plus carbon dioxide, appears to be relatively independent of the nature of the ester. For example, for poly(methylmethacrylate), where the ester group is located in the polymer side chain, the total yield of carbon monoxide plus carbon dioxide is approximately 2 (*4*), and similar yields have also been reported (*5*) for several low molecular weight esters. This suggests that all esters degrade via similar mechanisms.

The observation that the ester group undergoes significant degradation resulting in chain scission is consistent with our previous observation that chain scission radicals are formed, as well as other radicals, when the polymer is irradiated at 77K (*1*). The G-value for total radical formation for the homopolymer was approximately 1.9 at 77K, which includes contributions from abstraction and ionic radicals, as well as the chain scission radicals. This G-value is of a similar magnitude to G(-COO) of 1.7. However, it has been suggested previously (*6*) that ester groups may degrade via both free radical and excited state pathways. Thus it may be expected that the yield of scission radicals would be less than G(-COO).

The G-value for hydrogen for irradiation of the homopolymer is comparable with that for carbon monoxide and carbon dioxide, and it increased slightly with the inclusion of the hydroxyvalerate unit in the copolymer. The observation of radicals formed by the loss of hydrogen from the polymers during irradiation (*1*) is consistent with formation of a significant yield of hydrogen gas. To the extent that the hydrogen yield can be used as an indicator of the tendency of a polymer to undergo crosslinking, this increase may imply a greater propensity for crosslinking in the copolymer.

The most likely source of the low molecular weight alkanes is scission of the polymer side chains. The higher yields of ethane, propane and butane and the smaller yield of methane for the copolymer compared to the homopolymer supports is thesis. The total G(alkane products) increases from 6.44×10^{-2} for the homopolymer to 9.66×10^{-2} for the copolymer containing 0.263 mole fraction hydroxyvalerate. This increase could be attributed several factors, including (i) a decrease in the crystallinity of the copolymer compared to the homopolymer, (ii) a higher probability for scission of the ethyl compared to the methyl side chain, or (iii) the different glass transition temperatures of the polymers.

A higher probability for scission of an ethyl compared to a methyl side chain has been reported for the radiolysis of copolymers of ethene with propene or 1-butene (5). Here the probability of scission of an ethyl side chain was found to be 3.3 times that for scission of a methyl side chain. If it is assumed that the differences between the yields for the alkanes observed for the two poly(hydroxybutyrate)s can be attributed solely to an enhanced probability for scission of the ethyl compared to the methyl side chain, the ethyl group would need to be approximately 2.9 times more sensitive to side chain scission. This close agreement between the relative scission probabilities for methyl and ethyl side chains in the polyolefines and the polyesters supports the proposal that the alkanes observed in this work arise predominantly as a result of side chain scission.

Molecular Weight Changes. The G-value for scission, G(S), and crosslinking, G(X), can be obtained from studies of the variations in the molecular weight averages M_N and M_W with irradiation dose. Plots of $1 / M_N$ and $1 / M_W$ versus the absorbed dose are linear if the scission is random and the dispersity index of the polymer is equal to 2. Typical plots for the homopolymer are shown in Figure 2. The slopes of these plots are related to G(S) and G(X) through the following equations:

$$(1/M_N)_D = (1/M_N)_O + 1.037 \times 10^7 \, (G(S) - G(X)) \times D \qquad (1)$$

$$(1/M_W)_D = (1/M_W)_O + 0.519 \times 10^7 \, (G(S) - 4G(X)) \times D \qquad (2)$$

in which the absorbed dose, D, is expressed in kGy (7).

The G-values obtained for scission and crosslinking for gamma radiolysis of the homopolymer at 303K were 1.3 ± 0.1 and 0.0 ± 0.1 respectively. A G(S) of 1.3 compares closely with the value of G(-COO) = 1.7 obtained from the volatile product analysis. These values of G(S) and G(X) can be compared with values reported previously for similar polymers. Collet et al. (8) have studied the molecular weight changes in poly(lactic acid) and its copolymers with glycolic acid. For poly(lactic acid) they reported that G(S) = 1.26 and G(X) = 0.53 and that these values were both slightly higher in the copolymer. Pittman et al. (9) studied the molecular weight changes in poly(hydroxyisobutyric acid). They have reported values of G(S) = 0.37 and G(X) = 0.07 for poly(hydroxyisobutyric acid). Thus, the G-value for scission for poly(lactic) acid and poly(hydroxybutyrate) are similar in magnitude, suggesting that scission is controlled by the radiation sensitivity of the ester linkage in these two polymers. On the other hand, the G-value for crosslinking varies for these two polymers, suggesting that crosslinking is sensitive to the structure of the hydrocarbon moiety. For poly(hydroxybutyrate) and poly(hydroxyisobutyric acid) G(S) is greater than 4 times G(X), so that these polymers undergo nett scission. On the other hand, for poly(lactic acid) Collet et al. (8) report that G(S) is less than 4 times G(X), so that it will form a gel.

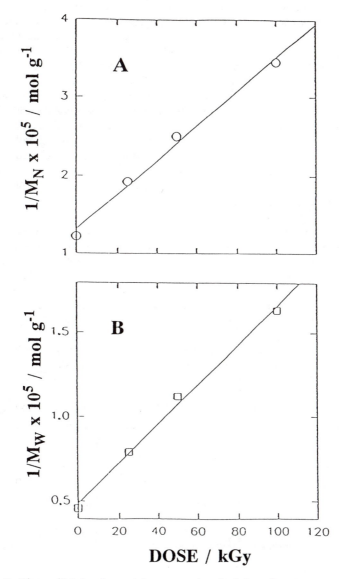

Figure 2: Plots of Molecular weight versus absorbed dose for gamma radiolysis of poly(hydroxybutyrate) at 303K. (A) $1/M_N$ and (B) $1/M_W$.

The molecular weight changes which result from electron beam irradiation of the polymers yielded the values for G(S) and G(X) given in Table II. The value of G(S) for the homopolymer was significantly higher than that found for the gamma radiolysis at 303K. The reason for this discrepency is not clear, but it could lie in the fact that at 303K the homopolymer is close to its glass transition temperature (10). In the study of the radical decay on annealing of the homopolymer (1), the radical stability was found to decrease significantly at temperatures slightly higher than 303K. The higher glass transition temperature of the homopolymer compared to the copolymers was also shown to result in a much higher G-value for radical formation in the homopolymer at 303K, G(R·) = 1.6 (1). If the high dose rates used in the electron beam irradiations caused a small localized increase in the temperature of the sample during radiolysis, such that the polymer chains became significantly more mobile, the G-value for scission might be higher than expected on the basis of the gamma radiolysis study at 303K. A temperature increase in the sample during electron beam irradiation would also explain why the values of G(S) for the copolymers are higher than those expected on the basis of the values for G(-COO) obtained from volatile product analysis following their gamma radiolysis.

Analysis of the changes in the molecular weight of poly(hydroxybutyrate) films following electron beam radiolysis yielded G(X) = 0.0, which is consistent with the observation for the gamma radiolysis of this polymer. However, incorporation of the ethyl side chain in the copolymers was shown to result in crosslinking as well as scission. This observation is consistent with studies of D'Alelio et al. (3) on other polyesters containing methyl and ethyl side chains, which have indicated that an ethyl branch increases the tendency towards crosslinking.

Table II. Scission and crosslinking yields for electron beam irradiation of poly(hydroxybutyrate-co-hydroxyvalerate)

MOLE FRACTION HYDROXYVALERATE	G(S)	G(X)
0.0	2.2 ± 0.3	0.0 ± 0.2
0.184	2.7 ± 0.03	0.3 ± 0.2
0.263	3.4 ± 0.3	0.5 ± 0.3

Chemical Changes. NMR provides a useful tool with which to study the structural changes in a polymer after irradiation. The ^{1}H NMR spectrum of the homopolymer after gamma radiolysis is shown in Figures 3 and 4, while the ^{13}C NMR spectrum of the homopolymer after radiolysis is shown in Figure 5. The major new peaks which appear in the proton and carbon spectra have been assigned on the basis of information available from the literature (3, 11-15) and on the new chemical structures (see Table

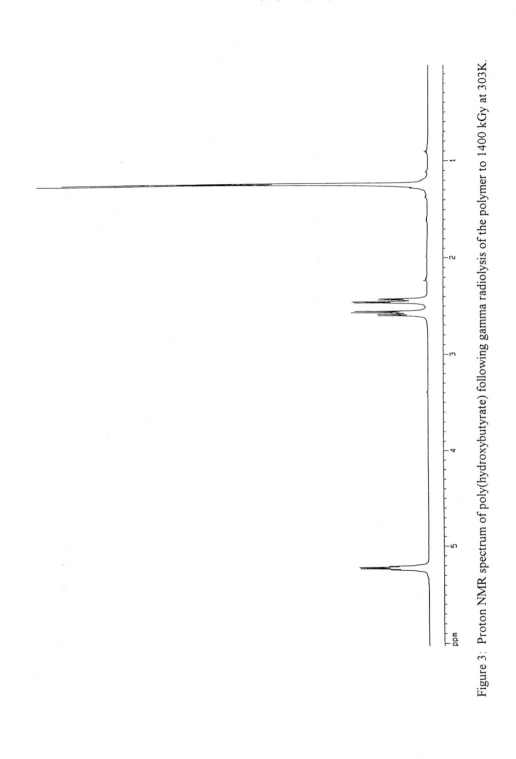

Figure 3: Proton NMR spectrum of poly(hydroxybutyrate) following gamma radiolysis of the polymer to 1400 kGy at 303K.

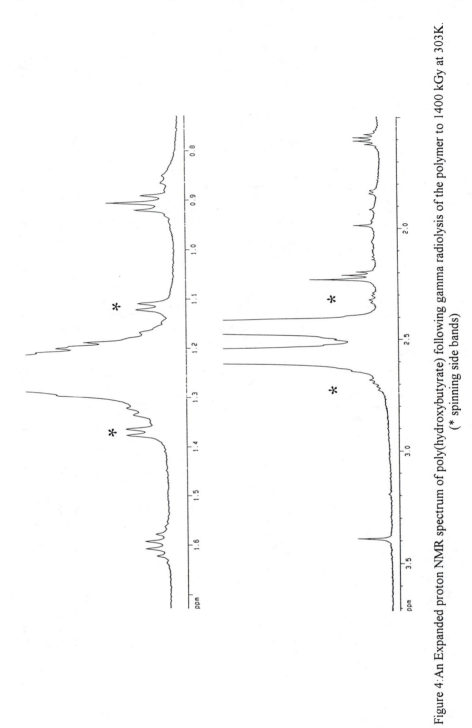

Figure 4:An Expanded proton NMR spectrum of poly(hydroxybutyrate) following gamma radiolysis of the polymer to 1400 kGy at 303K.
(* spinning side bands)

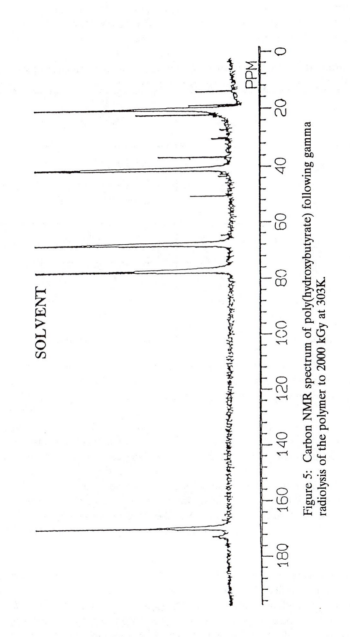

Figure 5: Carbon NMR spectrum of poly(hydroxybutyrate) following gamma radiolysis of the polymer to 2000 kGy at 303K.

III) which may be predict using the results obtained from the ESR study, the volatile product analysis and the molecular weight changes.   The chemical shifts for the expected new structures have been predicted from NMR data for model compounds (*16,17*).   The assignments were further confirmed by undertaking ^{1}H homodecoupling, ^{13}C DEPT and 2D COSY analyses.

Table III.   Possible chain-end structures formed on radiolysis of poly(hydroxybutyrate) at 303K

DESIGNATION	STRUCTURE
I	$CH_3-CH_2-CH_2-CO-$
II	$(CH_3)_2-CH-O-$
III	CH_3-O-
IV	$CH_3-CH(OH)-$
V	$OCH-CH_2-$

Scission of the backbone ester group with subsequent loss of carbon dioxide would be expected to produce the saturated end groups designated by structures I and II in Table III. After irradiation of poly(hydroxybutyrate), new resonances appeared in the proton spectrum at 2.22 ppm (triplet), 1.60 ppm (multiplet) and 0.90 ppm (triplet). The 2D COSY spectrum in Figure 7 indicates that the protons at 1.60 ppm are coupled both to those at 2.22 ppm and to those at 0.90 ppm.   New resonances appeared in the carbon spectrum at 36 ppm and 19.2 ppm due to methylene carbons, and at 13 ppm due to a methyl carbon.   In addition, a new carbonyl carbon resonance appeared at approximately 172 ppm.   The chemical shifts for these resonances in the proton and carbon spectra agree closely with those predicted for structure I.   The relative areas of the three proton resonances were found to be in close agreement with the expected peak intensities of 2:2:3.   Thus, there is very strong evidence for the formation of chain-end structure I.

The new resonance at 1.25 ppm, which is on the up-field side of the main-chain methyl proton peak, can be assigned as a methyl resonance, while the broad peak at 5.0 ppm is associated with a methine proton.   The 2D COSY spectrum in Figure 7 shows that these two proton resonances are coupled.   In addition, a series of proton decoupling experiments were undertaken at approximately 1.25 ppm, which caused the broad multiplet at 5.0 ppm to collapse to a singlet.   A new peak arises in the carbon spectrum at 22 ppm, which the DEPT experiment identified as a methyl carbon resonance.

Comparison of the peaks in the proton and carbon spectra with predicted chemical shifts indicate that these peaks are associated with structure II. It is believed that the methine carbon resonance of structure II is not resolved in the carbon spectrum and that it lies beneath the resonance of the main-chain methine carbon. Thus the NMR results show very strong evidence for the formation of the new chain end II.

In addition to the major new resonances which are associated with the chain-end structures I and II, a number of other less intense peaks arise in the proton spectrum of the irradiated polymer in the region 1.5 to 7 ppm. Most of these minor resonances have not been assigned. However, there are two new peaks in the proton spectrum at 3.4 and 2.0 ppm following irradiation. The peak at 2.0 ppm overlaps one of the methylene proton resonances associated with structure I. The 2D COSY spectrum in Figure 6 shows that these two peaks are coupled, and the proton detected one dimensional HMQC NMR spectrum shown in Figure 7 clearly demonstrates that both of these protons are attached to carbon atoms. The multiple pulses required for the HMQC experiment lead to J-modulation, and a spectrum which cannot be conventionally phased, but the spectrum is helpful in confirming the assignment of the peaks to protons attached to carbon rather than oxygen. The methine resonance at 5.0 ppm associated with structure II is also clearly shown in this spectrum.

A proton with a chemical shift of 3.4 ppm which is attached to a carbon atom is most likely to be associated with a methyl ester or methyl ether end-group, as shown in structure III. This assignment is supported by the observation of a resonance at 51 ppm in the carbon spectrum. The peak at 2.0 ppm is difficult to assign unequivocally, but it could arise from a methyl group attached to an in-chain double bond, as in the structure $-C(CH_3)=CH-$. The new resonances found in the region 5 to 7 ppm are most probably associated with protons attached to unsaturated carbon atoms, which would be expected to be found in the irradiated polymer. The weak coupling between the peaks at 2.0 and 3.4 ppm suggests that these groups are adjacent.

Scission of the ester backbone with loss of carbon monoxide could lead to the formation of an alcohol chain end, which has been designated as structure IV in Table III. In the proton spectrum there are two very small broad peaks at about 4.2 ppm, which could be associated with an alcohol. The methyl protons of the alcohol structure IV would not be resolved from the methyl protons in the polymer, so identification on this basis is not possible. However, in the carbon spectrum obtained after irradiation, the small new carbon resonance at 64 ppm could be due to a methine carbon adjacent to an alcohol group, and the small resonances at 24 ppm and 42 ppm could arise from the adjacent methyl and methylene carbon atoms associated with this chain end. Thus there is some evidence which could support the formation of a new alcohol chain end following rupture of the ester group, but it is somewhat uncertain.

Sevilla et al. (*11*) considered the possibility of hydrogen addition to the carbonyl radical anion, resulting in the formation a new alcohol end-group, structure IV, and a corresponding aldehyde end-group, Structure V. There is possible evidence for the formation of small amounts an alcohol end-group, and the corresponding aldehyde end-group formed by the reaction reported by Sevilla et al. should have a proton peak

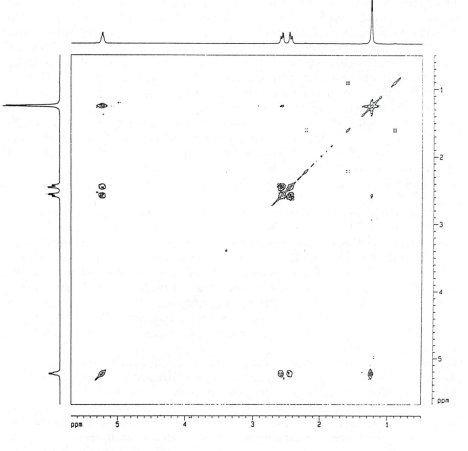

Figure 6: The 45 ° COSY spectrum of poly(hydroxybutyrate) following
radiolysis of the polymer to 1400 kGy at 303K.

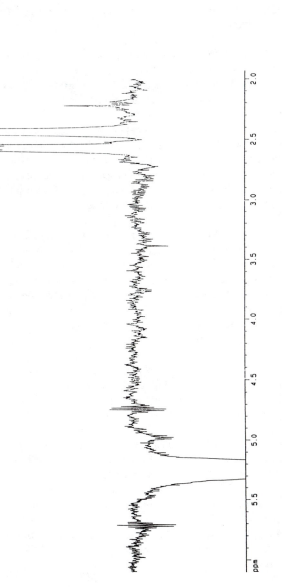

Figure 7: The HMQC spectrum of poly(hydroxybutyrate) following radiolysis of the polymer to 1400 kGy at 303K.

at approximately 9.7 ppm and a carbon peak at about 204 ppm. After irradiation, small resonance peaks were observed at these positions in the two spectra. Thus there is evidence of a reaction between the radical anion and the adjacent methylene group in these polyesters.

In addition to the small aldehyde peak observed in the proton spectrum, a second small proton peak was observed at 12 ppm, indicative of a free carboxyl group. This peak must arise from scission of the O-C bond in the backbone of the polymer, followed by hydrogen atom addition or abstraction.

Since the major new resonances in the proton spectrum arise from structures I and II, the yield of new end-groups represented by sturcture I can be used to obtain an approximate estimate of G(S). A value of G(S) of 1.4 has been calculated from the yield of this new end group, based upon the proposed proton NMR assignments presented above. This value compares very favourably with the value of G(S) = 1.3 calculated from molecular weight studies.

Degradation Mechanism. The mechanism of the high energy radiation degradation of the polymers of hydroxybutyrate and hydroxyvalerate has been shown to be similar to that which has been reported for other esters. The primary reactions which take place on irradiation of the polymer are molecular ionization, radical formation and excitation. Carbon monoxide and dioxide have been observed as the major volatile products of radiolysis at 303K, along with hydrogen, indicating that scission occurs at the ester groups with elimination of these gases.

Molecular weight studies confirm that chain scission is the predominant reaction in the polymers. There was no evidence for crosslinking in poly(hydroxybutyrate), but the value of G(S) was found to be greater for the higher dose rate electron beam irradiation. Some crosslinking was observed for electron beam irradiation of the copolymers, with G(X) increasing with valerate content.

The neutral polymer radicals may decay to yield new saturated and unsaturated groups in the polymer. Evidence for the formation of new saturated end-groups, resulting from cleavage and breakdown of the main-chain ester linkages, was found in NMR studies of the polymer after irradiation. Evidence for chain unsaturation was also found in the proton spectra of the irradiated polymer.

Acknowledgements. The authors wish to thank the Australian Research Council and the Australian Institute for Nuclear Science and Engineering for financial support for this research.

References

1.   Carswell-Pomerantz, T.G.; Hill, D.J.T.; O'Donnell, J.H. and Pomery, P.J.
        Polym. Degrad. and Stab., in press.
2.   Bowmer, T.N., Ph.D. Thesis, The University of Queensland, **1980**.

3.   D'Alelio, G.F.; Haberli, R.; Pezdirtz, G.F. *J. Macromol. Sci., Chem.*, **1968,** *A2*, 501.

4.   Carswell, T.G., Ph.D. Thesis, The University of Queensland, **1991.**

5.   Ho, S.Y., Ph.D. Thesis, The University of Queensland, **1985.**

6.   Calvert, J.G.; Pitts, J.N. *Photochemistry*, Wiley, New York, NY, **1966,** pp 429-433.

7.   Saito, A. *The Radiation Chemistry of Macromolecules*, Ed Dole, M , Academic Press, New York, NY, **1972,** 1, 223.

8.   Collett, J.H.; Lim, L.Y.; Gould, P.L. *Polym. Prep.*, **1989,** *30*, 468.

9.   Pittman, Jr, C.U.; Iqbal, M.; Chen, C.Y.; Helbert, J.N. *J. Polym. Sci., Polym. Chem. Ed.*, **1978,** *16*, 2721.

10.   Marchessault, R.H.; Monasterios, C.J.; Morin, F.G. *Pacific Polym. Prepr.*, **1989,** *1*, 449.

11.   Sevilla, M.D.; Morehouse, K.M.; Swarts, S. *J. Phys. Chem.*, **1981,** 85, 923 .

12.   Hudson, R.L.; Williams, F. *J. Phys. Chem.*, **1978,** *82*, 967.

13.   Ayscough, P.B.; Oversby, J.P. *J. Chem. Soc., Faraday Trans.*, **1972,** *68*, 1153.

14.   Nakajima, Y.; Sato, S.;Shida, S. *Bu l l. Chem. Soc. Japan*, **1969,** *42*, 2132.

15.   Josephson, E.S.; Peterson, M.S. *Preversation of Food by Ionizing Radiation*, CRC Press Inc., Boca Rotan, Fl, **1983**.

16.   Bremser, W.; Ernst, L.; Facheyer, W.; Gerhards, R.; Hardt, A.; Lewis, P.M.E. *Carbon-13 NMR Spectral Data*, Wernheim, New York, NY, **1987**.

17.   Pauchet, C.J.; Campbell, J. *The Aldrich Library of NMR Spectra*, Aldrich Chem. Comp., Milwaukee, Wi, **1974**.

RECEIVED October 5, 1995

Chapter 3

Radiation-Induced Energy and Charge Transport in Polystyrene

Laser Photolysis and Pulse Radiolysis Comparative Study

Guohong Zhang and J. K. Thomas[1]

Department of Chemistry and Biochemistry, University of Notre Dame, Notre Dame, IN 46556

The radiation induced processes in polystyrene films have been studied by laser photolysis and pulse radiolysis using transient spectroscopic techniques. Transport of the singlet excitation in polystyrene was found to be temperature dependent, and the average excitation migration rate constant was estimated to be ~5.0x10^{11} M^{-1}s^{-1} at 293 K and 0.5~1.0x10^{11} M^{-1}s^{-1} at 80 K. The concentration of excimer forming sites in room temperature equilibrated polystyrene was measured to be around 0.1 M by competitive quenching kinetics. Efficient generation of excited states, cation and anion radicals of solutes was observed in electron irradiated polystyrene films, concomitant with the quenching of the polymer singlet excited state. The yields of pyrene transients at [Py]=100 mM were measured: G(^{1}Py*)~0.6, G(^{3}Py*)~1.0, G(Py$^+$·)~1.6, and G(Py$^-$·)~0.8. Fast kinetic measurements resolved the concentration dependent formation rates of solute excited states and cation radicals that was within ~2 ns. Migration of the positive hole in the polymer matrix was proposed to account for the production of solute excited states via geminate recombination with solute anions and the generation of solute cations by charge transfer to solutes at encounter. The positive charge migration coefficient was estimated from the bimolecular reaction rate to be Λ_h~3x10^{-5} cm^2/s. Simulation of the experimental data shows that the thermalization length of excess electrons is 56 Å in solid polystyrene, and that ~70% of the polystyrene excimer is derived from the subsequent geminate ion recombination.

Polymeric solids have been widely used as resist, imaging, and coating materials in industrial applications and commercial products. The performance of this category of polymers is essentially determined by their response to radiation interaction. Because of their technological importance, radiation effects on polymeric materials have been the focus of recent studies (*1-3*). Processes induced by low energy photons in aromatic polymers have been well established over the past two decades (*4,5*), while the modes of interaction between radiation and solid polymers and the early processes leading to excited and ionic intermediates are still poorly understood (*6*).

[1]Corresponding author

0097–6156/96/0620–0028$13.75/0

Early work on polystyrene by pulse radiolysis experiments identified the production of solute excited states and solute cation radicals (*7,8*). It was observed that the triplet excited states of solutes were formed with the nanosecond pulse, independent of the lowest singlet excited states of the solutes, indicating that the intersystem crossing from the solute S_1 states is not the major channel for the triplet production. Ho and Siegal attributed the solute triplet state formation to triplet energy transfer from the polymer matrix (*7*). Nanosecond pulse radiolysis of solute free polystyrene films observed polystyrene excimer, polystyrene dimer cation, and polystyrene triplet (*9*), which again confirmed the capability of excited state and charge generation in aromatic polymers. Recent work from our laboratory (*6,10*) clearly shown the production of pyrene excited states (S_1 and T_1), and ion radicals in pyrene doped polystyrene films. The ion yield was measured to be $G_{ion}=2\sim3$ and the yield of excited states was $G_T+G_S=1\sim2$, indicating a very efficient charge generation in the nonpolar polystyrene matrix. In contrast, previous work in benzene and toluene solutions observed large yield of excited states $G_T+G_S=4.5\sim5.4$ but little ions $G_{ion}\sim0.1$ (*11*). Picosecond pulse radiolysis experiments in liquid benzene and toluene shown very fast (~7 ps) singlet excited state formation and subsequent energy transfer to solutes by molecular diffusion (*12*). Short electron thermalization length and very rapid geminate ion recombination were suggested to be responsible for the nearly complete energy conversion into excited states (*13*). From liquid monomer model compounds such as benzene and toluene to solid polystyrene, an abrupt change in radiation products was observed. This is difficult to explain by the popular radiation chemistry theory which states that the radiolysis initially generates ions and thermalized electrons, and subsequent rapid geminate recombination produces excited states in nonpolar hydrocarbon liquids such as benzene and cyclohexane (*14*). Even when the rigidity of the glassy environment and therefore the low mobilities of the excited and charged species trapped in the polymer matrix were taken into consideration, the mechanism still remains uncertain. The fast formation of solute excited states might be derived from: 1) energy transfer from the polymer excited states, 2) geminate charge recombination which is not limited by the slow mass diffusion in the polymer matrix, 3) direct excitation by subexcitation electrons; while solute cations may be produced via: 1) charge transfer from the ionized polymer, 2) ionization by the superexcited states or the plasmon excitation of the polymer. It is also not clear what makes polystyrene quite different from its model compounds in the condensed phases, and whether or not the polymer carbon-carbon backbone is involved in the energy deposition processes.

Comparative studies have been performed using pulse radiolysis and laser photolysis techniques to examine the response of polymeric materials to excitation at different energy levels and search for the mechanism underlying the unique production of excited states and ion radicals in aromatic polymers upon high energy irradiation. This paper reports some of the results on the transport of the singlet excitation of polystyrene in photolysis, and the early ionic processes in radiolysis of the polymer which lead to the formation of excited states and ion radicals.

Experimental

Chemicals.   Polystyrene (PS, M.W.~280,000), poly(benzyl methacrylate) (M.W.~40,000), and polyvinylbenzyl chloride (PVBC) were obtained from Aldrich. Poly(2-phenylethyl methacrylate) solution (25% solid in toluene) was supplied by the Scientific Polymer Product. All the above polymers were further purified by precipitation from ethanol-benzene solution 2~3 times. Pyrene and bromopyrene were purified by passing their benzene solutions through activated silica column and recrystalized several times from ethanol-benzene solution. Other molecular probes like biphenyl, naphthalene, 4-bromobiphenyl (BrPh2), 1,4-dibromonaphthalene (DBN), benzophenone, anthracene (An), 9,10-diphenylanthracene (DPA), and ferrocene (Fc) were used as obtained from Aldrich.

Sample Preparation. Polymer films were made by casting the mixed toluene or 1,2-dichloroethane solutions of polymers and dopants on quartz plates. The same thickness of polystyrene films, which is required in two photon fluorescence measurements, was carefully controlled by casting a definite amount of polymer solution over the same plate area. Solvents were slowly evaporated under watch glasses. The polymer samples were then air dried over night and vacuum pumped at room temperature for a day. Dried PS and PVBC films can be easily peeled off from the plates. For radiolysis experiments, the film thickness is normally about 100~200μm. Thin films with thicknesses of 10~20μm were used in photolysis measurements. Film samples in air were used in most of the measurements. Electron scavenging experiments were performed in a vacuum chamber attachment where vacuum pumped film samples could be exposed to a controlled pressure of gas or vapor of a scavenger. Homogeneous distribution of dopant molecules in polystyrene films was examined by fluorescence spectroscopy: 1) using the pyrene excimer as an indicator of molecular aggregation, and 2) using the non-radiative energy transfer quenching of pyrene fluorescence as a spectroscopic ruler.

Instrumentation. UV-visible absorption spectra of polymer films were taken on a Varian Cary 3 UV-VIS Spectrophotometer. Steady state fluorescence emission of polystyrene and polymethacrylates were measured at both room temperature and 77 K on a Perkin-Elmer MPF-44B Fluorescence Spectrometer. Impurities and solutes in polymer films were checked by both absorption and emission spectroscopy. In energy transfer studies, absorption spectra of acceptors (used as quenchers) and fluorescence spectra of donor chromophores (for example, cumene) were measured for the Forster spectral overlap calculation (*15*).

Nanosecond fluorescence emission was measured in two different schemes. The fourth harmonic at 266 nm (~0.1 mJ/pulse) and the second harmonic at 532 nm (~10 mJ/pulse) from a Continuum mode-locked picosecond Nd:YAG laser (PY61-10) were used for excitation in one photon and two photon fluorescence experiments. The pulse width is ~20 ps or ~35 ps from the cross correlation measurements depending on the saturable absorbers used. The fluorescence from polymer samples was detected by a Hamamatsu microchannel plate PMT (R1644U) with a rise time ~150 ps. The signals were amplified by Tektronix 7A29 amplifier (bandwidth ~1 GHz) and digitized by a Tektronix 7912AD programmable digitizer. The rise time of the nanosecond measurement system is ~500 ps. Two photon excitation by using the high power green light from the picosecond laser was utilized to selectively excite polystyrene instead of the included solutes. The nanosecond fluorescence of polymer films at 77 K was measured by fixing the film samples in a quartz dewar filled with liquid nitrogen.

In picosecond fluorescence measurements, a Hamamatsu phototube R1328U-02 (rise time ~60 ps) was used as the detector, and signals were sampled by a Tektronix communication signal analyzer CSA 803 with a 20 GHz sampling head SD-26 (rise time ~17.5 ps). With a laser pulse width of ~35 ps, the pulse response of the system (FWHM) is ~100 ps, and the rise time of the step response is ~110 ps. For measurements at low temperatures, samples were cooled down in a quartz dewar by a stream of cold nitrogen which could be chilled to as low as 80K after passing through a copper coil immersed in liquid nitrogen. The sample temperature was controlled by adjusting the nitrogen flow rate.

Picosecond transient absorption measurements were performed by the pump-probe method in a double beam setup. When using the laser pulses at 355 nm or 532 nm as the probe beam, the noise level is as low as 0.002 absorbance unit. Normally 100 shots were averaged at each time delay. Using a 20 ps pulse, processes as fast as ~10 ps can be measured by reconvolution fitting. The intersystem crossing time of benzophenone in cyclohexane was measured to be ~30 ps, which is in agreement with the early work in benzene solution (*16*).

High energy radiation experiments were performed on a Field Emission Febetron (Model 706) which delivers a short pulse (FWHM ~2 ns) of fast electrons with the output energy of 0.4~0.6 MeV and a dose of ~200 krad per pulse. Transient absorption measurements on microsecond timescale were made in a conventional setup using a Hamamatsu 1P-28 PMT, a Tektronix 7A16 amplifier (bandwidth ~100 MHz), and a Tektronix 7912AD programmable digitizer. The response of the entire measuring system is ~5 ns. Transient absorption spectra at different time after the electron pulse were obtained by assembling the measurements at different wavelengths with a bandwidth of 3~4 nm. Spectra were corrected by monitoring the output fluctuation of electron pulse using a separate transient absorption line. Absorption measurements within 10 ns were made with an ITT F-4014 biplanar vacuum photodiode (rise time ~90 ps at 3 kV). The system response was checked by monitoring the absorption of the solvated electrons in water. Using a reconvolution algorithm, processes as fast as 400~500 ps can be resolved from these fast transient measurements. Nanosecond fluorescence measurements in radiolysis were made with the same microchannel plate PMT setup as in photolysis.

G-value Measurement. In radiation chemistry, absolute transient yields were measured in G values (number of transient species generated per 100 eV energy loss). The dose per pulse was determined relatively using the absorbance of the hydrated electrons at 600 nm at the end of the electron pulse in a 2 mm water cell. The transient absorbance of a film sample at the end of the pulse at a specific wavelength under the same irradiation conditions was measured with four layers of thick films. Based on the same dose absorbed by water and polymer films, G value of the measured transient was calculated by the following equation.

$$G(Tr) = \frac{A(Tr;\lambda)}{A(e_s^-;600nm)} \cdot \frac{\varepsilon(e_s^-;600nm)}{\varepsilon(Tr;\lambda)} \cdot G(e_s^-) \qquad (1)$$

where, $G(Tr)$: G value of the transient to be measured;

$G(e_s^-)$: G value of the solvated electron in water, taken as 3.0;

$\varepsilon(e_s^-; 600 nm)$: the extinction coefficient of the solvated electron at 600 nm,

$\varepsilon = 12,000$ $M^{-1}cm^{-1}$ was taken from literature;

$A(e_s^-; 600 nm)$: the absorbance of the hydrated electron at 600 nm measured at the end of the pulse;

$A(Tr; \lambda)$: the absorbance of the transient species at wavelength λ measured at the end of the pulse;

$\varepsilon(Tr; \lambda)$: the extinction coefficient of the transient at wavelength λ, which was taken as the same value measured in liquid hydrocarbons.

Data Processing. Data from kinetic measurements with timescales much longer than the 2 ns pulse width were analyzed with different decay functions for rate parameters by the Marquardt nonlinear least square fit (*17*). For measurements on the timescale which is comparable to the instrument response (~2ns), the transient signals were simulated by reconvoluting the model kinetics with the pulse response of the measurement system.

If the measured transient signal is $\tilde{S}(t)$, the system response function is $F(t)$, impulsive kinetics $S(t)$ can be obtained by reconvoluting model kinetics to simulate the transient signal $\tilde{S}(t)$ by the following convolution algorithsm.

$$\tilde{S}(t) = F(t) \otimes S(t) = \int_0^t F(\tau) \cdot S(\tau) d\tau \tag{2}$$

A simple biexponential function form was used for the model kinetics $S(t)$.

$$S(t) = S_0 \left\{ \exp(-k_2 t) - \exp(-k_1 t) \right\} \tag{3}$$

where, k_1 is the formation rate of the transient, k_2 is its decay rate which can be measured on a longer timescale in the same experiment. In some cases, the actual formation kinetics is not first order, and k_1 is the average formation rate of the transient.

Results

1. Singlet Excitation Transport in Laser Photolysis of Polystyrene

Steady State Measurements. The absorption spectrum of a thin PS film exhibits the same absorption band around 258 nm with the same vibrational structure as its model compound cumene. A spectral broadening of the S_1-S_0 transition of the phenyl chromophores was clearly observed in solid polystyrene when compared to a cyclohexane solution of cumene.

The fluorescence emission of PS was measured at room temperature (T=293 K) and at 77 K over the wavelength range 270 nm to 420 nm, with front surface excitation centered at 260 nm (bandwidth ~4 nm). Two types of fluorescence were observed in this work, as reported in the literature (*18*): monomeric emission near 285 nm and excimer emission at 330 nm. The term "monomeric" is used here to refer to the repeating unit of a polymer. Figure 1 shows the room temperature fluorescence spectra of thin films of three polymers containing phenyl chromophores. The spectra were adjusted to the same scale to show the dramatic spectral change from poly(2-phenylethyl methacrylate) to polystyrene. The spectrum of poly(2-phenylethyl methacrylate) film shows almost purely monomeric emission at 285 nm, while that of poly(benzyl methacrylate) film indicates an increased shoulder around 320 nm. PS film, however, exhibits a broad excimer band peaked at 330 nm with little monmeric emission around 285 nm. Even at 77 K, only a weak monomeric shoulder was observed, and the excimer still accounts for most of the PS fluorescence.

Nanosecond Fluorescence Measurements. Time resolved experiments were performed to measure the dynamics of excitation transport via excimer formation and fluorescence quenching by doped acceptors. Nanosecond fluorescence measurements were made by exciting a PS film at 266 nm with a Nd:YAG laser and monitoring the monomeric emission at λ=285 nm and the excimer emission at λ=330 nm. The emission bandwidth was controlled to less than 3 nm to clearly resolve the two types of fluorescence. The monomeric emission of PS at 293 K shows a very fast decay with a lifetime less than 200 ps, which is limited by the instrument resolution (Figure 2A). Also shown in Figure 2A is the decay profile of the PS monomeric emission measured at 77 K. A much slower but highly nonexponential decay was observed

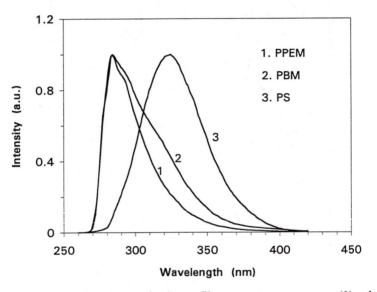

Figure 1. Fluorescence spectra of polymer films at room temperature: (1) poly(2-phenylethyl methacrylate); (2) poly(benzyl methacrylate); (3) polystyrene.

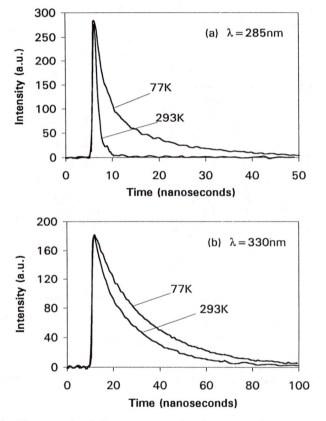

Figure 2. Time resolved fluorescence of polystyrene film: (a) monomeric emission decay at 293K and 77K; (b) excimer emission decay at 293K and 77K.

over several nanoseconds. Figure 2B shows the excimer decay traces measured at 293 K and 77 K. At both temperatures, excimer formation is faster than the instrument response (~500 ps). Excimer decay at 77 K can be fitted to a single exponential function with a lifetime of ~24 ns, while the decay at 293 K is nonexponential and was fitted to a biexponential function with a fast component (lifetime ~8 ns) and a slow component (lifetime ~24 ns). The slow component matches the normal excimer decay rate, and the fast component is attributed to S_1-S_1 annihilation which is absent at low temperature. Increasing the intensity of the excitation beam leads to a relative increase of the fast component and thus a faster decay of the PS excimer.

Picosecond Fluorescence Measurements. The rapid excimer formation in PS film was measured at different temperatures by picosecond fluorescence spectroscopy. The formation of the excimer emission around 330 nm was monitored at the rising edge following the excitation pulse. The two traces measured at 293 K and 80 K are dramatically different (Figure 3). Excimer formation is almost instantaneous at 293 K but exhibits a slower nonexponential growth within 1 ns at 80 K. An excimer formation rate of 5.0×10^{10} s^{-1} was obtained from the reconvolution fit of the signal at 298 K, indicating that, in PS film, the excitation initially produced at a monomeric site is trapped at a nearby excimer forming site (EFS) within ~20 ps at room temperature. The trapping of the excitation is slowed down at low temperatures and shows nonexponential formation kinetics.

Picosecond Transient Absorption. The previous work in pulse radiolysis of polystyrene films (*6,9*), and laser photolysis of PS in solution and in solid films (*19,20*) observed a short-lived, broad absorption around 530 nm (c.f. Figure 6), which was assigned to the S_1-S_n transition of the PS excimer, based on the well established excimer bands of benzene and its derivatives (*21*). Picosecond transient absorption experiments were performed to resolve the formation of the PS excimer by exciting the film sample at 266 nm and then monitoring the excimer absorption at 532 nm. Figure 4 shows that the PS excimer forms within 10~20 ps at room temperature, which is in agreement with the picosecond fluorescence measurement and the earlier work by Mataga (*20*).

Two Photon Fluorescence. The concentration of EFS traps and the rate constant of excitation migration can be measured via competition kinetics by doping the PS films with different concentrations of quenchers (acceptors) which compete with the excimer forming sites for the singlet excitation. Certain experimental problems arise in this sort of quenching studies due to the direct excitation of the quencher molecules embedded in PS matrix and the absorption of the excimer emission by the dopants. This situation requires many corrections in the data analysis (*18*). In the present study, selective two photon excitation of polystyrene at 532 nm, instead of one photon excitation at 266 nm, was made by taking advantage of the high peak power of the picosecond laser. The quencher molecules were selected by the following requirements: 1) no absorption or very weak absorption around 330 nm region, and 2) absolutely no absorption at 532 nm. Figure 5 shows the static decrease of the initial intensity of the PS excimer with increasing quencher concentration for both DBN and $BrPh_2$, clearly indicating a competition between the EFS and $BrPh_2$ (or DBN) for singlet excitation moving among phenyl chromophores.

$$^{1}PS^{*} + EFS \xrightarrow{\ k_{trap}\ } {}^{1}PS_{2}^{*} \qquad (4)$$

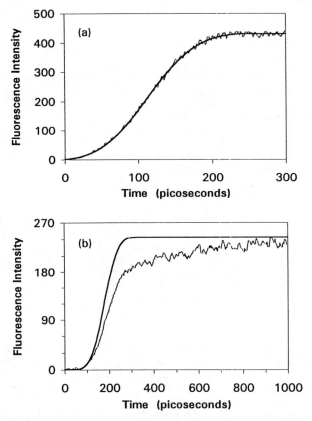

Figure 3. Polystyrene excimer formation by picosecond fluorescence measurements: (a) at 293 K, the solid line shows the reconvolution fit with a formation rate $k=5.0 \times 10^{10}$ s^{-1}; (b) at 80 K, the solid line shows the step response function.

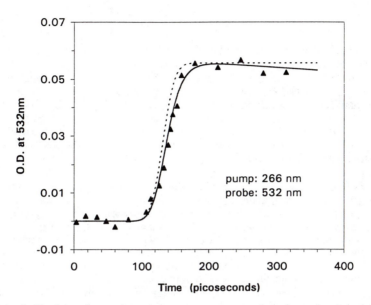

Figure 4. Excimer formation at room temperature by picosecond transient absorption measurement. The solid line is the reconvolution fit with a formation rate $k=7.0\times10^{10}$ s^{-1}, and the dash line shows the step response function.

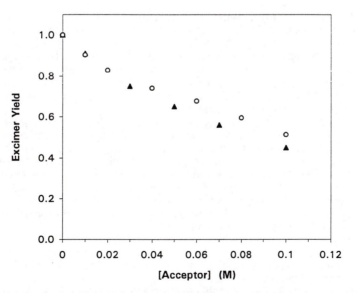

Figure 5. Static quenching of the excimer yield at the end of the laser pulse with increasing concentration of acceptors in two photon fluorescence: open circle: 4-bromobiphenyl (BrPh$_2$), black triangle: 1,4-dibromonaphthalene (DBN).

$$^1PS^* + BrPh_2 \xrightarrow{k_q} {}^1BrPh_2^* \tag{5}$$

The small difference between the DBN quenching curve and $BrPh_2$ curve might be explained by noticing a larger transfer radius from the phenyl excited state (S_1) to DBN ($R_0=18.1$ Å) than to $BrPh_2$ ($R_0=16.7$ Å). Little dynamic quenching of PS excimer was observed for either of the quenchers used, which is expected as the critical radii for energy transfer from PS excimer to $BrPh_2$ ($R_0=8.2$ Å) and to DBN ($R_0=13$ Å) are quite short.

Assuming that the precursor quenching of the PS excimer is due to a one step Forster energy transfer from an excited phenyl chromophore to a nearby $BrPh_2$, a calculation based on the Forster theory shows that even at a concentration of $BrPh_2$ as high as 0.1 M, the rate of energy transfer via dipole-dipole interaction is only $7.2 \times 10^7 s^{-1}$ on average. This leaves almost no chance for the very short lived phenyl excitation ($\tau \sim 20$ ps) to transfer its energy to $BrPh_2$ before it is trapped by the EFS. The fact that the precursor quenching process occurs within 20 ps suggests that the excitation on the phenyl chromophores should move around to give very fast trapping at EFS and competitive quenching by doped acceptors. At quencher concentration of $\sim$100 mM, the excimer yield is quenched to half of that in pure PS film. Using the homogeneous Stern-Volmer kinetics to describe the transport and trapping of excitation in polystyrene, the concentration of excimer forming sites is estimated to be around 100 mM, and the excitation migration rate is estimated to be 5.0×10^{11} $M^{-1}s^{-1}$ at 293 K and $0.5 \sim 1.0 \times 10^{11}$ $M^{-1}s^{-1}$ at 77K if the concentration of EFS is assumed to be the same at low temperatures.

2. Excitation and Charge Generation in Radiolysis of Polystyrene

Excited States of Polystyrene. The fluorescence emission of a PS film on irradiation with a pulsed electron beam is mainly excimeric. Monomeric emission around 285 nm is so weak that the measurement was interfered by the Cerenkov light induced by the fast electron beam passing through the film sample. The nonexponential decay of the excimer emission at 330 nm was fitted to a biexponential function, with a decay rate of 4.5×10^7 s^{-1} (lifetime $\sim$22 ns) for the slow component, in agreement with the photolysis measurements. A fast decay component is also present due to the excimer S_1-S_1 annihilation.

The transient absorption spectra of PS taken at the end of the pulse (t$\sim$5 ns) and 400 ns after the pulse are shown in Figure 6. The broad absorption band from 400 nm to 600 nm with a maximum around 535 nm is the well assigned excimer absorption. It also exhibits a nonexponential decay, which well matches the emission decay around 330 nm. The absorption below 400 nm was overlapped by the polystyrene triplet (lifetime $\sim$100 ns) (19), the biradicals from triplet reaction with neighboring phenyl chromophores (21), and the long-lived benzyl radicals due to the α-hydrogen abstraction from the main chain (22). The spectrum taken 400 ns after the pulse resembles that of the benzyl radical (23).

Addition of aromatic solutes leads to quenching of both the fluorescence and the transient absorption of the excimer. Figure 7A illustrates the typical decay traces of the excimer emission of PS films doped with different amounts of 1,4-dibromonaphthalene (DBN). Little dynamic quenching was observed as in photolysis. Static quenching of both emission and absorption of the PS excimer by DBN have the same dependence on concentration, which again confirms the assignment of the emission at 330 nm and the absorption at 535 nm to the PS excimer (Figure 7B). The quencher concentration at half quenching ($[Q]_{1/2}$) of PS excimer is

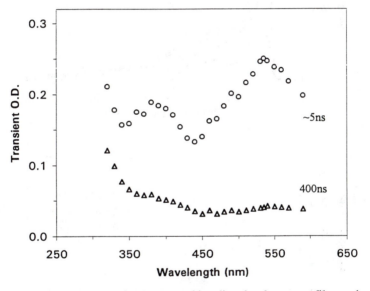

Figure 6. Transient absorption spectra of irradiated polystyrene film at the end of pulse (t~5 ns, cicles) and 400 ns after the pulse (triangles).

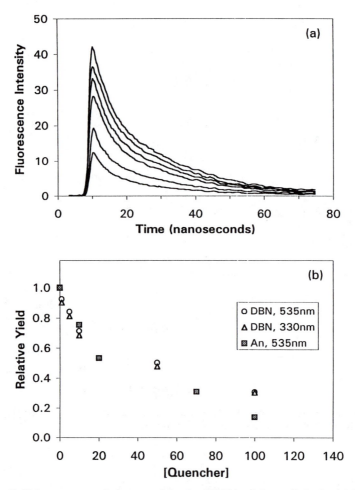

Figure 7. Polystyrene excimer quenching by DBN in pulse radiolysis: (a) Time-resolved fluorescence quenching measurement with increasing DBN concentration: (1) 0 mM, (2) 1 mM, (3) 5 mM, (4) 10 mM, (5) 50 mM, (6) 100 mM; (b) Static quenching pattern from both fluorescence (λ=330 nm) and transient absorption (λ=535 nm) measurements.

found to be ~40 mM, which is significantly less than the $[Q]_{1/2}$~100 mM in laser photolysis. More pronounced excimer quenching by anthracene at high concentrations is due to the dynamic quenching via the Forster energy transfer.

Energy Transfer to Solutes. Pyrene (Py) was used to probe the excited state formation and ion production in this work. In our experiments on Py/PS films, all the pyrene transient species, $^1Py^*$, $^3Py^*$, $Py^{+\cdot}$, and $Py^{-\cdot}$, were observed by transient absorption spectroscopy. Figure 8 shows the transient spectra of a pyrene doped PS film with [Py]=10 mM at ~10 ns and 2μs after the pulse in vacuum. The spectrum taken at t=2 μs clearly shows the long lived absorption bands due to $^3Py^*$ (420 nm, 520 nm), $Py^{+\cdot}$ (455 nm), and $Py^{-\cdot}$ (490 nm), after the fast decay of the PS excimer around 535 nm and that of $^1Py^*$ which absorbs strongly at 370nm and 470nm. The yields of all the transient species, measured in G values, increase with increasing pyrene concentration and show a trend of saturation at high concentrations (Figure 9). The growth pattern of the pyrene singlet excited state $^1Py^*$ vs concentration was also confirmed by fluorescence measurements. At the highest solute concentration [Py]=100 mM, $G(^1Py^*)$~0.6, $G(^3Py^*)$~1.0, $G(Py^{+\cdot})$~1.6, $G(Py^{-\cdot})$~0.8. It can be easily seen that more than 80% of the pyrene triplet state was produced independent of the pyrene S_1 state. Both the pyrene cation and the pyrene anion are trapped in the PS matrix and live up to hundreds of microseconds in vacuum.

The singlet excited states (S_1) and the triplet states (T_1) of other solutes were also produced in irradiated polystyrene films. The solute fluorescence measurements showed a similar yield-concentration growth pattern for anthracene, and DPA doped PS samples. The triplet states of benzophenone, naphthalene, anthracene, and bromopyrene were all observed to obey a similar growth pattern as that of the pyrene triplet. The anion radical of biphenyl (λ=410nm) and the cation radical of naphthalene (λ=690nm) were readily observed in irradiated polystyrene films containing these solutes.

Charge Scavenging. Absorption of electron scavengers, such as methylene chloride, chloroform, and 1,2-dichloroethane, into the pure PS film leads to a decreased excimer yield. Both singlet excited states and triplet states of various solutes are statically quenched by the adsorbed electron scavengers. These electron scavengers also decrease the pyrene anion radical yield and make it decay faster. The initial yield of the pyrene cation is little affected, but a dynamic quenching was observed due to the plasticization of PS by adsorbed scavengers. Vacuum pumped samples were also compared to samples exposed to air and oxygen. In the presence of oxygen, both $Py^{-\cdot}$ and $^3Py^*$ show only a slight static quenching but a marked dynamic quenching due to the following reactions.

$$O_2 + e^- \longrightarrow O_2^- \qquad (6)$$

$$O_2 + Py^- \longrightarrow O_2^- + Py \qquad (7)$$

$$O_2 + {}^3Py^* \longrightarrow O_2(^1\Delta) + Py \qquad (8)$$

Similar reactions as 6 and 7 also account for the significant quenching of excited states and anion radicals by the electron scavengers absorbed in polystyrene.

Polyvinylbenzyl chloride (PVBC) gives another example of electron scavenging. PVBC, which is structurally similar to polystyrene, has a very high

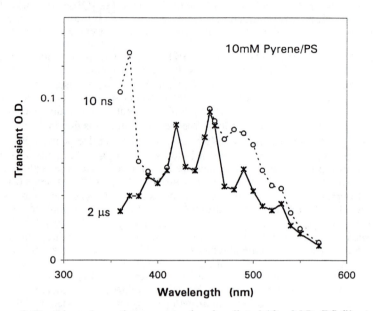

Figure 8. Transient absorption spectra of an irradiated 10 mM Py/PS film taken at 10 ns (broken line) and 2 µs after the pulse (solid line).

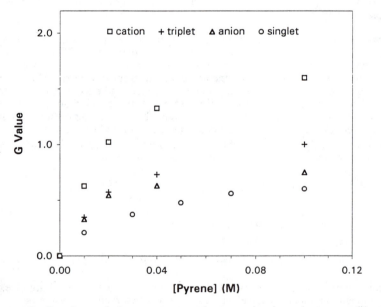

Figure 9. G values of the pyrene transient species in polystyrene films at different pyrene concentrations.

concentration (7 M) of chloride groups as the built-in electron scavenging centers. Irradiation of a pyrene doped PVBC film produces a simple transient absorption spectrum which can only be attributed to the pyrene cation radical (Figure 10). Neither ^{1}Py* nor ^{3}Py* was observed in transient absorption and fluorescence measurements, indicating that the direct excitation of solutes by subexcitation electrons is negligible. Nearly all of the excess electrons are trapped in the PVBC matrix due to the very efficient electron detachment reaction.

$$e^- + RCl \longrightarrow R \cdot + Cl^- \qquad (9)$$

The above electron scavenging studies suggest that excited states produced in polystyrene and other related polymers are mainly derived from ionic processes, and the excess electron is the negative precursor.

Ferrocene was used as a hole scavenger in pyrene doped polystyrene films due to its low ionization potential (6.81 eV) and its low reactivity with excess electrons. Figure 11 illustrates the static quenching effect of ferrocene on ^{1}Py*, ^{3}Py*, and Py$^+\cdot$, which results from its reaction with the positive precursor of these transient species, i.e. the positive hole in the polymer matrix.

$$PS^+(h^+) + Fc \longrightarrow Fc^+ \qquad (10)$$

Only a minimal decrease was observed for the pyrene anion radical with the increased doping of ferrocene.

Fast Kinetic Measurements. Fast kinetic measurements within the electron pulse were attempted to resolve the formation processes of the transients produced in PS. A reconvolution fit of the time resolved trace of polystyrene excimer absorption at 535 nm found that formation of the polystyrene excimer is not instantaneous but has a definite rate of ~1.0×10^9 s^{-1} at room temperature (Figure 12). The ~1 ns formation time is not comparable to the ~20 ps formation time in laser photolysis, indicating that recombination of the charged primary precursors is the rate limiting step in the excimer formation in radiolysis.

The formation rate of the solute singlet excited state S$_1$ was measured by monitoring the fluorescence of pyrene and DPA in irradiated PS films. For a 10 mM Py/PS sample, the ^{1}Py* was found to be formed within ~1.5 ns, while for pyrene concentration above 30 mM, ^{1}Py* is formed within the instrument resolution (~500 ps).

The broad absorption of the PS excimer around 535 nm interferes with transient absorption measurements on the pyrene anion radical at concentrations below 20 mM. Reliable measurements were made with pyrene concentrations larger than 20 mM, and Py$^-\cdot$ was found to be formed within the instrument resolution. This defines the lower limit of the rate constant of 1.0×10^{11} M^{-1}s^{-1} for the electron-pyrene reaction in polystyrene. Pyrene doped PVBC films were used to remove the interference from the polymer background absorption. The formation rate of the pyrene cation radical was measured to be 8.0×10^8 s^{-1} for 5 mM Py/PVBC and 1.0×10^9 s^{-1} for 10 mM Py/PVBC (Figure 13), suggesting that the positive charge transfer from the hole on the polymer matrix to pyrene proceeds with a concentration dependent rate.

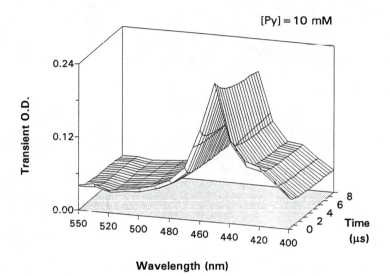

Figure 10. Transient absorption spectrum of an irradiated PVBC film doped with pyrene only shows the production of the pyrene cation radical.

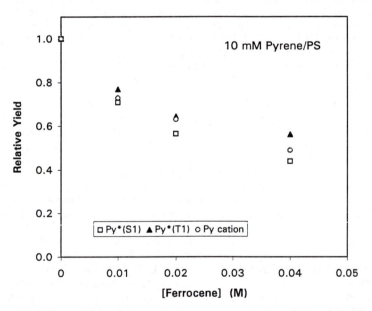

Figure 11. Effect of ferrocene as a hole scavenger on the yields of the pyrene transient species in 10 mM Py/PS films.

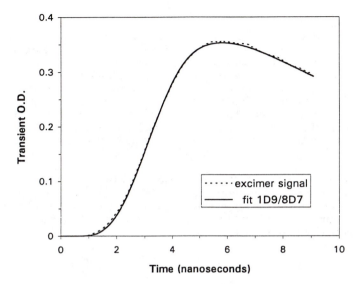

Figure 12. Measurement of PS excimer formation rate at $\lambda=535$ nm. The reconvolution fit gives the following rate parameters: formation rate $k_1=1.0\times10^9$ s^{-1}, decay rate $k_2=8.0\times10^7$ s^{-1} within the first 4 ns.

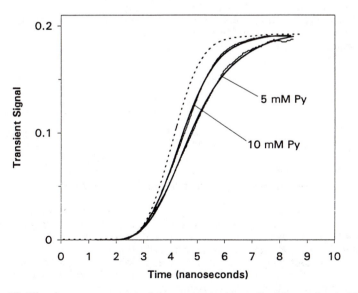

Figure 13. Kinetic measurements of the pyrene cation radical formation in PVBC films at different pyrene concentrations. Formation rate $k=8.0\times10^8$ s^{-1} for 5 mM Py/PVBC, and $k=1.0\times10^9$ s^{-1} for 10 mM Py/PVBC. The dash line shows the step response function.

Discussion

Ionic Precursors of Excited States. The excited states of polystyrene can be derived from either direct excitation including the internal conversion from some higher excited states or geminate recombination between the excess electron and the positive hole initially produced in the polymer matrix. The electron scavenging effects on the PS excimer, the static excimer quenching by aromatic dopants, and the concommitant formation of ion radicals and excited states of the doped aromatics clearly demonstrate that the majority of the polymer S_1 state is derived from ion neutralization. The ~1 ns time constant for such an electron-hole recombination suggests a high mobility of the excess electrons in polystyrene. If we assume that the polymer has the same negative electron affinity as its model compounds benzene and toluene (24), the excess electrons in polystyrene are only trapped to a shallow depth, which can be described as the quasi-free electrons in a two-state model. These quasi-free electrons have a mobility on the order of 10^{-4} cm^2/s in polystyrene (see Modeling and Simulation). The geminate electron-hole recombination within 1 ns results in formation of a phenyl S_1 state, further trapping of the phenyl excitation at EFS within ~20 ps leads to the PS excimer.

$$PS^+(h^+) + e^- \longrightarrow {}^1PS^*(S_1) \xrightarrow{\text{EFS}} {}^1PS_2^*(S_1) \qquad (11)$$

A separate study on the polyvinylnaphthalene triplet state shown that the monomeric T_1 state is also derived from such a geminate recombination. The mobile nature of the positive precursor, as will be discussed later in this paper, suggests that the positive hole is not in a dimeric form. The trapping of the mobile hole on dimeric sites leads to the formation of the PS dimer cation (9).

It has been well established that in benzene and toluene solutions, solute excited states are produced by diffusion controlled energy transfer from the solvent excited states (12).

$$^1M^* + S \longrightarrow M + {}^1S^* \qquad (12)$$

$$^1S^* \longrightarrow {}^3S^* \qquad \text{(intersystem crossing)} \qquad (13)$$

$$^3M^* + S \longrightarrow M + {}^3S^* \qquad (14)$$

Here, M stands for benzene or toluene molecules, while S stands for solutes. The bimolecular reaction rate for singlet energy transfer is on the order of $\sim 10^{10}$ M^{-1}s^{-1} (12).

Previous work suggested energy transfer as the major channel for solute excitation in polystyrene by making analogy to the molecular liquids of the model compounds (7). However, molecular diffusion is too slow to account for the fast formation of solute excited states in solid polystyrene. Excitation energy can be transferred to solutes by either the long range donor-acceptor interaction or the excitation migration among the donors over certain distance to acceptors. Laser photolysis studies of the excitation dynamics in polystyrene clearly shows that: 1) direct long range energy transfer from the excited phenyl (S_1) to nearby aromatic solutes via dipole-dipole interaction is negligible; 2) energy migration to solutes is not efficient due to the excitation trapping by ~100 mM EFS. Another source for the production of solute singlet excited states is the non-radiative energy transfer from the PS excimer to solutes, which leads to the observed dynamic quenching of the excimer over ~10 ns in anthracene doped PS. The formation of solute excited states from

~1.5ns to within ~400 ps in radiolysis of polystyrene films can not be explained by any of these non-ionic energy transfer modes.

For the formation of the excited states of a solute (S) via ion recombination, there exist several possible reaction pathways.

1) recombination between the solute cation and the excess electron:

$$S^+ + e^-_{quasi-free} \xrightarrow{\quad} {}^{1,3}S^*$$ (15)

2) recombination between the solute cation and the solute anion:

$$S^+ + S^- \xrightarrow{\quad} {}^{1,3}S^*$$ (16)

3) recombination between the positive hole and the solute anion:

$$PS^+(h^+) + S^- \xrightarrow{\quad} {}^{1,3}S^*$$ (17)

The excess electron in solute doped PS films, for example, 10 mM Py/PS, has a lifetime that is shorter than ~1 ns, due to the geminate electron-hole recombination and the electron scavenging by the solute. However, the solute cation radical was not readily formed until $t \sim 1$ ns when all excess electrons have disappeared in recombination and scavenging reactions. For reaction 15, there is a mismatch in the time domain between the solute cation and the excess electron. The benzophenone triplet formation gives another strong argument against the solute cation-electron recombination mechanism. Since benzophenone has a higher ionization potential (~9.10 eV) than the phenyl chromorphores (~8.70 eV for cumene), the benzophenone cation radical, even if generated in some way, will disappear by transferring the positive charge to polystyrene before reacting with the excess electron.

Numerical simulation shows that geminate recombination between S^+ and S^- via electron tunneling is very inefficient. As we have seen from the pyrene cation decay kinetics, reaction 16 proceeds over hundreds of microseconds via the slow diffusive motion of the ion radicals that are trapped in the polymer matrix. The same argument applies to reaction 17. The fast formation of solute excited states within 2 ns suggests that the positive charge (hole) in polystyrene must be moving among the monomeric sites until it reacts with a solute anion at encounter. The monotonic growth of solute excited states with solute concentration results from the increasing production of solute anions and the increased amount of the positive holes free from recombination with the excess electrons. The concentration dependent rate of formation of the S_1 states of pyrene and DPA indicates that the positive charge migration is the rate determine process.

Formation of Solute Ion Radicals. Just the same as in the radiation chemistry of molecular liquids, formation of anion radicals of aromatic solutes in polystyrene results from the scavenging of excess electrons by the solute molecules which are randomly distributed in the polystyrene matrix. The reaction rate between the quasi-free electrons and pyrene in PS is on the order of $\sim 10^{11}$ $M^{-1}s^{-1}$, based on the estimated mobility of the excess electrons in polystyrene $D_e \sim 10^{-4}$ cm^2/s from a numerical simulation of the electron-hole recombination kinetics. The observed growth pattern of solute anion yield vs solute concentration is typical of charge scavenging reactions: the larger the solute concentration, the higher the scavenging yield. The electron thermalization length was found to be ~56 Å from the simulation of the pyrene anion yield growth pattern (see Modeling and Simulation).

In contrast to the simple mechanism for the solute anion generation, the solute cation might be derived from two different types of processes: ionic and non-ionic.

1) positive charge transfer from the ionized polymer

$$PS^+(h^+) + S \longrightarrow S^+ \tag{18}$$

2) ionization by the superexcited states or the plasmon of the polymer

$$PS^{***} + S \longrightarrow S^+ + e^- \tag{19}$$

The probability of direct ionization of small concentrations of solute molecules by primary and secondary electrons is very low. Subexcitation electrons (energy below 3~4 eV) do not have enough energy to ionize the aromatic solutes used which have ionization potentials above ~6 eV.

Experiments were performed to simulate the ionization of solute molecules by the superexcited states via UV two photon ionization of perylene in polystyrene. Only a very small amount of perylene cations (Φ~0.01), after escaping the fast electron-cation geminate recombination, were detected as free ions on nanosecond timescale (25). The solute cation radicals with fairly high yields (G~1.6) are not generated by solute interaction with the short-lived, highly energetic non-ionic precursors, neither the superexcited states nor the plasmons of the polymer.

Therefore, solute cations are produced from reaction 18 between the positive hole initially produced in the polymer matrix and the solute molecules. Suppose that a solute cation is formed by a long distance electron transfer reaction between an immobile hole and a solute molecule. Fitting the pyrene cation yields in PS films using the Perrin model (26) gives a critical radius of 23.2 Å for such a charge transfer process. According to the electron tunneling theory (27), transfer of an electron from a solute to a positive hole takes at least 1 μs for a distance of 23.2 Å. The resolvable fast formation of the solute cation within ~2 ns does not favor the concept of much slower long distance electron tunneling. The movement of the positive charge (hole) is required here to account for the solute cation generation in polystyrene, the same as we found for the production of solute excited states.

$$(PS^+)_i + (PS)_j \longrightarrow (PS)_i + (PS^+)_j \tag{20}$$

Here, i and j indicate different sites in the polymer. Whether the hole is on the polymer backbone or on the phenyl groups is not known at this stage. More work are needed to elucidate the nature of the mobile hole in polystyrene and the molecular mechanism for positive charge migration.

Energy Deposition by Charge Transport. Based on all the experimental facts and the above arguments, a kinetic scheme is proposed to illustrate the energy deposition via different channels into the different products in polystyrene: the dimer cation of PS, the polymer excited states, solute excited states, solute anion radicals and solute cation radicals (Figure 14). It can be seen that hole transport plays a very important role in the excited state formation and the charge generation in polystyrene, following the initial action of high energy radiation.

$$PS \xrightarrow{e^-} PS^+ + e^- \qquad \text{(ionization of the polymer matrix)}$$

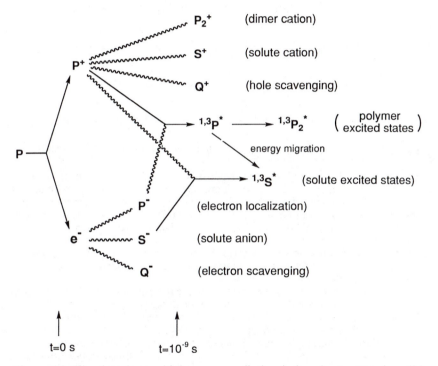

Figure 14. Kinetic scheme: high energy radiation induced processes in solid polystyrene (P: polymer, S: solutes, Q: scavengers). The time markers indicate the temporal evolution of the transient species and the chemical reactions from the very beginning of radiation action.

$$e^- + PS \longleftrightarrow PS^-$$
(quasi-free electrons in the polymer)

$$PS^+ + e^- \xrightarrow{\quad} {}^{1,3}PS^*$$
(electron-hole geminate recombination)

$${}^1PS^* \xrightarrow{\text{EFS}} {}^1PS_2^*$$
(excimer formation)

$$e^- + S \longrightarrow S^-$$
(electron scavenging)

$$(PS^+)_i + (PS)_j \longrightarrow (PS)_i + (PS^+)_j$$
(charge migration)

$$(PS^+)_j + PS_2 \longrightarrow PS_2^+$$
(dimer cation formation)

$$(PS^+)_j + S \longrightarrow S^+$$
(positive charge transfer to solute)

$$(PS^+)_j + S^- \xrightarrow{\quad} {}^{1,3}S^*$$
(geminate hole-anion recombination)

where, PS is polystyrene, PS_2 is a dimer site in the polymer, and S is a solute.

Positive holes and excess electrons are first produced in irradiated polystyrene. The excess electrons, which are shallowly trapped in the polymer matrix after thermalization and exist as quasi-free electrons, either recombine with geminate holes within ~1 ns to form the polymer excited states, or react with solute molecules leading to formation of solute anion radicals. The positive holes free from the recombination with the geminate electrons, migrate in the polymer matrix until they are trapped by the dimeric sites to form the dimer cations, or trapped by solutes to give solute cation radicals, or recombine with geminate solute anions from electron scavenging to produce solute excited states.

The electron scavengers used in the experiments compete with solute molecules for the excess electrons, leading to the decreased yields of the PS excimer, solute anions and solute excited states. Hole scavengers such as ferrocene decrease the yields of solute cations and solute excited states by reacting with the mobile positive holes. If a homogeneous bimolecular reaction is assumed for the production of solute cations via the positive charge migration, the rate constant is estimated to be $3\sim5\times10^{10}$ M^{-1}s^{-1}. By assuming a hole-anion encounter radius of 6 Å, the positive charge migration constant is estimated, from the Einstein-Smoluchowski relation, to be around 3×10^{-5} cm^2/s. Such a fast hole transport process has not been observed in liquid benzene and toluene.

Modeling and Simulation. As a good approximation, we neglect the interaction among the ion pairs (electron-hole pairs) which are initially produced in microzones (spurs) by the high energy radiation, and treat them as isolated single pairs in a statistical ensemble. This single pair model has been used successfully in radiation chemistry (28,29). For a single geminate ion pair in which the charge carriers with opposite charges diffuse in a dielectric media under the influence of mutual Columbic interaction, the movement of the geminate ions in the presence of charge scavengers can be adequately described by the Debye-Smoluchowski equation (29).

Mathematical derivation and algorithm for numerical simulation are given elsewhere (*25*).

Simulation of the electron scavenging by pyrene reproduces the total pyrene anion yield growth pattern with the increasing pyrene concentration (Figure 15A). Here we assume that: 1) the reaction of pyrene with the quasi-free electrons in polystyrene is diffusion controlled; 2) all the excited states of pyrene come from secondary recombination; and 3) the total ionization yield in solid polystyrene is $G_i=4.3$ (*28*). The yield of the electron scavenging by pyrene in PS is

$$G_{total}(Py^-) = G(Py^-) + G(^1Py^*) + G(^3Py^*). \qquad (21)$$

The simulation well fits the experimental data by using the following parameters: electron mobility $D_e=1\times10^{-4}$ cm^2/sec, and initial charge separation $b_G=50$ Å. The free ion yield in PS film is thus calculated $G_{fi}=0.12$, and the electron thermalization distance is $L=56$ Å. A simulation with the second order exponential function gives similar results $b_E=18$ Å, $L=54$ Å.

By using the scavenging curve obtained from the pyrene anion simulation, the PS excimer quenching by DBN was simulated by a consecutive two step model.

$$PS^+ + e^- \longrightarrow {}^1PS^* \xrightarrow{\text{EFS}} {}^1PS_2^*$$
$$\downarrow \qquad\quad \downarrow \qquad\qquad\qquad (22)$$
$$DBN^- \qquad {}^1DBN^*$$

where, DBN can react with both precursors of the PS excimer, i.e., the quasi-free electron and the monomeric S_1 state. It is reasonable to assume that DBN has the same reactivity with the quasi-free electrons and thus the same electron scavenging pattern as pyrene. After the deduction of the excimer quenching via singlet excitation transport to DBN, a quenching pattern was obtained which is due to the electron scavenging reaction of DBN (Figure 15B). A lower efficiency of electron scavenging suggest that not all the monomeric S_1 state are derived from electron-hole recombination. A portion of the monomeric S_1 state might be derived from either direct excitation of the phenyl chromophores or electron-hole pairs with very short separation which cannot be adequately described by the Gaussian distribution used. The experimental data is reproduced only by assuming that 30% of the phenyl S_1 state is unscavengable. Therefore, ~70% of the excited states are produced from geminate recombination of the electron-hole pairs which are separated on an average by 56 Å.

Conclusion

It is shown in the present work that singlet excitation transport is the key photophysical process in solid polystyrene films, which results in a very fast excitation trapping at excimer forming sites and energy transfer to solutes. The low efficiency of photon harvesting by polystyrene is due to the high concentration of EFS, [EFS]~0.1 M. The nonexponential kinetics and temperature dependence of excitation transport originate from the site inhomogeniety, which will be discussed in a separate work (*30*).

Excimer formation was also observed in pulse radiolysis of polystyrene films and was shown to be derived mainly from electron-hole geminate recombination and subsequent excitation trapping at EFS. The inclusion of aromatic solutes leads to the

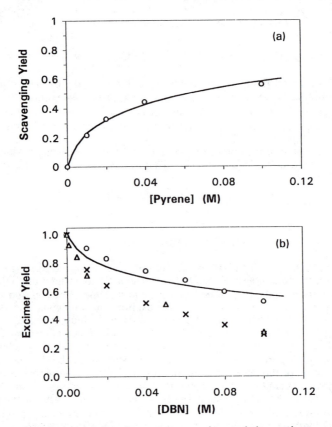

Figure 15. Simulation of the electron scavenging and the excimer quenching patterns observed in pulse radiolysis of polystyrene films: (a) fit pyrene anion yield by a Gaussian function with the initial charge separation $b_G = 50$ Å; (b) fit PS excimer quenching pattern, where the quenching of excimer via excitation trapping is shown in circles using laser photolysis data, the electron scavenging by DBN is given by the solid line, and the resultant quenching of excimer by the two step model (shown as crosses) well matches the experimental data (triangles).

static quenching of the PS excimer and the simultaneous formation of solute excited states, anion radicals and cation radicals. These solute transients are produced by reactions with the ionic precursors initially generated in polymer matrix by high energy radiation. The laser photolysis studies on the singlet excitation transport in polystyrene were successfully combined with the pulse radiolysis experiments to show that excitation transfer from the polymer to solutes is not the major channel for energy deposition. Fast positive charge transport was proposed to account for the observed solute excited state formation and the efficient generation of solute cation radicals. The migration constant of the mobile positive charge was estimated to be on the order of $\Lambda_h \sim 3 \times 10^{-5}$ cm^2/s. Numerical simulation of the polymer excited state formation and the solute anion production shows that the excess electrons have a lower mobility of $D_e \sim 1 \times 10^{-4}$ cm^2/s and a larger thermalization distance L=56 Å in solid polystyrene than in liquid benzene and toluene. About 70% of the polystyrene excimer is derived from such separated electron-hole pairs, while the other ~30% is from direct excitation and tightly bound geminate electron-hole pairs.

Comparative studies between laser photolysis and pulse radiolysis have shown that the much higher yield of charge generation in polystyrene (G_{ion}=2.4) than in its model compounds such as benzene and toluene (G_{ion}=0.1) results from a more efficient charge scavenging in the polymer due to: 1) a larger initial charge separation, i.e., 56 Å in polystyrene vs 42 Å in benzene (*28*); and 2) a fast positive charge transport to solutes in polystyrene, which has never been observed in aromatic liquids and solids.

Acknowledgment

We are grateful to the National Science Foundation for support of this work. G. Zhang would like to thank the Department of Chemistry and Biochemistry, University of Notre Dame, for awarding the Reilly Fellowship.

References

1. *The Effects of Radiation on High Technology Polymers* ; Reichmanis, E.; O'Donnell, J. H., Eds.; ACS Symposium Series 381; ACS: Washington, D.C., **1989**.
2. *Radiation Effects on Polymers* ; Clough, R. L.; Shalaby, S. W., Eds.; ACS Symposium Series 475; ACS: Washington, D.C., **1991**.
3. *Photophysics of Polymers* ; Hoyle, C. E.; Torkelson, J. M., Eds.; ACS Symposium Series 358; ACS: Washington, D.C., **1987**.
4. Guillet, J. *Polymer Photophysics and Photochemistry* ; Cambridge University Press: Cambridge, **1985**.
5. Rabek, J. F. *Mechanisms of Photophysical Processes and Photochemical Reactions in Polymers* ; John Wiley & Sons, Inc.: New York, **1991**.
6. Mezyk, S. P.; Yamamura, S.; Thomas, J. K. In *Radiation Effects on Polymers* ; Clough, R. L.; Shalaby, S. W., Eds.; ACS Symposium Series 475; ACS: Washington, D.C., **1991**, p.53.
7. Ho, S. K.; Siegal, S. *J. Phys. Chem.* **1967**, 71, 4527; Ho, S. K.; Siegal, S. *J. Chem. Phys.* **1969**, 50, 1142; Siegal. S.; Stewart, T. *J. Chem. Phys.* **1971**, 55, 1775.
8. Thomas, J. K. *J. Chem. Phys.* **1969**, 51, 770.
9. Tagawa, S. *Rad. Phys. Chem.* **1986**, 27, 455.
10. Zhang, G.; Thomas, J. K. *Polymer Preprints* **1994**, 35(2), 874.
11. Cooper, R.; Thomas, J. K. *J. Chem. Phys.* **1968**, 48, 5097.

12. Beck, G.; Thomas, J. K. *J. Phys. Chem.* **1972**, 76, 3856; Beck, G.; Ding, K.; Thomas, J. K. *J. Chem. Phys.* **1979**, 71, 2611.
13. Choi, H. T.; Hirayama, F.; Lipsky, S. *J. Phys. Chem.* **1984**, 88, 4246.
14. Thomas, J. K. *Chemistry of Excitation at Interfaces*; ACS monograph 118, Washington, D.C., **1984**.
15. Forster, Th. *Ann. Phys.* **1948**, 2, 55; *Discuss. Farad. Soc.* **1959**, 27, 7.
16. Hochstrasser, R. M.; Jutz, H.; Scott, G. W. *Chem. Phys. Letters* **1974**, 24, 162.
17. Bevington, P. R. *Data Reduction and Error Analysis for the Physical Sciences*; 2nd ed.; McGraw-Hill: New York, **1992**.
18. Coulter, D. R.; Gupta, A.; Yavrouian, A.; Scott, G. W.; O'Connor, D. B.; Vogl, O.; Li, S.-C. *Macromolecules* **1986**, 19, 1227; O'Connor, D. B.; Scott, G. W.; Coulter, D. R.; Yavrouian, A. *J. Phys. Chem.* **1991**, 95, 10252.
19. Tagawa, S.; Nakashima, N.; Yoshihara, K. *Macromolecules* **1984**, 17, 1167.
20. Miyasaka, H.; Fumihiko, I.; Mataga, N. *J. Phys. Chem.* **1988**, 92, 249; *Chem. Phys. Letters* **1993**, 202, 419.
21. Cooper, R; Thomas, J. K. *J. Chem. Phys.* **1968**, 48, 5097; Bensasson, R. V.; Richards, J. T.; Thomas, J. K. *Chem. Phys. Letters* **1971**, 9, 13.
22. Harah, L. A. *Molecular Crystals* **1969**, 9, 197.
23. Sehested, K.; Corfitzen, H.; Christensen, H. C.; Hart, E. J. *J. Phys. Chem.* **1975**, 79, 310.
24. Itoh, K.; Holroyd, R. *J. Phys. Chem.* **1990**, 94, 8850; *J. Phys. Chem.* **1990**, 94, 8854.
25. Zhang, G.; Thomas, J. K. *Polymer Preprints* **1994**, 35(2), 973; Zhang, G.; Thomas, J. K. In *Radiation Effects of Polymers: Chemical and Technological Aspects*; Shalaby, S. W.; Clough, R. L., Eds.; ACS Symposium Series, in press.
26. Birks, J. B. *Photophysics of Aromatic Molecules*; Wiley-Interscience: New York, **1970**, p.441.
27. Tachiya, M.; Mozumder, A. *Chem. Phys. Letters* **1974**, 28, 87; Tachiya, M.; Mozumder, A. *Chem. Phys. Letters* **1975**, 34, 77; Murata, S.; Tachiya, M. *Chem. Phys. Letters* **1992**, 194, 347.
28. Freeman, G. R. In *Kinetics of Nonhomogeneous Processes*; Freeman, G. R., Ed.; John Wiley & Sons, Inc.: New York, **1987**; Chapter 2, p.19.
29. Noolandi, J. In *Kinetics of Nonhomogeneous Processes*; Freeman, G. R., Ed.; John Wiley & Sons, Inc.: New York, **1987**; Chapter 9, p.465.
30. Zhang, G.; Thomas, J. K. *J. Phys. Chem.* **1995**, in press.

RECEIVED July 27, 1995

Chapter 4

Photoionization in Polymer Films

Guohong Zhang and J. K. Thomas[1]

Department of Chemistry and Biochemistry, University of Notre Dame, Notre Dame, IN 46556

Cation radicals of aromatic solutes have been produced in polymer films by UV two photon ionization. The yield of charge separation and the mechanism of charge neutralization were studied via transient absorption spectroscopy. The thermalization length of the photoejected electrons from perylene was measured to be L~36 Å in polystyrene. Trapping of the excess electrons by polymer matrices competes with the geminate electron-cation recombination within 1 ns, which leads to the observed charge generation. Among the polymers studied, polystyrene has the lowest reactivity with excess electrons and therefore the lowest yield of charge separation $\psi_{PS}=1.4\%$, while poly(vinylbenzyl chloride) scavenges nearly all the excess electrons and shows the highest charge separation yield $\psi_{PVBC}=1$. Subsequent recombination of charge carriers trapped in polymer matrices over a wide dynamic range is diffusion limited, which gradually evolves from geminate nature to homogeneous second order reaction. The mobilities of charge carriers in different polymers were measured in this work.

Photogeneration of charge carriers in polymers has been studied for the potential applications in developing polymer based photoconductive and Xerographic materials (1). The charge separation is obtained at low energy photon excitation by electron transfer from photoexcited donors to strong electron acceptors doped into a polymer matrix. High electric field is often used to help separate the product ion pairs. The efficiency of photogeneration is normally very low, with the quantum yields around $10^{-3}\sim10^{-2}$. Under the influence of external field, charge transport results from the repetitive electron transfer from neutral molecules to cation radicals for p-type conduction, and from anion radicals to neutral molecules for n-type conduction. Recent work in this laboratory has shown that efficient charge generation can be achieved in nonpolar polymers like polystyrene without any electron acceptor upon high energy radiation (2). Solute cation radicals have been produced with a fairly large yield (G~1.6 per 100eV energy loss); and very fast positive hole transport ($\Lambda_h\sim3\times10^{-5}$ cm^2/s) was proposed to be the major mechanism for efficient charge generation. The significant difference in ion production seems to indicate an excitation energy dependence of charge generation in

[1]Corresponding author

0097–6156/96/0620–0055$12.00/0

polymeric materials. It is the purpose of this work to examine the effects of medium energy radiation (hv=5~10 eV) on solid polymers.

Photoionization of molecules by dumping photon energy up to 7~10 eV in the liquid phase has been well documented (3,4). Different distribution functions were used to describe the initial position of the thermalized electron around the parent cation. Diffusion controlled geminate kinetics leads to neutralization of the primary charge carriers and the formation of free ions (5, 6). Both pulse radiolysis (7) and laser photolysis (3,4) have shown that the recombination of geminate electron-cation pairs in nonpolar liquids occurs on the picosecond timescale, and is attributed to the high mobility of the excess electrons under the influence of the Columbic interaction between the geminate electron and cation. The diffusivity of the shallowly trapped electrons in hydrocarbon liquids are normally on the order of 10^{-3} cm^2/s. In contrast, processes in polymers induced by radiation with energy between the high and the low extremes are not well understood. Many questions remain unanswered about charge generation, trapping and transport in polymers: Can we photoionize solute molecules in polymer matrices just like in liquid solutions? How is the initial charge separation? Are polymer matrices capable of trapping excess electrons? Could rigid polymers help separate and stabilize the charged species?

Recent work by Yamamoto and coworkers (8) observed that thermalized electrons produced by photoionization of aromatic solutes are well trapped in a bulk polymerized poly(methyl methacrylate) (PMMA) matrix. Long range electron tunneling from the trapping sites back to the parent cations was suggested to explain the very slow decay of the ion radicals in hours. Temperature dependent recombination rate was attributed to the chain relaxation effect on the electron transfer reaction in the PMMA matrix.

As part of the comparative studies between low energy photochemistry and high energy radiation chemistry of polymers, laser photolysis with two UV photons was performed to simulate the high energy events in polymers. Photoionization of solute molecules and subsequent electron trapping and charge recombination kinetics in different polymers were examined both experimentally and theoretically in the present work.

Experimental

Chemicals and Instruments. Polymers used in this work are polystyrene (PS, M.W.~280,000), poly(vinylbenzyl chloride) (PVBC), poly(benzyl methacrylate) (PBM), poly(methyl methacrylate) (PMMA, M.W.~25,000), and polycarbonate (PC) from Aldrich. Films of PS, PVBC, PBM, and PMMA with thicknesses around 200 μm were cast from their benzene solutions with specific amount of perylene. PC films were cast from a 1,2-dichloroethane (DCE) solution of the polymer with added perylene. The concentration of perylene in the polymer films were controlled to be around 5 mM. A Varian Cary 3 UV-Vis Spectrophotometer was used to check the perylene absorbance at 337 nm in the film samples for charge generation yield measurements. The 337 nm light from a Laser Photonics nitrogen gas laser (Model UV-24) with output energy of ~5 mJ per pulse and pulse width of ~10 ns was used to excite perylene in the polymer films. The formation of the perylene cation radical was monitored by the conventional transient absorption spectroscopy. Transient absorption spectra were collected by averaging 4~8 shots at each wavelength. Kinetic measurements were taken over a wide dynamic range, from 10^{-7} second to 10 seconds. Transient signals within 10 ms were digitized by a computer interfaced Tektronix 7912AD programmable digitizer, while signals on longer timescales were captured by Tektronix C1001 Video Camera and then digitized by the computer.

Theories and Simulation. For a quantitative understanding of the charge separation and recombination processes, the experimental measurements were simulated by available models for different mechanism: 1) diffusion controlled geminate recombination, as suggested by Noolandi and Hong (6); and 2) geminate recombination via electron tunneling, as described by Tachiya and Mozumder (9).

Geminate Recombination via Diffusion. The initial photoionization produces geminate electron-cation pairs well isolated from each other. We neglect the interaction among the ion pairs and treat them as isolated single pairs in a statistical ensemble. This single pair model has been used successfully in radiation chemistry (5, 6). For a single geminate ion pair in which the charge carriers with opposite charges diffuse in a dielectric media under the influence of mutual Columbic interaction, the movement of geminate ions can be adequately described by the Debye-Smoluchowski equation (6). In a spherically symmetrical system, the equation is written as follows.

$$\frac{\partial \rho}{\partial t} = D \left\{ \frac{\partial^2 \rho}{\partial^2 r} + \left(\frac{2}{r} + \frac{r_c}{r^2} \right) \frac{\partial \rho}{\partial r} \right\} \tag{1}$$

Distribution function $\rho(r,t)$ is the probability density to find the geminate ions separated by distance r at time t. In equation 1, D is the mutual diffusion constant of geminate ions, and r_c is the Onsager radius $r_c = \dfrac{e^2}{4\pi\varepsilon_0 \varepsilon k_B T}$.

The pair survival probability at time t is P(t).

$$P(t) = \int_0^\infty \rho(r,t) 4\pi r^2 dr \tag{2}$$

The distribution function at t=0 is defined in the initial condition.

$$\rho(r,0) = f(r) \tag{3}$$

For an ionization induced electron-cation geminate pair, different forms of f(r) have been used in the early work of radiation chemistry, including the exponential and the Gaussian functions. In the present work, a Gaussian function is used to describe the distribution of thermalized electrons.

$$f_G(r) \sim \exp(-\frac{r^2}{b_G^2}) \tag{4}$$

where, b_G is the distribution parameter which determines the most probable position of a thermalized electron around its parent cation.

Reaction between geminate ions at encounter is specified in the boundary condition at the reaction radius. For a diffusion controlled charge recombination, the probability of reaction at encounter is assumed to be unity.

$$\rho(r, t) = 0 \qquad\qquad \text{for } r \le R_r \tag{5}$$

where, R_r is the encounter radius for the recombination reaction.

Boundary condition at infinity is obvious.

$$\rho(r, t) = 0 \qquad\qquad \text{for } r \to \infty \qquad\qquad (6)$$

In the presence of charge scavengers, the scavenging reaction term was added to equation 1.

$$\frac{\partial \rho}{\partial t} = D\left\{ \frac{\partial^2 \rho}{\partial^2 r} + \left(\frac{2}{r} + \frac{r_c}{r^2} \right) \frac{\partial \rho}{\partial r} \right\} - k_s C_s \rho \qquad\qquad (7)$$

The bimolecular rate constant for a diffusion controlled scavenging reaction is given by the Smoluchowski relation.

$$k_s = 4\pi N_a R_s D \qquad\qquad (8)$$

where, R_s is the scavenging radius (cm), D is the mutual diffusion constant (cm^2/s), and $N_a=6.02 \times 10^{20}$.

Charge scavenging yield at different concentration was expressed by the following equation.

$$Y_s = \int_0^\infty dt \int_0^\infty dr \, 4\pi r^2 \left\{ -\left(\frac{\partial \rho}{\partial t} \right)_s \right\} = \int_0^\infty dt \, k_s \, C_s \, P(t) \qquad\qquad (9)$$

Using the relationship between the survival probability functions with and without a scavenger, i.e.,

$$P(t) = P_0(t) \exp\left\{ -k_s C_s t \right\} \qquad\qquad (10)$$

equation 9 can be rewritten as follows.

$$Y_s = C_s \int_0^\infty dt \, k(t) P_0(t) \exp\left\{ -k_s C_s t \right\} \qquad\qquad (11)$$

where, P(t) is the survival probability function at scavenger concentration C_s, and $P_0(t)$ is the survival probability function without any scavenger.

The forward difference method was used to numerically calculate $\rho(r,t)$ and P(t) at successive time (10). The space grid Δr and time step Δt are selected for the precision and the stability of calculation. The calculated survival probability P(t) is compared with the experimentally measured transient kinetics. Information about the initial charge distribution $f_G(r)$ and the diffusivities of charge carriers D in polymers are obtained from the simulation of ion recombination kinetics and charge scavenging patterns.

Geminate Recombination via Electron Tunneling. When charge carriers are trapped in a rigid environment, the mobilities of the charge carriers are negligibly small, geminate ion pairs recombine only via long distance electron transfer reaction. The survival probability of such trapped ion pairs is given by Tachiya (9).

$$P(t) = \int_0^\infty \exp\{-k(r)t\}\, f_G(r)\, 4\pi r^2\, dr \qquad (12)$$

k(r) is the rate of the electron transfer between an ion pair separated by distance r.

$$k(r) = v \exp(-r/a) \qquad (13)$$

where, v is the transfer frequency, and a is the overlap radius, and the initial distribution of ion pairs is defined by $f_G(r)$. It can be shown that a linear relationship exists between $\ln P(t)$ and $\ln(vt)$ over a broad range of timescales, which is characteristic of the geminate ion recombination via electron tunneling.

$$P(t) \propto (vt)^{-a/b_G} \qquad (14)$$

Comparison between the above two different kinetics over a wide dynamic range helps clarify the charge recombination mechanism in solid polymers.

Results and Discussion

Pyrene and perylene, which are readily ionized by absorption of two UV photons at 337 nm, were used as solute molecules in our experiments. The strong transient absorption by the pyrene cation radical at 455 nm and by the perylene cation radical at 540 nm were monitored to examine the yield of charge generation and the decay kinetics of the charge carriers. Perylene was especially favored due to: 1) its short-lived S_1 state and its low yield of triplet from intersystem crossing $\Phi = 0.02\sim0.05$; and 2) the well separated perylene triplet and perylene cation radical absorption spectra. Figure 1A shows the absorption spectra of perylene transients produced in a 1,2-dichloroethane (DCE) solution with a high laser intensity. DCE was used as a solvent to capture the photoejected electrons via the electron detachment reaction. The spectrum measured at 100 ns after the laser pulse in oxygenated 5 mM/DCE solution is purely due to the perylene cation radical, which has an absorption maximum at 545 nm. The spectrum measured under the identical excitation conditions but in a nitrogen bubbled DCE solution exhibits an additional weak absorption band on the blue side of the cation radical spectrum. After subtracting the cation radical spectrum, the difference spectrum matches well with the T_1-T_n absorption spectrum of the perylene triplet by comparing with the absorption spectrum of the perylene triplets produced in 5 mM perylene/benzene solution via benzophenone (20 mM) sensitization (Figure 1B). Obviously, there is no triplet absorption underneath the perylene cation radical absorption above 540 nm.

Perylene was ionized in polymer films by a resonant two photon process with absorbed energy added up to ~6.5 eV.

$$Pe \xrightarrow{h\nu} Pe^*(S_1) \xrightarrow{h\nu} Pe^+ + e^- \xrightarrow{\text{trapping}} Pe^+ + e_{tr}^- \qquad (15)$$

The second photon is absorbed by the S_1 state of perylene, which leads to electron ejection from perylene. Figure 2 illustrates the transient absorption spectrum of a 5 mM perylene doped PBM film measured at 20 ns after the laser pulse. Formation of the perylene cation was clearly seen, with a small amount of triplet also produced. Similar ionization and cation formation processes were observed for pyrene. A

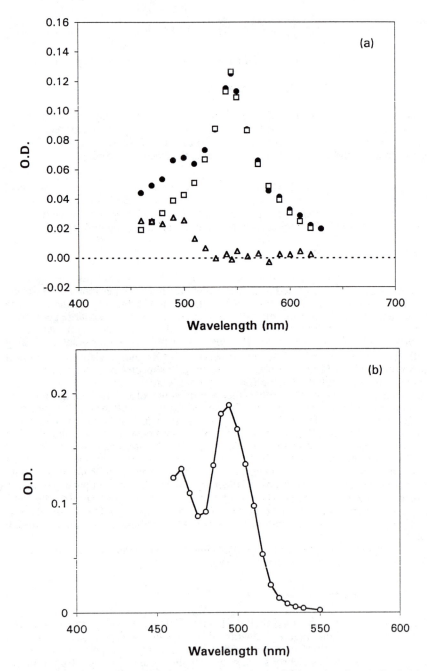

Figure 1. Transient absorption spectra of perylene triplet and cation radical. (a) Spectra in a 5 mM Pe/DCE solution: oxygenated (square), nitrogen bubbled (dot), and difference spectrum (triangle); (b) Benzophenone sensitized perylene triplet in a 5 mM Pe/benzene solution.

decay trace of the pyrene cation radical produced in a vacuum pumped PVBC film was shown in Figure 3. The monitoring wavelength is 455 nm where the pyrene triplet ($\Phi\sim0.3$) still has a significant absorption. The contribution of the pyrene triplet to the observed signal at 455 nm is also illustrated in a broken line in Figure 3. This spectral overlap makes it impossible to detect the low ion yield and analyze the ion recombination kinetics with pyrene as a solute. Most of our experiments were performed with perylene as the ionization probe.

Measurements of Charge Separation Yield. In order to compare the electron trapping in different polymers, charge separation yields were abstracted from the transient absorption measurements of each polymer on the basis of a quantitative description of two photon ionization processes.

The following coupled equations can be written for a transparent film sample of thickness d, containing a certain concentration (C_0) of solute which absorbs at the excitation wavelength with an extinction coefficient ε_0.

$$-\frac{\partial I(x,t)}{\partial x} = I(x,t)\left[C_0\varepsilon_0 + C_1(x,t)\varepsilon_1\right] \tag{16}$$

$$-\frac{\partial C_1(x,t)}{\partial t} = I(x,t)C_0\varepsilon_0/N \tag{17}$$

The laser pulse passing through the film is a function of depth and time, $I(x,t)$. $C_1(x,t)$ is the concentration of the solute intermediate, ε_1 is its extinction coefficient at the same excitation wavelength, and N is Avogadro number. The formation of the ionizing state is determined by the absorption of the second photon by the solute intermediate. Assume that the ionization yield of the ionizing state is 1, and that a fraction (ψ) of initially generated ion pairs escape the very fast electron-cation geminate recombination, leading to the observed charge separation. The end-of-pulse transient absorption of solute cation radicals at a specific wavelength with a concentration $C_2(x,t)$ and an extinction coefficient ε_2 at the monitoring wavelength is given by the Lambert-Beer's law.

$$A = \int_{-\infty}^{\infty}dt\int_0^d dx\, C_2(x,t)\varepsilon_2 \tag{18}$$

where, $C_2(x,t)$ is defined by the following equation.

$$-\frac{\partial C_2(x,t)}{\partial t} = \psi\, I(x,t)C_1(x,t)\varepsilon_1/N \tag{19}$$

The laser intensity inside the sample can be solved from equations 16 and 17.

$$I(x,t) = I_0(t)\exp\left\{-C_0\varepsilon_0 x\left[1 + \frac{\varepsilon_1 I_0\delta t}{N}\Phi(\frac{t-t_0}{\delta t})\right]\right\} \tag{20}$$

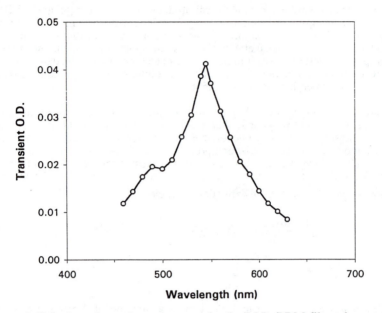

Figure 2. Transient absorption spectrum of a 5 mM Pe/PBM film taken at the end of the pulse (t=20 ns).

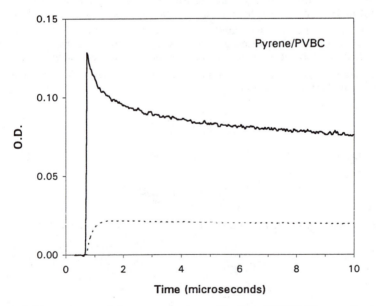

Figure 3. Decay trace of the pyrene cation radical in a PVBC film monitored at 455 nm. Pyrene triplet also contributes to the measured signal (broken line).

where, $I_0(t) = I(0,t) = \pi^{-1/2} I_0 \exp\left\{-\left(\dfrac{t-t_0}{\delta t}\right)^2\right\}$ is the laser pulse before entering

the sample, $\Phi(\dfrac{t-t_0}{\delta t}) = \pi^{-1/2} \int_{-\infty}^{t} \exp\left\{-\left(\dfrac{t-t_0}{\delta t}\right)^2\right\} dt$ is the integration of the

excitation pulse, and δt is the pulse width.

For the excitation conditions used in our experiments $\dfrac{\varepsilon_1 I_0 \delta t}{N} \ll 1$, $I(x,t)$ can

be simply written as follows.

$$I(x,t) = I_0(t)\exp\{-C_0\varepsilon_0 x\} \tag{21}$$

Integration of equation 18 gives the cation radical absorption, which was measured 20 ns after the laser pulse in the experiments.

$$A = \psi\left(\frac{I_0\delta t}{N}\right)^2 \varepsilon_1\varepsilon_2 \left[\frac{1-\exp(-2C_0\varepsilon_0 d)}{2}\right] \lim_{t\to\infty} \Phi(\frac{t-t_0}{\delta t}) \tag{22}$$

For the thin films used in our experiments, due to the low extinction coefficient of perylene at 337 nm ($\varepsilon \sim 1{,}370$ $M^{-1}cm^{-1}$), the cation radical signal is proportional to the perylene absorbance at 337 nm and the quantum yield of charge separation ψ in the polymer sample under measurement.

$$A \propto \psi(C_0\varepsilon_0 d) \tag{23}$$

The yield of charge separation was set to 1 in PVBC films due to complete scavenging of the excess electrons, i.e., $\psi_{PVBC}=1$. The charge separation yields in other polymers were measured relative to PVBC.

Electron Trapping in Polymers. The perylene cation radical was produced with different yields in the polymers used. The highest yield of the perylene cation, i.e., the highest charge separation efficiency, was observed in PVBC, while the lowest was measured in PS, with PC, PMMA, and PBM in between. This variation can be understood from the reactivity of polymers with the excess electrons. The quantum yield of charge separation ψ is determined by the competition between the geminate electron-cation recombination and the electron scavenging by polymer matrices.

Polystyrene. Low yield of the perylene cation $\psi_{PS}=1.4\%$ was observed in polystyrene, since polystyrene does not react with excess electrons, the same as its model compounds benzene and toluene. Previous work on liquid benzene and toluene (*11*) found that electron attachment to the benzene ring is not favored unless under very high pressures, which force the localization of excess electrons by formimg the benzene and toluene anions. Under normal conditions, excess electrons are only trapped to a very shallow depth in benzene and toluene with a diffusivity on the order of 10^{-3} cm^2/s (*11*). The same situation is expected for excess electrons in polystyrene. High energy radiation experiments have shown that the mobility of the quasi-free electrons in polystyrene is on the order of 10^{-4} cm^2/s (2). Geminate recombination between the perylene cation and the shallowly trapped excess electron occurs rapidly within ~1ns. The low ion yield observed in polystyrene is due to the fact that only a very small fraction of charge carriers escape the early recombination as free ions. Electron scavenging experiments were performed to help separate the geminate ion pairs. Different amounts of 2,4,6-trimethylbenzyl chloride (TMBC) were doped into polystyrene with perylene. The transient absorption signals monitored at 550 nm are shown in Figure 4A. The growth of the perylene cation with increasing TMBC concentration was simulated by the diffusion controlled geminate recombination model (Figure 4B). The parameter for the initial Gaussian distribution function was obtained by fitting the scavenging pattern, b_G~32 Å. The electrons are ejected 36 Å on average away from the geminate perylene cation radicals in two photon ionization at 337 nm, which well compares with the results of photoionization in liquid hydrocarbons (3,4). The ion yield in a 5 mM perylene/PS film is predicted to be ~1.2% from the above simulation, in agreement with 1.4% measured relative to the ion yield ψ_{PVBC}=1 in PVBC films.

Polyesters. Yields of charge separation in polyesters are ψ_{PC}=0.46 for polycarbonate, ψ_{PMMA}=0.32 for poly(methyl methacrylate), and ψ_{PBM}=0.14 for poly(benzyl methacrylate). The ester group in the monomeric units of PC, PMMA, and PBM has very high concentrations (6~8 M) in solid polymer films, and its reaction with the excess electrons leads to the formation of polymer anions and efficient charge separation. Electrons trapped in PMMA matrix have been studied using EPR spectroscopy by Tabata (12) and Yamamoto (13). A strong singlet peak was observed at the center of the EPR spectrum of irradiated PMMA at 77 K, which was assigned to an ester radical anion of PMMA ($PMMA \cdot^-$).

$$e^- + PMMA \longrightarrow PMMA \cdot^- \tag{24}$$

At 77 K, the $PMMA \cdot^-$ anion is not stable and thermal fragmentation results in side chain scission and leads to the formation of $P-COO^-$, and methyl radicals $\cdot CH_3$. At room temperature, the decarboxylation of $P-COO^-$ leads to the formation of $\cdot CO_2^-$ anion radical and main chain radical P$\cdot$ which further induces main chain scission (12).

$$PMMA \cdot^- \longrightarrow P-COO^- + \cdot CH_3 \longrightarrow P \cdot + \cdot CO_2^- + \cdot CH_3 \tag{25}$$

Similar processes occur in PBM after electron trapping by the ester groups. Since the reactivity of these polymers with electrons is lower than that of PVBC, more electrons recombine with the parent cations before being trapped by the polymers. The higher yield of charge generation observed in PMMA than in PBM might be attributed to the higher concentration of ester groups and the shorter Onsager length in PMMA (Table I).

PVBC. PVBC has a high concentration of chloride groups (7 M) which capture excess electrons efficiently via an electron detachment reaction. This corresponds to the high concentration limit of the electron scavenging in polystyrene by TMBC. Electrons trapped in PVBC matrix exist as chloride anion Cl^-.

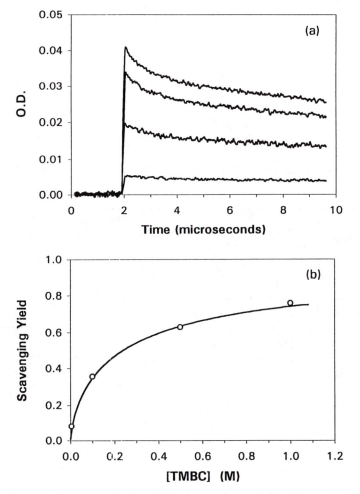

Figure 4. Electron scavenging by 2,4,6-trimethylbenzyl chloride in polystyrene. (a) Observation of increased signal level at 550 nm with increasing perylene concentration, [Pe]=0 M, 0.1 M, 0.5 M, and 1 M; (b) Growth of the perylene cation yield fitted to a geminate ion pair model gives b_G=32 Å.

Table I Charge Separation and Transport in Different Polymers

Polymers	Onsager Length (Å)	Yield of Charge Separation Ψ	Mobilities of Charge Carriers (10^{-12} cm^2/s)	Molecular Diffusion (10^{-13} cm^2/s)
PS	220	0.014		1.1
PMMA	155	0.32	7.0	8.8
PBM	174	0.14	220	46
PVBC	139	1	750	68

$$e^- + \rangle CHC_6H_4CH_2Cl \longrightarrow \rangle CHC_6H_4CH_2 \cdot + Cl^- \tag{26}$$

Nearly 100% charge separation in PVBC ($\psi_{PVBC}=1$) was confirmed by the electron scavenging experiments in polystyrene.

Charge Recombination: Mechanism and Kinetics. The effects of rigidity and temperature were examined in order to determine whether the recombination is diffusion controlled or it occurs via long distance electron tunneling. A dry perylene/PBM film was compared with a wet perylene/PBM film with ~5% toluene residue left. A faster cation decay was observed in the wet film than in the dry PBM film (Figure 5A), suggesting a mass movement limited reaction, i.e., diffusion controlled. Measurements on the cation decay kinetics were made at 240 K in comparison with the room temperature observation in perylene doped PVBC films (Figure 5B). With only a 50°C drop of temperature, ion recombination in the polymer is significantly slowed down, which is again indicative of a rigidity effect on a diffusion limited reaction.

Kinetic measurements over a wide range of time were made on different polymers. Figure 6 shows the semi-log plots of the perylene cation radical decay in PBM and PVBC with the timescale variation over seven orders of magnitude. The decay pattern generated by a computer simulation using the diffusion controlled geminate recombination model was given in Figure 7. The semi-log plot was drawn in a reduced timescale where $t_c = r_c^2/4D$. The decay pattern does not depend on how fast charge carriers move, whether in liquid solutions or in solid polymers. It is only a function of the initial charge separation $f_G(r)$ and the Onsager length r_c. The observed perylene cation decay curves agree with the theoretical pattern, which confirms that the charge recombination in solid polymers is diffusion controlled.

However, on a short timescale ($10^{-8} \sim 10^{-6}$ s) when there is no extensive molecular movement and charge carriers are relatively frozen, the cation radical decay can not be well described by the diffusion model. Trapped charges at short separation might recombine via electron tunneling. Ferrocene (Fc) doped perylene/PVBC films were examined for possible electron tunneling reactions. Figure 8 demonstrates the quenching of the perylene cation by ferrocene due to the following electron transfer reaction.

$$Pe^+ + Fc \longrightarrow Pe + Fc^+ \tag{27}$$

It was shown by the molecular mobility measurements (see below) that there is no mass diffusion involved on the microsecond timescale. The dynamic quenching of Pe^+ by ferrocene can only be explained by reaction 27. Pulse radiolysis experiments in PS and PVBC films have observed a similar pattern of the pyrene cation quenching by ferrocene within 10μs (*14*).

When the time window is shifted to the middle range of observation ($10^{-6} \sim 10^{-3}$ s), a linear relation of the semi-log plot exists for all the polymers studied (Figure 7), which is characteristic of the diffusion controlled geminate ion recombination kinetics (*15*).

$$\frac{C(t)}{C_0} = \text{cons} \tan t - k \log t \tag{28}$$

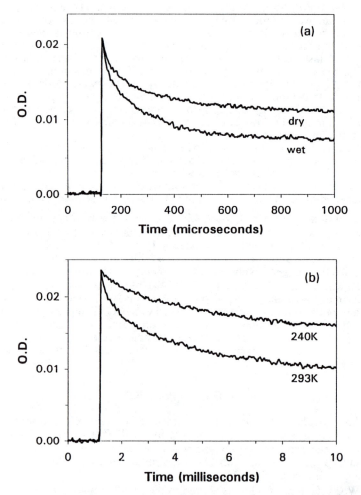

Figure 5. Rigidity and temperature effects on the decay kinetics of perylene cation radical (λ=550 nm): (a) in PBM, dry film vs film with 5% toluene residue; (b) in PVBC, 293 K vs 240 K.

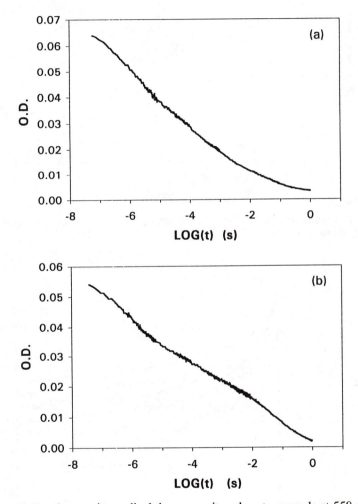

Figure 6. Perylene cation radical decay monitored up to seconds at 550 nm. (a) in PBM; (b) in PVBC.

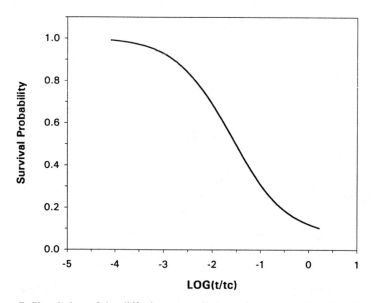

Figure 7. Simulation of the diffusion controlled geminate recombination kinetics. Parameters used for calculation: r_c=220 Å, b_G=50 Å, and R_r=4 Å.

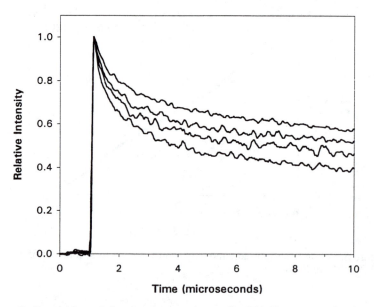

Figure 8. Quenching of the perylene cation radical by ferrocene via electron tunneling in 5 mM Pe/PVBC films. Ferrocene concentration is 0, 10, 20, and 40 mM for the decay traces from the top to the bottom.

The slope k is a function of the Onsager length r_c and the distribution parameter b_G. Similar slopes were observed for PMMA and PBM films, indicating very close b_G and r_c values. The smaller slopes for PVBC and PC are probably due to the shorter Onsager lengths in these polymers (see Table I). The relative shift of the semi-log curves along the time axis is related to the variation in the diffusion constants D of charge carriers in different polymers. Charge carriers moves 1~2 orders of magnitude faster in PVBC and PBM than in PMMA and PC.

The cation radical decay on the long timescale (Figure 9A) cannot be fitted to a geminate recombination which predicts a square root relation as follows.

$$\frac{C(t)}{C_0} = \rho_{fi}\left(1 + \frac{r_c}{\sqrt{\pi Dt}}\right) \tag{29}$$

where, C(t) is the concentration of the perylene cations. However, the long time decay can be fitted to a second order homogeneous ion recombination. Because the cation has the same concentration as the trapped electron, the A + B type charge neutralization reaction shows the same second order kinetics as an A + A type reaction.

$$\frac{1}{C(t)} = \frac{1}{C_0} + kt \tag{30}$$

For PVBC, a straight line is obtained on the 1/O.D.~t plot after ~100ms (Figure 9B), while for PMMA, a straight line is reached after ~1 s. This means that, after a long enough period, some of the ion pairs escape the geminate recombination and spread over the polymer matrices, and the recombination kinetics switches from the geminate to the homogeneous regime. Second order rate constants were measured to be $k{\sim}3.4{\times}10^5$ $M^{-1}s^{-1}$ for charge carriers in PVBC, $k{\sim}3.0{\times}10^3$ $M^{-1}s^{-1}$ for charged species in PMMA, and $k{\sim}9.3{\times}10^4$ $M^{-1}s^{-1}$ in PBM. The diffusion constants of the charge carriers were estimated to be $7.5{\times}10^{-10}$ cm^2/s in PVBC, $7.0{\times}10^{-12}$ cm^2/s in PMMA, and $2.2{\times}10^{-10}$ cm^2/s in PBM. Since the large perylene cation moves even slower in glassy polymers with $D=10^{-12}{\sim}10^{-13}$ cm^2/s as shown below, the measured mobilities of the charge carriers are mainly due to Cl^- in PVBC, and $\cdot CO_2^-$ in PMMA and PBM. The later confirms the chain scission reactions in PMMA and PBM after electron trapping.

Molecular Motion in Polymers. The movement of small molecules (such as oxygen) in glassy polymers have been well studied (16). The mobility of large molecules was measured in this work to compare with the charge carrier movement in different polymers. The pyrene triplet quenching by ferrocene was examined in PVBC, PMMA, PBM, and PS.

$$^3Py^* + Fc \longrightarrow Py + {}^3Fc^* \tag{31}$$

Nonexponential dynamic quenching was observed in all these films, with the quenching data of PMMA films illustrated in Figure 10. The pyrene triplet decay traces were fitted to a dispersive Gaussian kinetics (17), and the quenching rate constant in each polymer was obtained from measurements over different ferrocene concentrations. PMMA, for example, exhibits a second order quenching rate

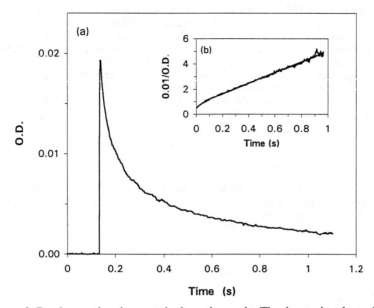

Figure 9. Perylene cation decay at the long timescale. The decay signal was fitted to second order kinetics (insert).

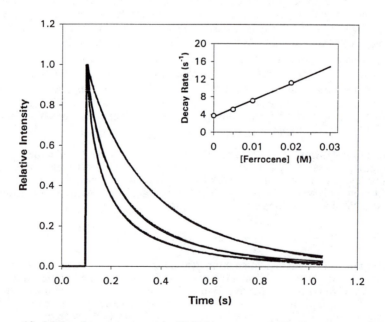

Figure 10. Molecular movement in PMMA measured via the pyrene triplet quenching by ferrocene. Insert shows that pyrene triplet decay rates increase linearly with ferrocene concentration.

constant k=370 $M^{-1}s^{-1}$, which corresponds to a diffusion constant D=8.8x10^{-13} cm^2/s. The results for different polymers were summarized in Table I.

Conclusion

Aromatic molecules are readily photoionized in solid polymers. Electrons can be ejected 36Å on average away from perylene molecules embedded in polystyrene. Subsequent electron trapping in polymer matrices leads to charge separation with the quantum yields varying from 0.014 in polystyrene, to 0.32 in PMMA, and up to nearly 1 in PVBC. The competition between electron-cation geminate recombination and electron scavenging was observed to be the same in polymer films as in liquid solutions. The rigidity of the polymer matrices helps stabilize the charged species up to a few hundred milliseconds. Charge neutralization occurs mainly via diffusion of charge carriers through polymer matrices. The low yield of charge generation in polystyrene indicates that solute cation radicals produced with a large yield (G~1.6) in high energy irradiation of polystyrene are not derived from direct ionization of solutes but from scavenging positive charges in the polymer matrix (2).

Acknowledgment

We are grateful to the National Science Foundation for support of this work. G. Zhang would like to thank the Department of Chemistry and Biochemistry, University of Notre Dame, for awarding the Reilly Fellowship.

References

1. Mort, J.; Pfister, G. In *Electronic Properties of Polymers* ; Mort, J.; Pfister, G., Eds.; John Wiley & Sons, Inc.: New York, **1982**; Chapter 6, p.215.
2. Zhang, G.; Thomas, J. K. *Polymer Preprints* **1994**, 35(2), 874; Zhang, G.; Thomas, J. K. In *Radiation Effects of Polymers: Chemical and Technological Aspects*; Shalaby, S. W.; Clough, R. L., Eds.; ACS Symposium Series, **1995**, in press.
3. Thomas, J. K.; Piciulo, P. L. *J. Chem. Phys.* **1978**, 68, 3260.
4. Scott, T. W.; Braun, C. L. *Chem. Phys. Letters* **1986**, 127, 501.
5. Freeman, G. R. In *Kinetics of Nonhomogeneous Processes;* Freeman, G. R., Ed.; John Wiley & Sons, Inc.: New York, **1987**; Chapter 2, p.19.
6. Hong, K. M.; Noolandi, J. *J. Chem. Phys.* **1978**, 68, 5163; Noolandi, J. In *Kinetics of Nonhomogeneous Processes*; Freeman, G. R., Ed.; John Wiley & Sons, Inc.: New York, **1987**; Chapter 9, p.465.
7. Beck, G.; Thomas, J. K. *J. Phys. Chem.* **1972**, 76, 3856.
8. Tsuchida, A.; Sakai, W.; Nakano, M.; Yoshida, M.; Yamamoto, M. *Chem. Phys. Letters* **1992**, 188, 254.
9. Tachiya, M.; Mozumder, A. *Chem. Phys. Letters* **1975**, 34, 77.
10. Hummel, A.; Infelta, P. P. *Chem. Phys. Letters* **1974**, 24, 559.
11. Itoh, K.; Holroyd, R. *J. Phys. Chem.* **1990**, 94, 8850; *J. Phys. Chem.* **1990**, 94, 8854.
12. Tabata, M.; Nilsson, G.; Lund, A. *J. Polym. Sci., Polym. Chem. Ed.* **1983**, 21, 3257.
13. Sakai, W.; Tsuchida, A.; Yamamoto, M.; Matsuyama, T.; Yamaoda, H. *Makromol. Chem., Rapid Commun.* **1994**, 15, 551.
14. Zhang, G.; Thomas, J. K. manuscript to be submitted to *J . Phys. Chem.*
15. Debye, P.; Edwards, J. O. *J. Chem. Phys.* **1952**, 20, 236.
16. Chu, D. Y.; Thomas, J. K.; Kuczynski, J. *Macromolecules* **1988**, 21, 2094.
17. Kransnansky, R.; Koike, K.; Thomas, J. K. *J. Phys. Chem.* **1990**, 94, 4521.

RECEIVED July 27, 1995

Chapter 5

Radiation Chemistry of Polybutene-1

W. Ken Busfield and Gregory S. Watson

School of Science, Griffith University, Nathan, Brisbane, Queensland 4111, Australia

Molecular weight changes and gel content measurements of moderately isotactic polybutene-1 following gamma irradiation at ambient temperature, show that both scission and crosslinking occur with G values of 3.2 and 1.0 respectively. ^{13}C NMR shows there is also a loss of stereoregularity. The G value for the loss of isotactic pentads is 13.9, considerably less than that for polypropylene. Thus the amount of temporary chain scission is significantly less during the irradiation of polybutene-1 than with polypropylene.

A recent investigation of the effect of high energy irradiation on the stereoregularity of isotactic polypropylene (1) showed dramatic effects. Gamma radiation produced a G value for the loss of isotactic pentad sequences of 94 at ambient temperature, and 220 if the irradiation was carried out above the melt temperature. With a knowledge of the mechanism, deduced from changes in the full pentad distribution, the ambient temperature value translates into a G value for configurational inversion of >21 $(100 \text{ eV})^{-1}$. This value is considerably higher than the G value for any other chemical event in polypropylene, e.g. G(crosslinking) and G(scission) have values of about 0.9 and G (H_2) = 2. The suggested explanation is that the irradiation promotes many more chain scission events than G (scission) suggests, but a large proportion are nullified by recombination in the cage. With stereoregular polymers however, there is sufficient time between scission and recombination for bond rotation and configurational inversion of the free radical end, leading to extensive loss of stereoregularity. In contrast, a previous investigation of the effect of irradiation on isotactic polymethylmethacrylate (2) showed that even a 2 MGy dose produced only minor changes in tacticity. A more recent study (3), however, has shown that irradiation of stereoregular polymethyl methacrylate at 80°C undergoes inversion with a G value of 18.6. The mechanism suggested contrasts with that for polypropylene in that the inversion process involves temporary scission of side groups $(COOCH_3)$ followed by cage recombination rather than main chain scission. Side group scission in polymethyl methacrylate is thought to be the precursor to main chain scission.

0097–6156/96/0620–0074$12.00/0

In this paper we report our preliminary findings on the influence of gamma irradiation on isotactic polybutene-1, a polymer of commercial interest which has a chemical structure similar to that of polypropylene. The irradiation chemistry of polybutene-1 has not been extensively studied previously. Mark and Flory (4) found that irradiation produced predominant crosslinking in isotactic films and that the extent of crosslinking was much reduced in atactic polymer.

Experimental

The sample of polybutene-1 was moderately isotactic (69.1% mm triads) and of relatively low MW (2.1×10^4). Gamma irradiations were carried out at the ANSTO facility in Sydney with the sample in sealed evacuated ampoules ($<10^{-4}$ torr) at a dose rate of 2.8 kGy h^{-1} at either ambient (solid sample) or 140°C (molten sample).

Gel contents were the fraction insoluble in refluxing xylene.

Viscosity average MW's were determined in Ubbelhode viscometers at 89°C with the sample dissolved in anisole. The Houwink constants used were (5) K = 111 x 10^{-3} ml g^{-1}; a = 0.50.

^{13}C NMR spectra were recorded at 100 °C on a Varian Unity 400 instrument at 100 MHz. Samples were dissolved in 9: 1 dichlorobenzene: deuterated nitrobenzene mixture. For samples gelling significantly in this mixture, the deuterated nitrobenzene was omitted and an external lock used instead. Pentad sequences were assigned according to Asakura *et al.* (6) and quantitative analysis was with a spectral simulation package assuming Lorentzian lineshapes.

Result and Discussion

Gel contents and viscosity average MW's measured following a range of doses at ambient temperature and 140°C (molten sample) are given in table 1. Gel contents are also plotted against dose in fig 1. Irradiation at 140°C with the polymer in the molten state has a similar gel dose to that following irradiation at ambient temperature with the polymer in the solid state. This is in marked contrast to the irradiation behaviour of polypropylene which formed no insoluble material when irradiated in the molten state up to 2 MGy.

At doses below the gel dose there is only a small increase in MW with dose. Assuming that the viscosity average is equal to the number average, a value for [G(S)-4G(X)] can be derived from the equation:

$$\frac{1}{\bar{M}_n} = \left(\frac{1}{\bar{M}_n}\right)_o + [G(S) - 4G(X)] \frac{R}{100 \, N_o}$$

where $G(S)$ = G(scission); $G(X)$ = G(crosslink); R = dose in eV g^{-1}; N_o = 6.02 x 10^{23}.

Table 1. Molecular Weight and Gel Content Data for Irradiated Polybutene-1

Dose kGy	Temp. °C	MW 10^4g mol^{-1}	Gel Content (%)	Dose kGy	Temp. °C	MW 10^4g mol^{-1}	Gel Content (%)
0	-	2.09	0	1100	amb	-	18.5
50	amb	2.30	0	1500	amb	-	25.5
100	amb	2.46	0	2000	amb	-	30.1
200	amb	a	0	500	140	-	0
500	amb	a	0	1000	140	-	24.0
750	amb	-	8.0				

Note: a = incompletely soluble in anisole.

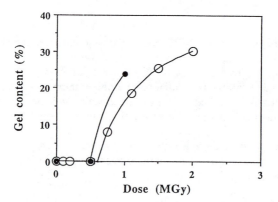

Figure 1. Influence of irradiation dose on gelation of polybutene-1 for irradiation at ambient temperature (open circles) and at 140°C (closed circles).

The slope of the graph for the soluble polymers shown in fig 2 gives the result:

$$G(S) - 4G(X) = -0.71 \pm 0.10$$

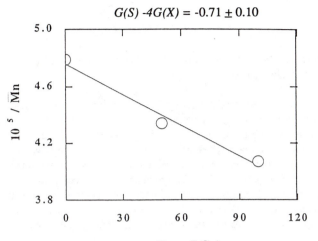

Dose (kGy)

Figure 2. Influence of irradiation dose on the molecular weight of polybutene-1 below the gel dose at ambient temperature.

At doses above the gel dose, the Charlesby-Pinner equation can be used to obtain individual G values.

$$S + S^{\frac{1}{2}} = \frac{G(S)}{2G(X)} + \frac{100 \, N_o}{R\bar{M}_w G(X)}$$

where S is the soluble fraction. The slope and intercept of the Charlesby-Pinner plot, shown in fig 3, give $G(S) = 3.0$; $G(X) = 1.3$.

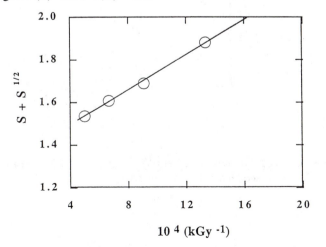

10^4 (kGy $^{-1}$)

Figure 3. Charlesby-Pinner plot for ambient temperature irradiation of polybutene-1. This equation is strictly only valid for amorphous polymers with the most probable MWD.

In view of the assumptions made in these analyses, the agreement is remarkably good. In summary:

$$G(S) = 3.2 \pm 0.5 \ and \ G(X) = 1.0 \pm 0.3.$$

In addition, the gel dose can be more precisely determined from the Charlesby-Pinner plot than that from fig 1; the value is 605 kGy.

The resolution required for determining stereoregularity in polymers can generally only be achieved with the polymer in solution. We have found, however, that as long as the polymer solvent system remains as a homogeneous swollen gel for relatively lightly crosslinked polymer, the resolution is adequate for stereoregularity analysis. This was the case for doses up to 1100 kGy at ambient temperature. When separation into two phases occurs, as with more highly crosslinked samples, then the spectrum is not necessarily representative of the total sample. The expanded regions of the spectra of polybutene-1 before and after 1100 kGy gamma irradiation are shown in fig 4. The relative increase in non-isotactic pentad sequences is easily seen, particularly in the side-chain methylene C resonance which

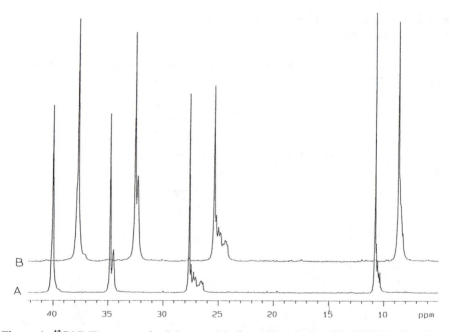

Figure 4. ^{13}C NMR spectra of polybutene-1 before (A) and after (B) 1100 kGy irradiation dose. Note spectrum B is displaced laterally 2 ppm for clarity.

is the most sensitive C atom in response to the stereoregularity of the adjacent main-chain C atom. The increase can also be seen in the methyl and backbone methine C atoms, but to a lesser extent. A quantitative analysis of the side-chain methylene C resonance based on the pentad sequence assignments given by Asakura *et al*. (6) has been carried out. The percentage loss of mmmm pentad sequences with irradiation dose is shown in fig 5. The

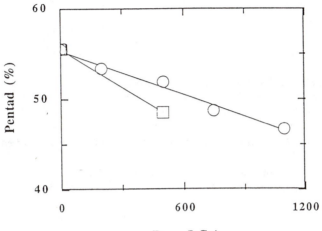

Figure 5. Influence of irradaiation dose on the fraction of mmmm pentad sequences in polybutene-1. Circles: ambient temperature; squares: 140°C irradiation.

slope of the graph representing the ambient temperature series gives a G value for the loss of isotactic pentads of 13.9. This value is significantly less than the value of 94 found for isotactic polypropylene (7) even allowing for the much lower starting isotactic pentad content (56 vs 88%). The one point determination of G(-mmmm) for the irradiation of polybutene-1 in the molten state gives the value 24. This again is very much less than the corresponding value of 283 for molten polypropylene (8).

The changes in concentration of the individual non-isotactic pentads as a function of dose, derived from the line shape analyses, are too scattered to be able to make any deductions about the mechanism of the irradiation induced loss of stereoregularity. Thus, it is not possible to estimate a G value for the inversion process as with polypropylene.

In conclusion MW measurement below the gel dose and gel contents of gamma irradiated polybutene-1 show that both scission and crosslinking occur with G values 3.2 and 1.0 respectively. These values are of the same order of magnitude as those for polypropylene. On the other hand the loss of isotactic pentad sequences from the moderately high isotactic polybutene-1 occurs with a G value of 13.9, which is considerably less than that for isotactic polypropylene (94). Thus, although G (-mmmm) indicates that some temporary chain scission occurs, i.e. chain scission followed by cage recombination, it is far less extensive than in polypropylene. The full sequence of events requires chain scission, inversion at the carbon free radical, bond rotation and finally radical recombination. Compared with polypropylene, one of these events must be impeded when the methyl is replaced by an ethyl group in polybutene-1, thus giving the free radicals a greater chance to escape the cage before recombination.

Acknowledgements: Assistance with NMR by Peta Harvey is greatly appreciated. Access to irradiation facilities at ANSTO and the assistance of AINSE is also acknowledged.

References

1.   W.K. Busfield and J.V. Hanna, *Polym. J.*, 1991, *23*, 1253.
2.   E. Thompson, *J. Polym. Sci., Letters Edit.*, 1965, *3*, 675.
3.   L. Dong, D.J.T. Hill, private communication.
4.   J.E. Mark and P.J. Flory, *J. Amer. Chem. Soc.*, 1965, *87*, 1423.
5.   J. Brandrup and E.H. Immergut, *Polymer Handbook*, J. Wiley; 3rd edition, New York, NY, 1989.
6.   T. Asakuro, M. Demura and K. Yamamoto, *Polymer*, 1987, *28*, 1037.
7.   W.K. Busfield, J.V. Hanna, J.H. O'Donnell and A.K. Whittaker, *Brit. Polym. J.*, 1987, *19*, 223.
8.   P.F. Barron, W.K. Busfield and J.V. Hanna, J. Polym. Sci., Polym Letters, 1988, *26*, 225.

RECEIVED October 5, 1995

Chapter 6

Anisotropic Properties in Ultrahigh-Molecular-Weight Polyethylene after Cobalt-60 Irradiation

J. V. Hamilton, K. W. Greer, P. Ostiguy, and P. N. Pai

Johnson & Johnson Professional, Inc., 325 Paramount Drive, Raynham, MA 02767

A series of studies was performed to investigate the factors contributing to dimensional changes in UHMWPE during the radiation sterilization process and determine if the dimensional changes correlate with the changes in the mechanical properties, morphology, and chemical structure of UHMWPE. Anisotropic dimensional changes were observed with radiation dose being the dominate factor studied. FTIR, DSC, density and tensile tests were performed to characterize these changes. Anisotropic property changes were observed in mechanical properties, morphology and chemical structure.

The effects of Co 60 radiation on polyethylene, and ultra-high molecular weight polyethylene (UHMWPE) in particular, have been studied extensively. The free radicals formed during irradiation have been postulated to be involved in a variety of reactions. Oxidative chain scission and crosslinking have been identified as the two predominant reactions. The free radicals are thought to preferentially react at the outer surfaces of the lamella crystals (1,2,3). Free radical-initiated oxidative chain scission reactions result in chain folding and an increase in the size of the lamella structure. These radiation-induced changes in crystallinity and crosslinking have been shown to increase some mechanical properties, such as elastic modulus and tensile strength, and at the same time decrease other properties, such as ultimate elongation (4,5). Chain scission has the additional effect of reducing the molecular weight of the polymer. The unique properties of UHMWPE, such as toughness and abrasion resistance, are a function of the extremely large molecular weights of greater than 1 million.

Some of the factors that have been shown to influence the extent and type of reactions that may occur include dose, dose rate and stress. As the dose increases, the radiation-induced changes in the material increase (4,5). The rate at which the dose is applied to the material has been shown to influence the magnitude of changes that will occur, with greater changes occurring at slower dose rates (6,7). Mechanical stress in

0097–6156/96/0620–0081$12.00/0

the polymer is known to accelerate radiation-induced degradation, although the mechanism is not well understood (8).

An underlying assumption in many studies is that the mechanical and physical properties of unirradiated UHMWPE are isotropic. Anisotropic mechanical properties prior to sterilization have been reported in the literature (9,10), however the effects of irradiation on the mechanical properties are typically treated as if the material is isotropic (4,5) . Furthermore, no one has reported any anisotropic changes in the mechanical properties of polyethylene due to radiation. Density changes in UHMWPE orthopaedic implants after Co 60 sterilization have been shown to vary through the thickness of the material with maximum values occurring below the outer surface, but variations in density across the outer surface have not been reported. Dimensional changes in UHMWPE after irradiation have been reported to be -0.1 to -0.15% (shrinkage) and independent of sample size and direction (5).

Current design of implants for total joint arthroplasty relies on the mating of UHMWPE components with metal components for proper function. Any dimensional mismatch in the mating of these components will result in relative motion upon the application of loads. This motion has been identified as a source of UHMWPE particulates (10). Wear particles have been related to adverse tissue reactions in bone, which can lead to bone resorption, implant loosening, and mechanical failure (11,12). Understanding and controlling the changes in UHMWPE after Co 60 sterilization of the implant is therefore critical to the successful functioning of total joint implants.

The intent of this series of studies was to investigate the factors contributing to the dimensional changes in UHMWPE during the radiation sterilization process and determine if changes also occur in the mechanical and physical properties. The initial study utilized a statistical design of experiments approach to determine which factors or interaction of factors contribute to the dimensional changes. The next study evaluated the effects of sample orientation within the bulk material on the dimensional changes to the material during sterilization. The final studies relate the dimensional changes in UHMWPE to the changes in mechanical, chemical and morphological properties, as determined by tensile tests, FTIR, DSC and density measurements.

EXPERIMENTAL

Materials. Two grades of ultra-high molecular weight polyethylene (UHMWPE), GUR 412 and GUR 415, were evaluated in these studies. The GUR 412 was produced by Hoechst in Oberhausen, Germany with a number average molecular weight of approximately 4 million. The powder was compression molded into sheets 1 m by 2 m by 60 mm thick by VTP, a division of Hoechst. The GUR 415 was produced by Hoechst Celanese in Bayport, TX, with a number average molecular weight of approximately 6 million. The powder was ram extruded into rods by Westlake Plastics of Lenni, PA.

Test sample orientation was dependent on the form of the bulk material. The longitudinal direction was defined as the direction in which the compaction pressure was applied to the powder during processing. For ram extruded rod, the longitudinal direction was parallel to the long axis of the rod, and the transverse direction was defined as the   radial direction. For compression molded sheet, the longitudinal direction was therefore through the thickness of the sheet, and the transverse direction was oriented along the length and width of the sheet.

METHODS

Process Variable Studies.   Shrinkage of UHMWPE components after sterilization was observed at Johnson and Johnson Professional Inc. It was hypothesized that a combination of residual thermal and mechanical stresses due to the powder consolidation process, and/or the machining processes, and the Co 60 sterilization process could result in these changes. A study was undertaken to identify which factors were contributing to the changes. A two-level factorial design was chosen to minimize the number of samples required and to evaluate variable interactions. The effect is, the difference between the sum of the measured value for each level, divided by the number of samples in a level. The upper and lower confidence levels are calculated from the product of the t-values for 95% confidence and the standard error based on unassigned factor effects added or subtracted to the effect respectively(13).

Cylindrical samples were machined from the bulk material. Two lines 90° apart were drawn across the top surface of each cylinder to identify the locations for the diameter and height measurements. The cylinders were conditioned for 24 hours at 21° C prior to all measurements. All measurements were taken with digital micrometer (Mityomo) with a resolution of 0.00127 mm. In each of these studies, the test variable was the change in volume normalized by the presterilized volume. All measurement were taken within seven days of irradiation

GUR 415 Process Variable Study.   In the GUR 415 process variable study, five factors were identified that could affect the dimensional changes of the extruded GUR 415 material. Each of the factors was evaluated at two levels, which required a total of 32 samples.

The first factor was the level of residual stress in the material produced during the extrusion process. The same extruder was used to produce rods of varying diameters. The rod diameter was determined by the size of the final die the material pushed through. The smaller the rod diameter, the greater the residual stresses. Rods of 50 mm and 68 mm were chosen to represent the range of rods used in the manufacture of the components in question.

The second factor was the creation of residual stresses in the parts during the machining process. Samples were machined at two levels. The high stress level was produced by machining at high speeds and feeds, and machining the most material from the rod, resulting in the smallest sample volume for both rod sizes. The low stress level was created by machining at low speeds and feeds and machining a minimal amount material from the rod, resulting in the largest sample volume for both rod sizes. Sample sizes and machining parameters are listed in Table I. Orientation of the sample was controlled, with the height in the longitudinal direction.

The third factor was the removal of the stresses created during the extrusion process through a thermal annealing cycle. The manufacturer anneals the rods at a proprietary elevated temperature to reduce the residual stresses prior to shipping. This was defined as the high residual stress level. A second annealing cycle at 100° C was done at Johnson & Johnson Professional-Inc. (JJPI) to create the low residual stress level.

The fourth factor was the sterilization dose rate. Ethicon (Somerville, NJ), was used to irradiate at a slow dose rate, 6.3 KGy/Hr, and Isomedix (Northboro, MA) was used to irradiate at a fast dose rate, 18.5 KGy/Hr.

Table I Dimensional anisotropy test sample size and machining parameters

Material	Shape	Description	Machined Cylinder Size (mm)		Machining Parameters	
				SMM[1]	Feed[2]	Tool radius
415	68 mm	Heavy cut	31.75 x 42.67 dia.	185	.5	1.5 mm
	dia.	Light cut	31.75 x 45.72 dia.	78	.2	.75 mm
	50 mm	Heavy cut	25.4 x 24.13 dia.	185	.5	1.5 mm
	dia.	Light cut	25.4 x 28.45 dia.	78	.2	.75 mm
412	sheet	Heavy cut	25.4 x 24.13 dia.	185	.5	1.5 mm
		Light cut	25.4 x 28.45 dia.	78	.2	.75 mm

1 SMM - Surface meters / min.
2. Feed - mm / rev

The fifth factor was sterilization dose. Samples were passed through the sterilizer either once or twice. Typical doses for a single pass and a double pass were 34 KGy and 68 KGy, respectively.

GUR 412 Process Variable Study. In the GUR 412 process variable study, three factors were identified that could affect the dimensional changes of molded GUR 412 material. Each of the factors was evaluated at two levels which required a total of eight samples. The three factors identified were dose, dose rate and machining. Stock size was excluded since the material is produced in only one size and annealing was excluded since historical information indicated that the material processor (VTP) did an adequate job at minimizing residual stresses in their process. Dose and dose rate parameters were identical to those used in the GUR 415 process variable study.

The third factor was the creation of residual stresses in the parts during the machining process. The samples were machined using the same machining parameters as the GUR 415 samples. Sample sizes are listed in Table I. Orientation of the samples relative to the longitudinal and transverse directions was not controlled during machining.

Dimensional anisotropy study. The dimensional anisotropy study measured the effects of sample orientation within the bulk material on dimensional changes in the material during sterilization. Cylindrical samples (49.3 mm x 25.4 mm diameter) were machined from either extruded GUR 415 (n=16) or compression molded GUR 412 (n=16). The orientation of the sample within the material was controlled, with cylinder height in the longitudinal direction for both GUR 412 sheet samples and GUR 415 rod samples. Dimensional changes were measured using the same equipment and techniques used for the process variable studies. Changes in the height and diameter of the cylinders were measured and normalized by the dimensions prior to sterilization. Samples were irradiated at Isomedix (Morton Grove, IL) at nominal doses of 28 and 50 KGy.

Mechanical Properties. The intent of the mechanical properties study was to evaluate the effect of Co 60 sterilization on the static tensile mechanical properties of compression molded GUR 412 and ram extruded GUR 415 and compare those results with the dimensional changes. Tensile samples (n=10 for each material and direction) were prepared from 3 mm thick slabs machined from both the rod and the compression molded materials. Longitudinal rod samples were taken from the center of the rod and transverse samples were taken from material immediately adjacent to the section of rod used for the longitudinal tests. The sheet samples were taken from the center of the sheet and the transverse samples were also taken from material immediately adjacent to the longitudinal material. ASTM D638 type V samples were chosen for this study so that tests could be performed in both directions. Tests were done on an Instron universal test system (Instron Canton MA Model 4204) at a cross head speed of 25 mm per min. with an extensometer (Instron Canton, MA, Model 2630-014) used to determine strain for modulus calculations. The load, cross head displacement and strain data were collected using a PC based data acquisition system (Cyborg, Newton, MA. and Labtech Notebook, Wilmington MA). Ultimate tensile strength, ultimate elongation, yield stress, and elastic modulus were subsequently calculated from this data.

Anisotropy for a property was defined as the longitudinal value divided by the transverse value for the same material at the same dose. A property was considered anisotropic when the student's t test showed a significant difference ($p < .05$) between directions. Comparisons were only made at the same dose level presterilization or post sterilization. Samples were sterilized to a dose of 40 KGy at Isomedix (Northboro, MA).

Chemical / Morphological changes. The intent of this series of experiments was to assess the radiation-induced chemical and morphological changes in FTIR, DSC and density measurements for both ram extruded and compression molded materials. Sample films of a nominal thickness of 250 microns were prepared using a JJPI designed skiving tool mounted in a milling machine (Bridgeport, Bridgeport, CT). The films were taken from the same locations and orientations as described for the mechanical property experiments. The films were irradiated to 40 KGy at Isomedix (Northboro, MA).

Fourier Transform Infrared spectroscopy. FTIR (Nicolet Madison, WI Magna 550) was performed on the films at five randomly selected locations from each film. Carbonyl area and IR crystallinity index were chosen to measure the extent of the oxidation reactions and the morphological changes respectively. The area under the curve from 1800 cm^{-1} to 1660 cm^{-1} was reported as the carbonyl area. The ratios of the peaks at 1894 cm^{-1} (crystalline band) and 1303 cm^{-1} (amorphous band) defined the IR crystallinity index (14). Each spectra consisted of 256 scans with a resolution of 4 cm^{-1}. The data was normalized for sample thickness using the peak at 4250 cm^{-1}.

Calorimetry. Differential scanning calorimetry was performed on a single sample from each film, using a Perkin Elmer DSC 7 to compare the crystallinty changes in the material . Samples weighed approximately 4 mg. and were heated at a rate of 10° C / min. in nitrogen. The heat of fusion was determined during the initial heating cycle. The percent crystallinity was determined using 288 Joules/gm for 100% crystalline UHMWPE (15).

Density. Density measurements were done on each of the films for both materials and directions. Density measurements were made using a density gradient column with a mixture of distilled water and isopropyl alcohol per ASTM D1505. A total of five measurements were made from random locations within each film. Accuracy of the column was .0002 gm./cc.

RESULTS

Process Variable Studies. The results for the GUR 415 study are shown in Table II. The influence a factor has on the variable being studied was determined by the magnitude of the effect and whether or not the confidence interval included 0. If the confidence interval did not include zero the factor had an effect on the response. The dominate factor identified was dose.

Table II GUR 415 rod process parameter results

	Rod Stock	Annealing	Machining	Dose rate	Dose
Effect	2.94	-1.44	0.81	6.06	13.69
Lower confidence level	1.52	-2.85	-0.6	4.65	12.27
Upper confidence level	4.35	-0.02	2.23	7.48	15.10

The results for the GUR 412 study are shown in Table III. Initial evaluation of the effect values and confidence intervals showed that none of these factors significantly effected volumetric changes. However, inspection of the data at the 68 KGy dose level (four samples) showed that the normalized volumetric changes could be grouped in two separate groups of two each, with average shrinkage values of -0.0022 and -0.0007 respectively. Due to the limited number of samples, no further statistical analysis was possible.

Table III GUR 412 sheet process parameter results

	Machining	Dose rate	Dose
Effect	4	-0.5	-1
Lower confidence level	23.05	18.55	18.05
Upper confidence level	-15.05	-19.55	-20.05

Dimensional Anisotropy Studies. Changes in height and diameter after sterilization were normalized by the respective dimension prior to sterilization. Figure 1 shows the effect of dose on the dimensional changes for both materials at doses of 28 and 50 KGy. Table IV shows lists the comparisons that had significant differences ($p < .05$) at each of these dose levels.

Table IV. Dimensional change comparisons ($p < .05$)

Dose (KGy)	Comparisons
28	Rod Height vs. Plate Height
	Plate Dia. vs. Plate Height
50	All comparisons

The effect of dose on dimensional change rate is shown in Figure 2. Dimensional change rate is defined as the difference of normalized dimensional change divided by the dose required to produce that change. Dose ranges evaluated are 0 to 28 and 28 to 50 KGy. The dimensional change rate of the GUR 412 sheet material in the pressing direction was significantly lower ($p < .05$) than the other three measurements for the 0 to 28 KGy range. In the 28 to 50 KGy, range all rates were equivalent.

An evaluation for anisotropic response within a group of samples was performed. Samples with the same material, dose and direction were grouped together. Comparisons of the diameter and height measurements taken 90° apart were made to determine if the changes in the dimensions were nonuniform. No significant differences ($p < .05$) were identified for materials, direction, or dose.

Mechanical Properties. Results for the GUR 415 and GUR 412 tensile tests are shown in Figures 3 and 4 respectively. All properties of the extruded GUR 415 were isotropic prior to sterilization. After irradiation, ultimate tensile strength and ultimate elongation were both anisotropic. Modulus and yield strength remained isotropic. Sheet GUR 412 initially exhibited isotropic modulus, yield strength and ultimate elongation. After sterilization, only modulus was isotropic. All comparisons were statistically significant ($p < .05$).

FTIR. Results of the FTIR scans are shown in Table V. All values were calculated from the absorbance mode. The change in carbonyl area was calculated by taking the area at 40 KGy and subtracting the same area for non-irradiated samples. There were no significant differences ($p < .05$) between directions for either material or between materials.

The crystallinity index values also showed no evidence of anisotropy either prior to or after irradiation ($p < .05$).

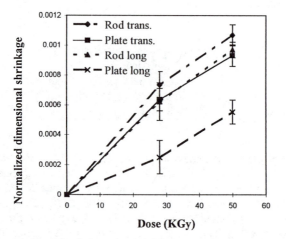

Figure 1. Normalized dimensional shrinkage vs. Dose. Normalized dimensional shrinkage = |(dimension irradiated - dimension non-irradiated)| / dimension non-irradiated. Error bars = standard deviation

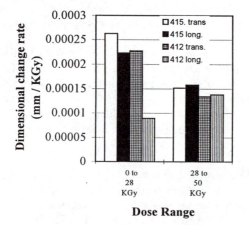

Figure 2. Dimensional change rate vs. Dose range. Dimensional change rate = |(Dimension @ final dose - Dimension @ initial dose)| / (final dose - initial dose)

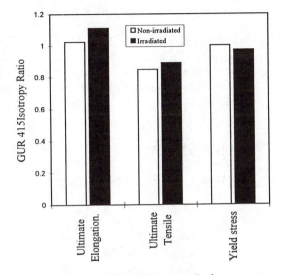

Figure 3. GUR 415 Isotropy Ratio.
Isotropy ratio = longitudinal value / Transverse value

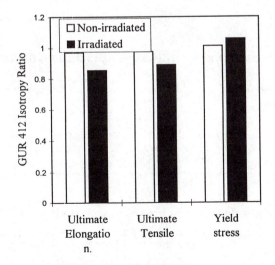

Figure 4. GUR 412 Isotropy Ratio.
Isotropy ratio = longitudinal value / Transverse value

Calorimetery. DSC results are shown in Table VI. No statistical analysis was possible since a single sample for each condition was evaluated. In general, there was an increase in crystallinity after sterilization and the extruded GUR 415 material shows a greater increase in crystallinity than the sheet GUR 412.

Density. Density measurements are shown in Table VII. Prior to sterilization both materials were anisotropic (p < .05). After sterilization, the sheet GUR 412 became isotropic and the extruded GUR 415 continued to be anisotropic. The anisotropy in the latter material was reversed, however, with the transverse density becoming greater than the longitudinal density.

Table V. FTIR scans of GUR 415 and GUR 412.

	Dose (KGy)		412 long.	412 trans.	415 long.	415 trans.
Crystallinity Index	0	Mean	0.348	0.348	0.316	0.3170
(1894 cm^{-1} / 1303 cm$^{-1)}$)		Std. Dev.	0.007	0.003	0.003	0.002
	40	Mean	0.399	0.399	0.3831	0.379
		Std. Dev.	0.006	0.004	0.010	0.004
Carbonyl Area		Mean	9.70	9.16	10.69	11.31
(1800 cm^{-1} to 1660 cm$^{-1)}$)		Std. Dev.	0.29	0.63	0.25	0.25

Table VI. DSC results of non irradiated and irradiated GUR 412 and GUR 415 (J/gm)

Dose (KGy)	412 long.	412 trans.	415 long.	415 trans.
0	51.19	51.07	48.58	48.56
40	58.41	58.16	58.35	57.94
delta	7.22	7.10	9.78	9.38

Table VII. Density values for non irradiated and irradiated GUR 412 and GUR 415 (gm/cc).

Dose (KGy)		412 long.	412 trans.	415 long.	415 trans.
0	Mean	0.9308	0.9316	0.9291	0.9286
	Std. Dev.	0.0003	0.0002	0.0001	0.0001
40	Mean	0.9400	0.9400	0.9389	0.9397
	Std. Dev.	0.0004	0.0004	0.0006	0.0004

DISCUSSION

The use of statistical design of experiments allowed for the identification of the variables that contribute to the dimensional changes observed in UHMWPE components. The variables identified were thermal and mechanical induced stresses in the material, dose and dose rate. The extruded GUR 415 results indicated that dose was the dominate factor in the range of doses studied. Dose rate also contributes to the changes but not to the same degree that dose does. In contrast to these results, the sheet GUR 412 had no dominate factor affecting the volumetric changes. However the assumption of isotropic material response for both materials was incorrect and caused problems with interpreting the results of the experiment. Inspection of the data gave a hint of an anisotropic response in the material, when the samples could be divided into two groups with much different responses. This can be explained by the fact that the orientation of the sheet samples with regard to the processing direction had not been controlled.

The dimensional studies using samples with controlled orientation showed that the dimensional changes in GUR 412 sheet material are anisotropic at a dose of 28 KGy and continue to be anisotropic at 50 KGy. The dimensional changes were greater in the transverse direction than in the longitudinal direction. The mechanical properties also followed these trends, where the ultimate tensile strength and elongation decreased after sterilization in the transverse direction; however, there were no significant changes in these properties in the longitudinal direction. These results also explains why there was an increase in the isotropy ratio at 40 KGy for the ultimate tensile strength and ultimate elongation values of the 412 sheet material.

The GUR 415 rod material did not respond the same way to sterilization as the GUR 412 sheet material. The dimensional changes are isotropic after exposure to 28 KGy, but with the transverse changes being greater than the longitudinal, though not statistically significantly different. However, at a dose of 50 KGy the dimensional changes became anisotropic. The mechanical properties prior to sterilization were all isotropic. After irradiation to a dose of 40 KGy, both the longitudinal and transverse ultimate mechanical properties decreased with the longitudinal properties decreasing more than transverse properties resulting in isotropy ratios less than one for both ultimate tensile strength and ultimate elongation. This was in contrast to the dimensional changes where the transverse dimension decreased more than the longitudinal and also the behavior of the GUR 412 material where the transverse response was always greater than the longitudinal.

The GUR 415 rod material had a more random structure and lower crystallinity (48% vs. 51%) and maintained isotropic properties at the lower dose. However as the dose increased the structure rearranged itself such that the anisotropy developed with greater changes in the transverse direction than the longitudinal direction. This may be a tendency for the material to return to the anisotropic condition as has been observed in the melt (16). Of interest is that in both the GUR 412 sheet and GUR 415 rod material the transverse dimensional changes are greater than the longitudinal changes.

The differences in the rates at which the dimensional changes occurred from 0 to 28 KGy may be partially explained if the relative reactivity of the polymer chains immediately adjacent to the crystalline region is taken into account. The atoms immediately outside the crystalline regions are more reactive to absorbed oxygen (1,8) and may favor the growth of the crystalline region, through a chain scission reaction, in the direction already established. The ram extruded material has no preferred orientation initially and shrinks uniformly until a preferred orientation develops. The orientation may be determined by the anisotropic nature of the UHMWPE melt.

Prediction of anisotropic responses with either FTIR or DSC in the bulk material of either form of UHMWPE was not successful. Statistical comparisons of the longitudinal and transverse values for FTIR crystallinty index, and changes in carbonyl area at the dose levels evaluated, could not determine any anisotropy. Comparisons of the percent crystallinity values from DSC were inconclusive showing an increase in crystallinity with dose for both materials and directions. The GUR 412 sheet results were not predictive of the anisotropic dimensional changes. The increase in crystallinity measured each direction was almost identical unlike the dimensional changes.

Anisotropic property measurements from density values were able to show anisotropy in 412 at 0 KGy and 415 at both 0 and 40 KGy. GUR 412 sheet material irradiated at 40 KGy was determined to be isotropic. Mechanical properties for the GUR 412 sheet material showed an increase in anisotropic properties with dose not a decrease. The GUR 415 rod material showed no signs of anisotropic properties prior to sterilization unlike the density, which was initially anisotropic.

In summary, dose was the dominate factor in producing shrinkage in both GUR 415 rod and GUR 412 sheet. These changes became anisotropic in both materials at a 50 KGy dose. A similar response was also seen for the mechanical properties. Changes in dimensions were not predictive of mechanical properties. No correlation could be made between changes in the morphology or chemical structure and changes in the dimensional or mechanical properties.

Acknowledgements
We would like to thank Dr. M.B. Schmidt, L. Macdonald, M.Gorhan and F.Senkel of Johnson & Johnson Professional Inc. for their assistance in these studies and editing of the manuscript.

Literature Cited

1. Shinde, A. ; Salovey, A. *J. Polym. Sci. , Poly. Phys. Ed.* 1985 23 1681-1689
2. Shimade, S.; Hori, Y.; Kashiwabara, H. *J. Polymer* 1977 18, 19
3. Kashiwabara, H.; Shimade, S.; Hori, Y. *Radiat. Phys. Chem.*, 1991, 1, 43
4. Zhao Y.; Luo, Y.; Jiang, B. *J. Appl. Ply. Sci.*, 1993, 50, pp. 1797-1801.
5. Streicher R., *Beta-gamma* 1/89 34-43.
6. Kurth, M.; Eyerer, P. In *Ultra-High Molecular Weight Polyethylene as Biomaterial in Orthopedic Surgery*; Willert,H.-G.; Buchorn, G.H.; Eyerer, P. Eds. ; Hogrefe & Huber, Toronto, 1991, pp. 82-87.
7. Ishigaki,I.; Yoshii, F.; Makuuchi, K.; Tamura,N. In *Sterilization of Medical Poducts , Vol.V* , Morrissey, R.F.; Prokopenko, Y.I. eds.; Polyscience, Morin Heights Canada 1991, pp. 308-321.
8. O'Donnell, J.H. In *The Effects Of Radiation On High-Technology Polymers, ACS symposium series 381*; O'Donnell Ed.; American Chemical Society, Washington D.C., 1989, pp. 1-13.
9. Rimnac, C.M.; Wright, T.M. In *Ultra-High Molecular Weight Polyethylene as Biomaterial in Orthopedic Surgery*; Willert,H.-G.; Buchorn, G.H.; Eyerer, P. Eds. ; Hogrefe & Huber, Toronto, 1991, pp. 28-31

10. Eyerer,P.;   Rodlaver,   D.;Siegmann,   A.   In   *Ultra-High Molecular Weight Polyethylene as Biomaterial in Orthopedic Surgery*; Willert,H.-G.; Buchorn, G.H.; Eyerer, P. Eds. ; Hogrefe & Huber, Toronto, 1991, pp. 236-239.
11. Gross, T.P.; Lennox, D.W. 1992, *J.Bone Joint Surg.* [Am]74, 1095-1101.
12. Harris, W.H.; McGann, W.A. 1986 *J.Bone Joint Surg.* [Am] 68, 1064-1066.
13. *Strategy of Experimentation* , Edition 4.1, 1988, E.I. du Pont de Nemours and Company.
14. Nagy, E.V.; Li, S. *10th Trans. Soc. Biomat.*, 1990, 109
15. Wundlerich, B.; Cormier, C. *J. Polym. Sci.*, 1967, A-2, 5, 987.
16. Zacharides, A. E.; Logan, J. A. . *J. Polym. Sci. , Poly. Phys. Ed.* 1983, 21, 821-830.

RECEIVED July 25, 1995

MODELING AND MONITORING OF RADIATION EVENTS

Chapter 7

Computer Simulation of Polymer Network Formation by Radiation Cross-Linking

Eric S. Castner and Vassilios Galiatsatos

Maurice Morton Institute of Polymer Science, University of Akron, Akron, OH 44325–3909

Computer simulation is used to analyze network formation, specifically network structures and weight fraction of gel, and chain degradation. These divisions represent the possible responses of a polymer material to high energy irradiation. Where validation with experimental systems cannot be performed, comparison is made with theoretical approaches to modeling polymer irradiation. The systems addressed are those of *cis*-1,4-polybutadiene, *cis*-1,4-polyisoprene, and polyisobutylene. The methodology employed is that of a coarse-grained simulation. Validation of simulation results is facilitated by the addition of a method to relate extent of reaction to dose.

Computer simulation is a predictive tool, the results of which generally give insight to material properties. Detailed investigation into the topology of the post treatment gel is not attainable through current experimental methods due to its inherent insolubility. As a result current methods rely on theoretical treatment(1,2). Current work is based on the computer simulation algorithm developed by Leung and Eichinger(3-5) which is described herein. This algorithm has an advantage over other methods in that it is able to monitor the gelation process and separate the soluble portion (sol) from the gel. Through connectivity analysis, the topology of the network can also be described with great detail. Structures such as dangling chain ends and intramolecular loops which are inherent to the process of gelation are therefore able to be identified. This is particularly applicable to those systems which experience crosslinking and chain scission simultaneously as is the case in irradiation processes.

Due to the detail with which the algorithm monitors the system, a separation between effective and ineffective elastic material can be made. This separation, and the subsequent incorporation of values for effective material into theories of rubber-like elasticity(6-9) make the prediction of physical properties all the more quantitatively accurate. Validation of this model addresses network structures and its comparison with theory(1,2) and gel formation as a function of dose with respect to both bulk and dilute solutions.

When modeling systems of such large molecular complexity and size as that found in polymeric systems, an approach other than fully atomistic simulations must be employed due to the large computational effort otherwise required. In order to avoid the expected large number of calculations, a more coarse-grained approach is taken whereby chemical and structural detailed is maintained while removing calculations which involve unreactive repeat units. This approach makes use of the methodology of graph theory(10). The coarse-graining lies in the treatment of those repeat units which are not involved in crosslinking or chain degradation as in the case of polymers exposed to high energy irradiation. Another coarse-grained simulation to address random crosslinking has been developed by Grest and Kremer(11). Here, molecular dynamics is performed on systems of greater detail where small segments of the polymer chain are modeled as point particles.

Application of Graph Theory to Irradiation Crosslinking

For future results and discussion, and to impart a clear picture of the algorithm in order to convey its strengths and weaknesses, a brief explanation of the language of graph theory(10) is necessary. Graph theory has been used for many years by mathematicians, physicists, and engineers(10). Graph theory lends itself to the description of matters involving connectivity. The language of graph theory as it applies to that of polymer networks, or large connected systems for that matter, consists of labeling junction points and the connections between them. For the polymeric network undergoing exposure to high energy irradiation, the sites which absorb the energy and become reactive are identified as the junction points or vertices, whether they participate in crosslinking or chain scission. The portion of the chain between these active points are termed edges. The sets of μ vertices and ν edges may then be defined as $V=\{v_1, v_2, ..., v_\mu\}$ and $E=\{e_1, e_2, ..., e_\nu\}$ respectively. Every distinct pair of vertices is connected by at least one unique edge. A typical polymer chain therefore, is constructed with μ vertices and $(\mu+1)$ edges. The connectivity, and thus the network description, is accomplished by relating the vertices which are related by their connections through edges. This construction appears in its simplest form as a connectivity matrix. The matrix is $2 \times \nu$ and has the form

$$\begin{bmatrix} i_1, i_2, i_3, ..., i_\nu \\ j_1, j_2, j_3, ..., j_\nu \end{bmatrix}$$

where vertex i_1 is connected to j_1, and so on. The connectivity of the vertex determines its chemistry in the network structure. The structures of junction, backbone, and chain end are described by vertices connected to four edges, two edges, and one edge respectively. A graph then is a structure defining the various connections in the network. A graph may include circuits whereby there are multiple paths to selected junctions. A graph however may also be void of such circuits, in which case there exists a single unique path to every junction in the graph. This structure void of circuits is defined as a spanning tree. The number of vertices in a spanning tree is given by μ and thus the number of edges (ν) is μ-1. If the number of circuits in the graph increase while keeping the number of vertices constant, then the number of edges in the graph must also increase at the same time. The cycle rank (ξ)(6) is described as the difference between the number of edges in a system and that required for a spanning tree of the same number of vertices ($\xi = \nu - \mu + 1$). The term describes the degree of circuit formation. The value of cycle rank is therefore proportional to crosslink density. The parameter of cycle rank has been incorporated into the theories of rubber-like elasticity by Flory and Erman(6-9).

The limitation of a coarse-grained simulation such as this where the unreactive material is generalized into vectors between active sites, is the lack of correlations between polymer chains. Polymer chains absent of correlations (no excluded volume) are defined as "phantom" chains. Just as the name suggests, phantom chains, are void of structure and are free to move within the system unabated by other structures within the system. This property of the phantom chain however does not lend itself to the calculation of dynamic moduli. When a material is deformed in a real elastomeric system the modulus results from the contribution of both the chemical crosslinks and entanglements(12-14) - trapped or otherwise. In a network comprised of phantom chains, entanglements of any type are non-existent and any measurement of the modulus is solely the response of the chemical crosslinks. As a result, the calculation of the modulus of a polymer network comprised of phantom chains represents a static system and not one under deformation.

Simulation Input

The simulation described herein requires as input basic parameters which describe the polymer system in question. Inherent to the particular type polymer being examined and required for the simulation are the repeat unit molecular weight (M_0), bond length of a single repeat unit (l), characteristic ratios at various degrees of polymerizations (C_n), polymer density (ρ), and the Charlesby-Pinner ratio (S). The Charlesby-Pinner ratio is defined as the probability of an excited site to participate in crosslinking (q) as opposed to participating in chain scission (p). The bond length and characteristic ratios are required as input in order to correctly distribute the chains in the box.

After the above has been completed, the remaining input are terms which go to describe the particular system. These are the polymer number average molecular weight (M_n), molecular weight between crosslinks (M_c), volume fraction of polymer in the system (V_p), and the desired number of primary chains (N_p). The simulation is capable of simulating dilute polymer systems by reducing the volume fraction of polymer from 1.00 for a bulk system. If V_p is less than 1.00 then the size of the box is increased accordingly while maintaining the same number of polymer components. The M_c as defined in the input is a measure of the dose to which the sample is exposed. As the M_c decreases there are created more active radical sites along the backbone of the constituent polymer chains. This would be expected with a greater dose exposure. In fact, the dependence of M_c on dose has been established(15,16).

With regards to the current simulation input, there exist ranges for the number average molecular weight, the number of polymer chains, and the molecular weight between crosslinks within which the simulation performs satisfactorily well. With respect to these values, one of the limiting parameters is the number of active sites. The computational time and the size of several matrices (e.g. the matrix of active site coordinates) are linearly dependent on the number of active sites in the system. The number of active sites in the system increases with both an increase in the molecular weight and number of primary polymer chains or a decrease in the molecular weight between crosslinks. These currently represent a somewhat upper limiting case with respect to the activity and size of the system. The lower limiting case is that of enough chains so that the system is sufficiently large that erroneous effects which may arise from the limited size of the system are not encountered.

Algorithm

The simulation at present only addresses amorphous polymer systems. A consequence of this is the relaxation processes and molecular motion that the chains may undergo. The problem that is imparted with a dynamic system as it applies to molecular modeling is the constant change in the position of the corresponding atoms over time and the relocation of active sites. Given the above description of the system as a coarse grained model this presents a problem. The effects of the random relaxation motions of an amorphous polymer on the crosslinking process however has been examined(17) and has been shown to have no impact on the statistics of crosslinking. The conclusion is that the chains in the system are under no bias and therefore completely random motions of the segments in the chain results equally in local increases and decreases in proximity of active sites. The result is an overall averaging and reproduction of crosslinking statistics.

Once the input data is read into the code, the simulation goes about constructing the desired system. Calculations are made with respect to the total number of active sites and the average number of active sites per chain from the input data. The dimension of the cubical simulation box which is to contain the chains is calculated with the following equation.

$$L = (M_n N_p / \rho N_A V_p) \tag{1}$$

The variable names are the same as above with ρ being the polymer density and N_A Avogadro's number. Addition of the chains into the box is begun by determining whether or not a chain end is an active site. This end is placed at random in the simulation box. The body of the chain is constructed in parts. From the input data, the number of repeat units ($n = M_c/M_o$) between active sites is known and the flexibility of the chain is reflected in the characteristic ratio (C_n). The distance of the succeeding active site is then located and placed within the box in an off-lattice method according to a gaussian distribution with one dimensional-variance of the form

$$\sigma^2 = C_n n l^2 / 3 \tag{2}$$

Once the succeeding active site is located, the determination of whether it acts as a crosslinking site or a scission site is made according to the Charlesby-Pinner ratio of probabilities (q/p). This process continues for the remaining chain segments recording their location and to which chain they belong. The above is repeated for the remaining chains.

At this point the desired system has been in effect generated for the fact that the active sites for the component chains have been created and distributed accordingly. The gelation process proceeds by placing the first active site in the box in question and determining whether it is a crosslink site or scission site. If the active site was defined to be that of the crosslinking type then a sphere of a given radius around the site is examined. The radius of this sphere is termed the capture radius and represents the distance the active site may diffuse through the system in a given amount of time. If another active site of the crosslink type is found in the region of the sphere then radical combination is assumed to have occurred creating a tetrafunctional junction. If there are two such active sites in the sphere, the active site which is closest to that under question is determined to have reacted with the site. The radius of this sphere is examined for all the crosslinkers in the system and the corresponding reactions (crosslinking and chain scission) are allowed to take place. The algorithm returns to the first active site, increases the capture radius, and performs the possible reactions. The increase in the capture radius is analogous to the passage of time as this possible diffusing distance is increased. This process is repeated for all active sites until the entire array of all capture radii has been examined. In order to characterize the system more easily, when two radical intermediates combine to form a tetrafunctional crosslink, the two vertices participating in the reaction have their labels "equivalenced". What is meant by "equivalencing" is that the vertex with the higher numeration is replaced with that of the lower. In that way the two vertices are now one as in the combination of two radical sites to yield one crosslink.

After each entry in the array of capture radii is examined for all the active sites, the connectivity table which has been created to this point is submitted to a graph theoretical algorithm(18). Here the structure of the network is discerned form the connectivity of vertices. Because the connectivity table includes all of the active sites in the system, the algorithm is capable of examining both the structure of the gel particle and the remaining structures comprising the soluble portion. Before the simulation begins the gel particle at any point during the gelation process is defined as the single largest particle. The component referred to as the "gel particle" will most likely change several times until an extent of reaction is met where there is a greater difference in the distribution of sizes for the components.

The detail with which the simulation addresses a real system has been discussed above. With the examples shown herein, there are certain material properties and reactions which have been omitted for simplicity. The first is the different possible morphologies which may exist in a polymer. Polyethylene for example may have various degrees of crystallinity dependent upon the temperature and its thermal history. A partially crystalline material will have two distinct types of chain conformations: those of folded lamella and those of random amorphous chains. This poses a challenge for the simulation due to different affinities for radical formation in the different morphologies(19-21). The simulation is currently constructed to examine amorphous polymeric material (e.g. PE above its T_m, 140° C) in that the chain segments, and therefore the chains themselves are generated with a gaussian distribution and the affinity for crosslink formation is homogenous throughout the system.

The presence of strictly tetrafunctional crosslinks is the second simplification. The possibility exists for the chain end to be excited and made reactive. In this event, the subsequent junction which is formed has a functionality of three. The effect of these junctions is diminished with the size of the system and the molecular weight of the sample polymer. Therefore, this assumption is made with the expectation that the topology will be little effected.

Simulation Output

The exposure of a polymeric material to high energy irradiation results in two competing chemical changes. These changes consist of crosslinking and chain degradation by scission. In the case of polymer crosslinking, a detailed three dimensional network is created. The description of the topology of the network is complicated by its resistance to analytical techniques. The resistance lies in the fact that the polymeric gel is not soluble.

The ability of the simulation to record the weight fraction of the material which is included in the gel allows for comparison of gel fraction curves with those of experiment. By varying the Charlesby-Pinner ratio this algorithm has been shown to give excellent agreement with experiment.(22,23) Relating the output from simulation however is less straight-forward with respect to the dosage that the material is exposed to. For example, when a polyethylene sample is exposed to high energy irradiation, an allylic hydrogen radical is displaced from the backbone (Figure.1). This occurrence leads to a radical being created at that point. The greater the exposure to irradiation, the greater the number of radical sites, and therefore, greater crosslinking. In order to compare the dosage a material receives with information from simulation, the dosage from simulation is related to the number of active sites reacted and normalized for the radiation required to induce gel (R_{gel}). The classification of the gel point is defined as the point where a single particle in the simulation box is composed of a minimum of forty chains. This definition although arbitrary, is only sensitive at very low weight fractions of gel.

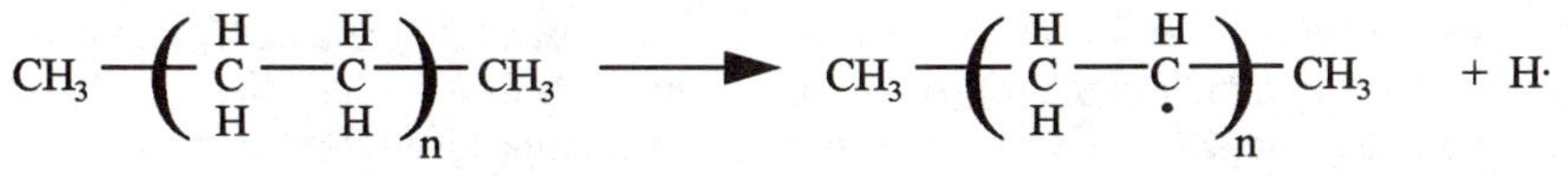

$$CH_3 \left(\begin{array}{cc} H & H \\ | & | \\ C & C \\ | & | \\ H & H \end{array} \right)_n CH_3 \longrightarrow CH_3 \left(\begin{array}{cc} H & H \\ | & | \\ C & C \\ | & \bullet \\ H & \end{array} \right)_n CH_3 + H\cdot$$

Figure 1. Mechanism of formation of a radical site during the irradiation process. A carbon atom along the backbone of a chain becomes excited and releases an allylic hydrogen.

The weight fraction of gel at a given exposure is certainly useful in characterizing the gelation of a material. However it is the description of the network topology which allows for calculation of static properties and gives insight to possible network structures otherwise not attainable by analytical methods. One of the strengths of the simulation described herein is the ability to identify a large variety of loop structures and dangling chain ends. With the detailed identification of network topology comes an ability to discern elastically effective material from that which is not. This ability makes the calculation of network properties all that more accurate. Crosslinks which are not effective cannot sustain stress because they are connected to the network structure by a single junction. They include those crosslinks involved in intramolecular loops and dangling chain ends. Dangling chain ends have been quantified by Flory(24). Intramolecular loops result from the radical combination between active sites on the same chain. However, not all intramolecular loops are totally ineffective. An intramolecular loop need only have a single elastically effective chain connected anywhere along its length between the junction in order to make the entire loop elastically effective. In addition, if an effective chain passes through the loop, the restriction created produces a physical crosslink termed a trapped entanglement.

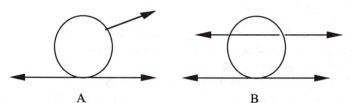

A   B

Figure 2. Representative sketch of (A) an intramolecular loop which is made elastically effective by the addition of an effective chain somewhere along the length of the loop. The appended chain, in order to be effective, must be connected in the network. (B) an intramolecular loop made effective by it acting as a trapped entanglement. This physical crosslink results from the presence of an effective chain passing through the loop structure.

Earlier results of Tonelli and Helfand(1,2) for the fraction of intramolecular loops formed during random crosslinking show that the average length between radical sites which combine to form these loops is quite small. This result reflects the probability for local active sites to interact. The percentage of crosslinks forming intramolecular loops was calculated to be 10-20%. In the case of solution cure however, the value rises to 20-70%. This is to be expected since the increase in intramolecular spacing brought about by an increase in dilution of the system increases the likelihood for intramolecular reactions. Because the amount of polymer involved in intramolecular loops is small, it is believed that the number of intramolecular loops made effective by possessing an appended effective chain is quite small. The additional consequence of the short length of these loops is the improbability for the presence of trapped entanglements.

The expression below for the number of elastically effective chains (ν) derived by both Bueche(25) and Mullins(26) was applied to test the accuracy of the simulation to discern effective material.

$$\nu = 2C - 2\rho/M_n + 2\varepsilon(1 - \rho/Cm_n) \tag{3}$$

where C is the density of chemical crosslinks, ρ is the polymer density, M_n is the number average molecular weight of the polymer, and 2ε is the maximum contribution that the entanglements may make toward the number of effective chains. The first two terms on the right side of the equation represent the contribution of the chemical crosslinks and the correction for the chain ends respectively.

The reason for making the distinction between effective and ineffective is its impact on physical properties. Earlier theories of rubber-like elasticity(27-31) either did not address network defects or concluded that their number was sufficiently low to be dismissed. The accounting of elastically ineffective material has since been made.(1,2,6-9,12-14) The number of intramolecular loops has been shown in the study above to be more than sufficiently large for recognition.

Results and Discussion

The simulation neglects correlations between the polymer chains. Given that the systems examined thus far are homopolymer melts, the lack of correlations should not have any effect on the resultant structure. This however, would not be the case were blends, block copolymers, or polymeric systems with additives being examined. In either of these situations, correlations would quite probably introduce heterogeneous crosslink density and a greater number of intramolecular loops. With regards to the simulation performed, the capture radii, the number of primary chains ($N_p = 800$) and a Poisson molecular weight distribution were used consistently for all the simulations. Where available, the simulation input reproduced that of experiment. It should also be mentioned that the results reported here are the outcome of a single simulation run. The reason being that fluctuations in critical observables is negligible.

Crosslinking. The first step in analysis of the algorithm involved simulation of cis-1,4-polybutadiene(32) and comparison of the weight fraction of gel versus radiation dose with that of experiment. The algorithm has been validated for the prediction of weight fraction of gel however, previous results(22,23) have been with respect to degree of crosslinking and not dose. The expression for absorbed dose derived from the simulation output gave excellent agreement with experimental results in a plot of weight fraction gel versus R/R_{gel} (Figure 3). Included in this analysis is the dependence of gelation on the ratio of scission sites to crosslink sites (p/q). At the various ratios, the simulation was never more than 4% in disagreement. Any disparity between the two results is quite probably due to the difference in molecular weight distribution: the gelation of a material is more sensitive to higher molecular weight polymer. As expected, the weight fraction gel increased with a decrease in p/q and achieved nearly 100% gel at $p/q = 0.1$. From the output produced it is also possible to calculate the critical exponents associated with gelation, although not performed here. The results of such an investigation would be expected to yield non-mean field exponents due to the presence of loop structures.

Network Topology. Validation of the simulation's prediction of network structure is made by comparison with results of Tonelli and Helfand(1,2) for cured *cis*-1,4-polyisoprene. The agreement between simulation and theory is very good for both

EFFECTS OF VARYING CROSSLINK/SCISSION RATIO
cis -1,4-polybutadiene

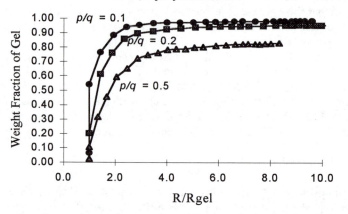

Figure 3. Comparison of experimental(31) and simulated gelation for polybutadiene samples with various scission to crosslink ratios (*p/q*). The x-axis is the radiation dose (R) normalized with respect to the radiation dose for incipient gel formation (R_{gel}). The polybutadiene samples had M_n 110,500 g/mol and all produced a R_{gel} of 2.00.

the total fraction of ineffective material (Figure 4) and the fraction of polymer in dangling chain ends (Figure 5). These results were encouraging however, not totally unexpected. The methodology employed by Tonelli and Helfand assumes that a site on a polymer chain is active by either irradiation from a high energy source or has reacted with a crosslinker, whereby the end of the crosslinker is now active. The densities for both repeat units on the same chain and on other chains within a specific volume are calculated. These densities then lead to the prediction of intramolecular loops and dangling chain ends. The application of graph theory to network formation results in the simulation also being dependent on the spatial distribution of the chains. Therefore, the algorithm employed is a manifestation of the densities calculated in the Tonelli and Helfand approach.

In both cases, the fraction of ineffective material is very high at low extents of reaction. When crosslinking begins, the amount of material in dangling chain ends and intramolecular loops is large. As crosslinking continues, the length of dangling chain ends decreases by continued incorporation into the network. The number of loops which are elastically ineffective decrease as a result of further crosslinking at points along the length of the chain. At the beginning of gelation an intramolecular loop has its greatest length. With continued reaction, portions of the loop react with the gel and change an otherwise elastically ineffective loop into one which is partially effective.

The comparison of the fraction of material in intramolecular loops is not made due to the discrepancy between the values. The difference is believed to be stem from the fact that Tonelli and Helfand employ a junction bridge of 6Å. This bridge reflects the distance from which the active site may react with another repeat unit. The

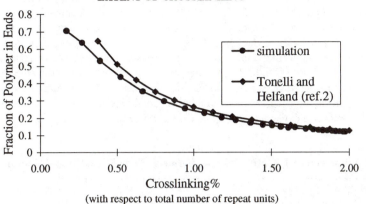

FRACTION OF POLYMER IN ENDS WITH RESPECT TO EXTENT OF CROSSLINKING

Figure 4.   Comparison of the total weight fraction of wasted material between that of theory(2) and simulation. The system examined was *cis*-1,4-polyisoprene of 68,000 g/mol molecular weight cured to 2% with respect to total number of repeat units.

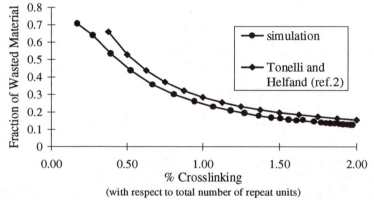

TOTAL FRACTION OF WASTED MATERIAL WITH RESPECT TO EXTENT OF CROSSLINKING

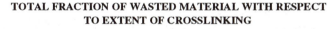

Figure 5.   Comparison of weight fraction of wasted material in dangling chain ends between that of theory(2) and simulation. The system examined was *cis*-1,4-polyisoprene of 68,000 g/mol molecular weight cured to 2% with respect to total number of repeat units.

simulation is different in this respect, for there are no distance constraints on possible reactions. So long as two active sites are within the examined capture radius, the reaction is assumed to have occurred. The result in a lower fraction of material in intramolecular loops from simulation.

The number of effective chains calculated by the simulation for various polybutadiene systems(33,34) is compared with predicted values from the derived equation of Bueche(25) and Mullins(26) (equation 3). The results (Table I) for the bulk systems were quite good with an increase in deviation being found for greater dilutions. The reason for the large difference in the dilute solutions is the increased number of intramolecular loops. Figure 6 shows the dependence of intramolecular loops on the degree of dilution. As the system becomes more dilute, the density of other chains in the vicinity of the active site decreases. This results in an increased number of intramolecular reactions and consequently more ineffective loop structures. Similar results and explanation are given in reference 2.

TABLE I
Comparison of Effective Chains Between Theory(24,25) and Simulation

sample	M_n g/mol	M_c g/mol	dilution (bulk=1.00)	v_{sim} $\times 10^{-2}$ nm	v_{calc} $\times 10^{-2}$ nm	%deviation
G1(32)	500,000	11,500	1.00	3.3	3.8	11
			0.80	2.5	2.9	13
			0.60	1.8	2.0	14
			0.40	1.0	1.2	17
			0.20	0.4	0.5	25
G2(32)	500,000	7,500	1.00	5.5	6.1	11
			0.80	4.2	4.8	12
			0.60	3.0	3.4	13
			0.40	1.8	2.1	16
			0.20	0.7	0.9	24
G2(33)	500,000	6,500	1.00	3.2	3.4	5.2
G2(33)	500,000	13,600	1.00	2.8	3.0	7.1
G2(32)	500,000	7,600	1.00	5.5	5.8	5.4

Degradation. In addition to crosslinking, polymers may also degrade when exposed to high energy irradiation. In order to validate this response polyisobutylene(35), a degrading polymer, was simulated for an exposure of 30 Mrad. The result of the simulation (Figure 7) showed very good agreement for the molecular weight of the constituent chains at higher doses. The reason for this being the molecular weight distribution difference between simulation and experiment. It can be seen in the low absorbed dose region that the molecular weight of the polyisobutylene of experiment was higher than that of experiment. The simulation was performed with the number average molecular weight as the only molecular weight input. The polymer chains in the simulation had a Poisson distribution which does not reflect that of experiment. As the radiation process continues however, the results of the two methods become closer. This is due to the larger polymer chains of experiment having a greater probability of chain scission as a result of their greater length. Given a sufficient dose, the molecular weight of the two systems should tend towards a common value as seen in Figure 7.

FORMATION OF SINGLE EDGE LOOPS IN GEL

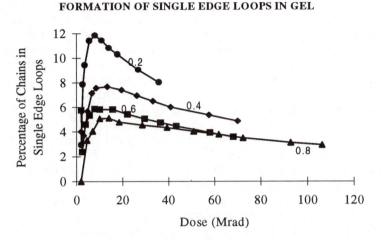

Figure 6. Dependence of intramolecular loops on the dilution of the system. The gelation of several poly(dimethylsiloxane) solutions were simulated and the percentage of intramolecular loops recorded. The inset values represent the volume fraction of polymer in the system.

DEGRADATION OF POLYISOBUTYLENE

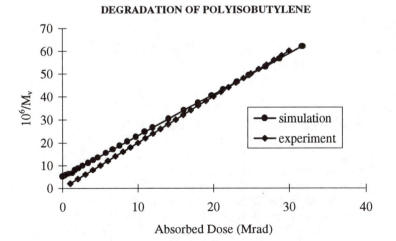

Figure 7. Comparison of the molecular weight of polyisobutylene during the irradiation process between experiment(34) and simulation.

Conclusions

The study performed is believed to have covered the pertinent aspects of irradiation curing. Presented first was gelation with respect to dose and examination of the topology of the network. Comparison of the gel curves of both experiment and simulation was the least detailed of the validation tests. It showed quite good agreement for all scission to crosslink ratios. This however, is not a definitive statement on the accuracy of the algorithm but did support the methodology for deriving the radiation dose from experiment. The true test and the strength of the algorithm is its ability to describe the topology of the resultant network structure. Validation of this portion with theoretical results for effective chains, dangling chain ends, and total weight fraction of elastically ineffective material was made. These results were in excellent agreement with those of theory. The only complication arising from the large molecular weight chains of a broad molecular weight distribution. The algorithm was then tested for a system which degrades upon irradiation (scission to crosslink ratio less than one) as a last test of the algorithm. This too showed very good agreement, particularly at higher irradiation doses where the effects of the higher molecular weight chains are less.

The algorithm has stood up to the challenges of various systems here however, the accuracy with which validation may be performed on a simulation algorithm is dependent on the wealth of experimental results. The simulation described here, although quite accurate in its prediction of gelation and degradation, lends itself to few comparisons with literature results. This however, is a point which is currently being addressed. The greatest error, albeit small, for the validations performed here is the molecular weight distributions. The ability to generate chains in a system with a broad molecular weight distribution would be the greatest step in bridging the gap between simulation and experiment.

Acknowledgments

The authors would like to express their gratitude to the Raychem Corp. for continuing financial support.

References

1.   Tonelli, A. E.; Helfand, E. *Macromolecules* **1974**, *7*, 59.
2.   Helfand, E.; Tonelli, A. E. *Macromolecules* **1974**, *7*, 82.
3.   Leung, Y.; Eichinger, B. E In *Characterization of Highly Cross-Linked Polymers*; Labana, S. S.; Dicckie, R. A., Eds; ACS Syposium Series 243; American Chemical Society:Washington, DC, 1984.
4.   Leung, Y.; Eichinger, B. E. *J. Chem. Phys.* **1984**, *80*, 3877.
5.   Leung, Y.; Eichinger, B. E. *J. Chem. Phys.* **1984**, *80*, 3885.
6.   Flory, P. J. *Proc. R. Soc. London. (A)* **1976**, *351*, 351.
7.   Flory, P. J. *Macromolecules* **1982**, *15*, 99.
8.   Erman, B.; Flory, P. J. *J. Chem. Phys* **1978**, *68*, 5363.
9.   Erman, B.; Flory, P. J. *J. Polym. Sci., Polym. Phys. Ed.* **1978**, *16*, 1115.

10.   Harary, F. *A Seminar on Graph Theory*; Hol, Rinehart and Winston, New York, 1967.
11.   Grest, G.; Kremer, K. *Macromolecules*, **1990**, *23*, 4994.
12.   Langley, N. R. *Macromolecules* **1968**, *1*, 348.
13.   Langley, N. R.; Ferry, J. D. *Macromolecules* **1968**, *1*, 353.
14.   Langley, N. R.; Polmanteer, K E. *J. Polym. Sci., Polym. Phys. Ed.* **1974**, *12*, 1023.
15.   Chapiro, A. *Radiation Chemistry of Polymeric Systems*; High Polymers Vol. XV John Wiley and Sons,New York, 1962.
16.   Charlesby, A *Atomics (London)* **1954**, *5*, 12.
17.   Neuburger, N. A.; Eichinger, B. E. *J. Chem. Phys.* **1985**, *83*, 884.
18.   Nijenhuis, A.; Wilf, H. S. *Combinatorial Algorithms Academic*: New York, 1975; Chapter 18.
19.   Basheer, R.; Dole, M. *J. Polym. Sci.* **1983**, *21*, 949.
20.   Mitsui, H.; Hosoi, F. *Polym. J.* **1973**, *4*, 79.
21.   Patel, G.N. *J. Polym. Sci.* **1975,** *13*, 339.
22.   Shy, L. Y.; Eichinger, B. E. *Macromolecules* **1986**, *19*, 2787.
23.   Galiatsatos, V.; Eichinger, B. E. *J. Polym. Sci., Polym. Phys. Ed.* **1988**, *26*, 595.
24.   Flory, P. *J. Chem Rev* **1944,** *35*, 57.
25.   Beuche, A.M. *J. Polym. Sci.* **1956**, *19*, 297.
26.   Moore, C.G.; Watson, W. F. *J. Polym. Sci.* **1956**, *19*, 237.
27.   Flory, P. *J. Am. Chem. Soc.* **1941**, *63*, 3083.
28.   Flory, P. *J. Am. Chem. Soc.* **1941**, *63*, 3091.
29.   Flory, P. *J. Am. Chem. Soc.* **1941**, *63*, 3096.
30.   Stockmayer, W.H. *J. Chem. Phys.* **1943**, *11*, 45.
31.   Stockmayer, W.H. *J. Chem. Phys.* **1943**, *12*, 125.
32.   Pearson, D. S.; Skutnik, B. J.; Bohm, G. G. A. *J. Polym. Sci.* **1974**, *12*, 925.
33.   Chiu, D. S.; Su, T.-K.; Mark, J. E. *Macromolecules* **1977**, *10*, 1110.
34.   Su, T.-K.; Mark, J. E. *Macromolecules* **1977**, *10*, 120.
35.   Wundrich, K. *Eur. Polym. J.* **1974**, *10*, 341.

RECEIVED October 5, 1995

Chapter 8

Low-Voltage Electron-Beam Simulation Using the Integrated Tiger Series Monte Carlo Code and Calibration Through Radiochromic Dosimetry

Douglas E. Weiss[1], Harvey W. Kalweit[1], and Ronald P. Kensek[2]

[1]Corporate Research Process Technologies Laboratory, 3M Company,
St. Paul, MN 55144–1000
[2]Simulation Technology Research Division, Sandia National Laboratories,
Albuquerque, NM 87185–1179

ITS(the Integrated Tiger Series) is a powerful, but user-friendly, software package permitting state-of-the-art modeling of electron and/or photon radiation effects. The programs provide Monte Carlo solution of linear time-independent coupled electron/photon radiation transport problems. A simple multi-layer slab model of a low voltage curtain electron beam was constructed using the ITS/TIGER code and it consistently accounted for about 71% of the actual dose delivered across the voltage range of 100 - 300 kV. The differences in calculated values was principally due to the 3D hibachi structure which effectively blocks 22% of the beam and could not be accounted for in a 1D slab geometry TIGER model. A 3D model was constructed using the ITS/ACCEPT Monte Carlo code to improve upon 1D slab geometry simulations. Faithful reproduction of the essential geometric elements involved, especially the window support structure, accounted for 86-99% of the dose detected by routine dosimetry.

Since numerical experiments tend to be cheaper, faster and easier than physical ones, computer modeling of electron beams can be very advantageous provided that the software is sufficiently easy to use, sufficiently flexible to model the important parts of the hardware, and sufficiently accurate for confidence in the results. Over the years, the U.S. national laboratories have developed powerful computer programs to solve problems involving the transport of electrons and photons. Examples of such codes which permit modeling of arbitrary geometries using the Monte Carlo technique are ETRAN(Electron TRANsport)(see Chapter 7 of Reference *1*), ITS 3.0(the Integrated TIGER Series)(*2*), EGS4(Electron-Gamma Shower)(*3*) and MCNP(Monte Carlo N-Particle)(*4*).

ETRAN was originally developed at the National Institute of Standards and Technology(formerly the National Bureau of Standards) to solve electron problems with energies up to a few MeV, gradually being extended up to a GeV, but restricted to a few materials in simple geometries. ITS, developed at Sandia National Laboratories, is based on the same ETRAN physics package, but extends to any number of materials in arbitrary geometries with a much more practical user interface. EGS was developed at the Stanford Linear Accelerator Center originally for high-energy(many GeV) cascades down to electron energies of 1 MeV, gradually being extended down to 10 keV. MCNP was developed at Los Alamos National Laboratory originally to treat photons and neutrons, with the electron physics of an earlier version of ITS added more recently.

Of course, different codes will have different strengths. To name only one of each: ETRAN and ITS-3.0 presently have the most sophisticated physics for low-energy electrons, EGS4 has an option for the most detailed treatment of boundary-crossing for electrons and MCNP has very powerful, built-in variance-reduction techniques for photon transport. However, selection of which code to chose is often more arbitrary: namely, whichever is perceived to be the most expedient for the user. Such was the case in our selection of ITS 3.0.

ITS 3.0, EGS4 and MCNP each enjoy a world-wide user base. In particular, ITS has been employed, for example, in the assessment of personnel radiation hazards, the disposal and cleanup of radioactive waste, the pasteurization and disinfestation of foodstuffs, the radiation vulnerability of satellite systems, electron-beam joining, the sterilization of hospital waste, radiation treatment planning, and the safety of nuclear power reactors. A comprehensive survey of ITS experimental benchmarks and engineering applications can also be found in Chapter 11 of Reference *1*. In this paper, we have employed the code for prediction of dose, depth/dose profiles and dose distributions in low voltage electron beams, now becoming more common place as a solventless industrial process technology.

There are many variables associated with electron beam processing, including all of the elements of process variability and gradient cure. Thin film radiochromic dosimeters are commonly used for dose measurement and calibration of electron beams, but often the results of dosimetry can not be correlated with the specific dose absorbed by thin coatings(2 microns) or dissimilar materials. A good coupled electron-photon transport simulation code can help address these types of problems. The specification of a cure gradient as a function of voltage, thickness, density and atomic number can be modeled. The effects of backscatter, air gaps and window thickness, and the effects of compositional variability on depth/dose penetration can also be screened.

For the past three decades, Monte Carlo calculations of low-energy beam dose distributions in various absorbing media have shown agreement(see Chapters 8 and 11 of Reference *1* and included references). Comparisons for simpler geometries are more definitive due to the greater fidelity of the simulation and often greater precision in the measurement. In particular, it has been shown that for degraded electron beams having primary energies in the 200-400 keV region and incident in air perpendicular to simple-geometry absorbers, there is reasonable agreement(generally within 5%) between experimental data provided by radiochromic film dose mapping and Monte Carlo calculations of dose distribution(*5*).

We have found the TIGER code of the Monte Carlo Integrated Tiger Series very useful for web product geometries because of the unlimited number of layers that can be modeled and the ability to independently subzone each layer(2). Since the web structure is large compared to dimensions in the beam direction, the one-dimensional nature of TIGER is not a factor limiting accuracy and the assumption of infinite boundaries is a reasonable one. Setup is simplified and run times shortened compared to more comprehensive(and complicated) multidimensional codes. Complex 3-dimensional geometries comprised of widely different materials can be handled using the features of the ITS/ACCEPT code, in which combinations of fundamental geometric shapes are described using Boolean Logic; different materials parameters can be assigned to any geometric zone. Using ACCEPT, a full scale model of an Energy Sciences, Inc (ESI) "Electrocurtain" electron beam system was constructed that includes reproduction of all major geometric features of the web chamber including metal plates, aluminum window clamp, titanium window, copper block with 'hibachi' support structure, nitrogen gap, lead sideplates and steel backplate. The hibachi structure required the greatest amount of coding and contains 112 ribbed slots which hold the titanium window against high vacuum. The electron source was modeled as a 2.54 cm by 40.64 cm rectangle to simulate the rectangular screen grid of the same dimensions at the cathode terminal of our electron beam. The source is assumed to be a planar wave based on the assumption that space-charge effects are not significant at the flux densities studied and that the thermal energies of the electrons as they reach the terminal grid are a small fraction of the accelerating potential. A multi-layered web target within the nitrogen inerted chamber was modeled to simulate a sandwich of materials consisting of a nylon FWT-60 dosimeter on top of a thick layer of paper moving on a biaxially oriented polyethyleneteraphthalate (PET) carrier web.

Overview of the Monte Carlo Method. The Monte Carlo method is based on statistical averaging of individual trials to estimate the desired numerical answer. Each trial is the history of the events during the passage of a single electron through the defined geometries, and the final result is the average of the passage of many electrons. Specific events which make up such histories are constructed by using random numbers to sample probability distribution functions to determine the type of interaction occurring and the interaction parameters. Such parameters may describe the motion of an electron moving in a given material at a specified energy, the generation of secondary particles, and the motion of these particles and any further cascade. The energy deposited in the interaction volume is then calculated and summed with the results of interactions of other electrons in that same volume to arrive at the total energy distribution for the mathematical "experiment." Clearly, the greater the number of electron histories, the lower the statistical uncertainty(and the longer the running time) of a calculation.

Due to the statistical nature of the results, any quantity of interest calculated by the code must have attached to it another quantity which represents the estimated statistical accuracy, typically expressed in terms of a standard deviation. In ITS, the total number of histories(source particles to be simulated) are divided up into an equal number of batches. In the present calculations, we typically ran 10 batches. The estimated mean value is simply the average of the mean values from each batch. The estimate of the standard deviation is the calculated deviation of this average value from each of the batch mean values. It is important to keep the code's estimate of the standard deviation a small fraction of the calculated quantity of interest. When this estimated fraction becomes large(>10%), the estimated fraction itself is unreliable and may be larger than what is indicated.

Another important aspect of electron/photon Monte Carlo is that the transport of photons is both faster and more accurate than that of electrons. The photons have a relatively large mean free path which means they typically pass through the material geometry of interest with few interactions(scattering events). The simulation can follow the photons from interaction site to interaction site accurately and efficiently. The electrons, however, are constantly interacting, changing direction and losing energy. It is completely impractical to follow the electrons from interaction point to interaction point. Therefore, the electrons are pushed a predetermined distance. The simulation then tries to account for the cumulative effect of all the multiple interactions which took place over this "step" length. This is termed the "condensed history" method.

Overview of the ITS Package. There are four essential elements of the ITS system; **(1) XDATA**, the atomic data file, **(2) XGEN**, the program file for generating cross sections, **(3) ITS**, the Monte Carlo program file and **(4) UPEML**, the update processor utility. The atomic data file,**(1)**, contains the data for generating cross sections for arbitrary homogeneous mixtures or compounds of the first 100 elements. The two program files,**(2)** and **(3)**, contain multiple machine versions of the multiple codes integrated in such a way as to take advantage of common coding. Their corresponding binary program libraries are input to the update processor,**(4)**, which selects a particular code for a particular machine and makes any modifications requested by the user(referred to as an 'update' or a 'correction' run). The output of the processor in this case is FORTRAN code that is ready for compilation.

Table I shows the eight member codes of the ITS Monte Carlo program file. From left to right, the codes grouped by column will be referred to as the standard codes, the P codes and the M codes, respectively. All member codes allow transport over the range of 1.0 GeV to 1.0 keV. All calculations in this work made use of the standard codes. These calculations take into account the primary electrons and all secondary radiations, including knock-on electrons from electron-impact ionization events, bremsstrahlung, Compton electrons, photoelectrons, and K-shell characteristic x-rays and Auger electrons resulting from electron and photon ionization events.

P Codes. Fluorescence and Auger processes in the standard codes are only allowed for the K shell of the highest atomic element in a given material. For some applications, for example, the calculation of energy spectra of low-energy escaping particles, it is desirable to have a more detailed model of the low-energy transport. In the P codes, a more elaborate ionization/relaxation model from the SANDYL(*6*) code was added to the standard codes.

M Codes. In the M codes, the collisional transport of the standard codes is combined with the transport in macroscopic electric and magnetic fields of arbitrary spatial dependence. This could be used for example, to model a magnetic field applied in an experiment to turn back electrons which would otherwise have escaped or in simulating a magnetic spectrometer.

To run the programs in the ITS package, the flow of input and output must be understood. Apart from the atomic data file, which is read by the cross-section generator, and the cross-section output file produced by the generator which is read by the Monte Carlo, there are two other types of input required for running either the cross-section generator code or the Monte Carlo code. First, there is the set of instructions to the processor that tells it how to produce a compile-ready FORTRAN code from either of the program libraries. Simple syntax allows the user the option to modify the source code via deletions, insertions and replacements of FORTRAN code. The second type of input is that required by the resulting executable code. For example, the problem materials and energy range must be specified for the cross-section generator code, and the problem geometry and the source distribution must be defined for the Monte Carlo code.

Operational Strategies. Real experiments are performed in the three dimensional world, but new users are urged to consider how the actual geometry may be simplified for modeling purposes. The input required to describe the TIGER geometry is quite straight-forward: simply specify the material, number of subzones and layer thickness in centimeters.

The input needed to describe arbitrary three-dimensional geometries requires some care and practice on the user's part. The ACCEPT code of ITS uses "combinatorial geometry"(see section 17.3 of Reference *1* for its three-dimensional input). In combinatorial geometry, geometrical regions are described as various logical combinations of a set of primitive body types(such as spheres, boxes, arbitrary polyhedra, ellipsoids, rectangular right cylinders, and truncated right cones). Logical combinations mean intersections(by an implied "AND"), unions(by an explicit "OR"), and negations. The escape zone is a region which completely surrounds the simulated universe that exists. When a particle enters the escape zone, the code considers it has escaped and will no longer continue to track it.

The ability to visualize the specified geometry is essential for verifying the setup of three-dimensional simulations. Appendix I in the ITS User Manual(*2*) describes how a user may interface the called FORTRAN subroutines within ITS with their local plotting package. Other users have employed a geometry-modeling code system such as SABRINA(*7*) to view their constructed geometry, others have written a translator(*8*) between AUTOCAD and ACCEPT for creating and/or displaying the specified geometry.

Low Voltage Electron Beam Simulation. Our interest in this simulation was to determine how well dosimetry calculations using a Monte Carlo code correlated with actual dosimetry data from an electron beam machine in use at 3M Company. A good correlation would lead to a calibrated model which could be used to simulate experiments quickly and in great detail with respect to energy deposition. Such a model could be used as a diagnostic tool and could extend the range of experimentation outside of the range of equipment capability.

A central issue addressed here is the effective beam stopping power of the hibachi support structure. This structure can not be included in a one dimensional ITS/TIGER model but a full ITS/ACCEPT model was constructed so that the results between the two codes could be compared.

Dosimetry Calibration and the "K" Curve. Calibration of the machine output to the measured dose is accomplished by way of the 'K' value or machine constant which adjusts the machine output to the dose observed. The machine constant(K) is also known as the cure yield and appears in the 'KIDS' equation, as it is commonly remembered, and shown in equation 1.(*9*) The K value changes as a function of voltage, especially below 175 kV. This is because the K value includes the effective stopping power of the window foil, air gap and other factors such as the window clamp support structure or "hibachi" and overspray. The shape of the K curve is mostly determined by the window foil and air gap since this has the most direct attenuating effect on the electrons and this effect is strongly influenced by voltage.

TIGER Calculated Dose. Since the TIGER model output is the average energy deposited per electron in a specified material layer, the dose delivered by an electron beam can be directly calculated without knowing "K". This is because the dose is directly related to the energy deposited per electron in a given layer times the current(number of electrons) divided by the area of the target material. There may still be a correction factor, k', to include, for example, 3-D effects that TIGER cannot model. The calculation is derived in the following manner:

Dose calculation for continuous operation of an electron beam:

$$\text{Dose} = \frac{K\ I}{S} = k' \times\ 100\ \text{Mrad}\ \times \frac{[I \times E]}{A} \qquad (1)$$

where: **K** = machine constant, **I** = current extracted from filament in mA, **S** = web speed in ft/min, **A** = area/sec = filament length[cm] x web speed [cm/sec], **E** = energy deposited per source electron(per cm^2) per gram = Mev -cm^2/g-electron, **k'** = correction factor
The 100 Mrad conversion factor is derived from the following relationship:

$$1\ \text{MeV-cm}^2/\text{g-electron} = 100\ \text{Mrad-cm}^2/\text{mA-sec}$$

The model k' value corrects the dose output calculated by the TIGER model to the dose experimentally observed for a specific machine and represents the lumping of the miscellaneous factors that are not a part of this model, such as the 'hibachi' structure. The average value of k' in the case modeled here was 0.71 and is mostly derived from the 22% blockage of the beam by the hibachi structure. Hence, there are other apparent losses in this particular piece of equipment as well, possibly a result of other 3D structures such as the aluminum window clamp that surrounds the window area.

Accept Calculated Dose. In ACCEPT, the dose is directly related to the energy deposited in a specific zone per electron times the current (number of electrons per unit time) divided by the total grams of the target material. The dose calculation for continuous operation of an electron beam is given by equation 2.

$$\text{Dose} = \frac{K * I}{S_{fpm}} = 100\ \text{Mrad}\ \times \left[\frac{E}{[d*T*W*L]}\right] \times \frac{[I*L]}{S_{cps}} \qquad (2)$$

TABLE I. Monte Carlo Member Codes of ITS

GEOMETRY	STANDARD CODES	ENHANCED PHYSICS	MACROSCOPIC FIELDS
1-D	TIGER	TIGERP	
2-D/3-D	CYLTRAN	CYLTRANP	CYLTRANM
3-D	ACCEPT	ACCEPTP	ACCEPTM

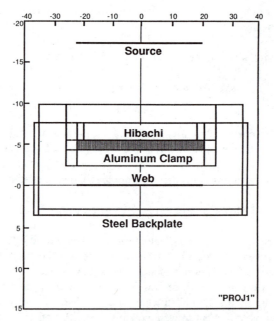

Figure 1: 3-D End View Plot of ACCEPT Electron Beam Model. (Reproduced with permission from reference 17. Copyright 1994 RadTech International North America.)

where: **K** = machine constant, **I** = current extracted from filament in mA, **S** = web speed in ft/min or cm/sec, **E** = energy deposited per source electron in modeled dosimeter = MeV/electron, **d** = density of dosimeter composition in g/cm^3, **T** = thickness of modeled dosimeter(cm), **W** = width of modeled dosimeter(cm), **L** = length of modeled dosimeter(cm)

The ACCEPT calculation differs from the TIGER calculation in several ways. Because the source is described in full detail in the input file, the filament length is already factored into the ACCEPT energy output and is therefore not included in equation 2 for area calculation; web speed alone is sufficient. ACCEPT does not calculate dose in MeV-cm^2/g-electron as does TIGER. Standard output is in MeV of energy deposited in a given subzone per source electron. Therefore, the energy(MeV) needs to be divided by the total grams in the modeled dosimeter zone to get dose per source electron. The total grams is the product of the density times the volume. Since we are converting the calculated dose, we need to use the volume of the dosimeter used in the model which is length(**L**) times width(**W**) times thickness(**T**).

By dividing by the total grams(density times volume), we have converted **E** into units of MeV/g-electron. Now we need to multiply by the number of electrons hitting the actual dosimeter chip when it has moved the length of our dosimeter used in the model. This is just the current **I** times an effective "time" which is the ratio of the model length **L** divided by the web speed **S**. Since **L** appears in both the numerator and the denominator, the calculation will be independent of the size of **L**(provided it is long enough to prevent energy from leaking out the ends).

Geometry Verification. Plots of the geometric construction are outputted by ACCEPT when the key word PLOT is used in the input file, followed by specification of scale and angles of perspective. These output files were plotted using GNUPLOT(*10*) and are shown in Figures 1 and 2. Figure 1 shows an "end view" cross-section of the electron beam. In this side view, the source appears as a line suspended in the vacuum above the window clamp/nitrogen-inerted chamber construction with the web suspended in the chamber. Note the individual rib slots that support the titanium window against the vacuum. Figure 2 shows a top view of the chamber construction; the ribbed hibachi structure is the most prominent feature. Within it is a rectangle which represents the grid source of electrons. The center strip in the web(vertical) direction represents a dosimeter that is 28 cm long x 10 cm wide. The dosimeters are actually only one cm square but a 10 cm width was used for calculation purposes to provide better statistics with fewer histories. This is possible because the model cross-web distribution is uniform. However, this was not true experimentally. The electron beam source non-uniformity was not accounted for in this Monte Carlo simulation model.

Experimental

Routine transfer dosimetry was performed using Far West Technologies FWT-60 nylon Radiochromic dosimeters(*11*). These dosimeters, like some others, are sensitive to humidity and temperature during exposure to radiation(*12-14*). They are also sensitive to light in general(*15*). Therefore, certain precautions were followed. The

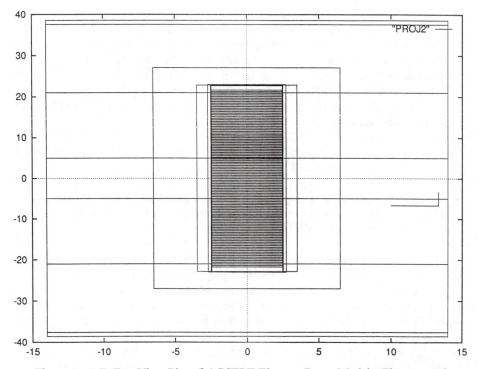

Figure 2: 3-D Top View Plot of ACCEPT Electron Beam Model. The array of vertical lines in the center of the figure is the hibachi structure that supports the window foil. The rectangular outline centered within this array describes the location of the grid source of electrons. The vertical column traversing the center of the figure is a dosimeter strip, the wider vertical outline is the web. Other rectangular shapes describe the window support structure and clamp. (Reproduced with permission from reference 17. Copyright 1994 RadTech International North America.)

dosimeters were kept in a chamber to maintain 52% relative humidity(R.H.) during storage. During handling and exposure, the ambient conditions were 50% R.H. with an air temperature of about 24° C. A total of 36 dosimeters were mounted by an edge onto a double thickness of index cards taped together to cover the full width of the web(30 cm). The dosimeters were spaced 2.5 cm apart and 3 deep across the 30 cm web width. The cards of dosimeters were passed through an ESI "Electrocurtain" model CB300 electron beam on a carrier web. The dosimeters were left uncovered during their passage through the electron beam at a web speed of 50 +/- 0.5 fpm with a current of 10 mA. The current was incremented an additional 0.1 to 0.8 mA as the voltage was increased from 100 to 300 kV to compensate for a certain amount of known current leakage in the system. The voltage was increased at increments of 25 kV from a voltage setting of 100 kV up to 300 kV. The momentary temperature increases during the irradiation do not affect the dosimetry readings at the dose rates studied(*14,15*). Except for the handling, exposure and reading of the dosimeters, the dosimeters were covered at all times to protect them from stray light. After allowing 24 hrs for the dosimeters to anneal at ambient temperature, the optical densities of the dosimeters were read on a FWT Radiachromic Reader at the 510 nm wavelength setting. The baseline absorption was subtracted from the readings and the corrected optical densities were converted to Mrads using a calibration curve established through NIST. Each of the optical densities recorded in Table II is an average of 36 individual dosimeter readings at each voltage. A series of 10 dosimeters were arranged closely together at the center of the web in the downweb direction at 200 kV to determine, as closely as possible, a standard deviation for the dosimetry measurement alone. The estimate of the standard deviation for the measurement was 2.5%. The estimate of the standard deviation for the dosimetry at each voltage, however, varied between 3 and 17%(Table II). This is the result of both variation in the measurement itself plus an additional variation due to some source non-uniformity that exists in this particular electron beam. Thus, the average value across the web is our best estimate of the beam output and is the number that Monte Carlo output would predict.

An additional experiment was run in the same manner using stacks of six dosimeters to obtain depth/dose information at several voltages(125, 175 and 300 kV) for the purposes of voltage calibration. The dosimeters from this lot were 45 microns thick and the stacks were run down the center of the web only. Because the dose is lower at the center than the average dose across the web, a k' value of 0.78 was used for TIGER calculations. These dosimeters were unstacked after exposure and, after a 24 hr period to anneal at room temperature, were individually read to get depth/dose information.

Monte Carlo Simulation

Monte Carlo simulation was modeled both as a one dimensional and as a three dimensional problem using the Integrated Tiger Series(ITS) codes(version 3). The code was compiled and run on an IBM Model 520 RS/6000 Workstation. A "correction" run was made to the source code to modify the source of radiation to a rectangle

TABLE II. Dosimetry Data and Simulation Results.

Voltage kV	Optical Density*	Dosimetry (Mrads)	Percent Standard Deviation	Machine Constant K	TIGER Output (Mrads)	TIGER k'	ACCEPT Output (Mrads)	ACCEPT vs Dosimetry % Deviation
100	0.0938	0.937	9.26	4.7	1.36	.68	0.976	+4
125	0.315	3.43	3.20	17	5.14	.66	3.91	+14
150	0.402	4.62	5.97	23	6.68	.68	5.09	+10
175	0.426	4.95	6.24	25	6.95	.71	5.13	+4
200	0.412	4.73	5.89	24	6.61	.71	4.79	+1
225	0.396	4.51	5.97	22	6.17	.73	4.48	-1
250	0.367	4.12	5.92	21	5.58	.73	4.24	+3
275	0.354	3.96	17.3	20	5.07	.76	3.91	-1
300	0.326	3.58	5.77	18	4.67	.75	3.66	+2

* average of 36 readings across web width, corrected for baseline

corresponding to the 2.54 cm x 40.64 cm grid source of electrons at the electron beam cathode.

The 1-D Monte Carlo simulation was modeled as a six layer problem using the TIGER code. The six layers, in order, are titanium(0.00127 cm)/nitrogen(4.25 cm)/ FWT nylon(0.00485 cm)/paper(0.040 cm)/PET carrier web(0.01016 cm)/steel(1 cm). A total of 20,000 histories in 10 batches were run at each voltage to ensure good statistics(1-2 % standard deviation). Each run took about 5-6 minutes. The 3-D Monte Carlo simulation was modeled using the ACCEPT code and included all of the essential elements of the electron beam. A total of 40,000 histories in 10 batches were run at each voltage to ensure good statistics(1-3 % standard deviation). Run time varied between 1.5 hours to 2.26 hours per run; lower voltages took more time. The default global cut-offs were used, they were 5% of maximum source energy for electrons and 10 keV for photons.

A summary of the data collected and model calculations are shown in Table II.

Results and Discussion

Monte Carlo Output vs Dosimetry. The model output calculated for the FWT-60 dosimeters were converted from Mev-cm2/g-electron to Mrads using the information in Table II and equation 1. A plot of the dosimetry data and also of the calculated doses(TIGER and ACCEPT) received by the dosimeter in units of Mrads as a function of voltage is shown in Figure 3. These curves all follow the same pattern as the "K" curve; increasing sharply up to 175 kV and then decreasing gradually as the voltage is increased to 300 kV. This is the predictable result of the energy peak moving from the window/gap area to the plane of the dosimeter(at 175 kV) and then moving beyond that plane into the substrate. The dose output of the models correlate very well with the shape of the "K" curve because the major attenuation factors of the window foil and air gap are included in both models.

The ACCEPT code overestimates the dosimetry measurements in the range of 125 - 150 kV by 10-14%, this is slightly outside of the average standard deviation of the measurements(7%). At other voltages, ACCEPT is within a few percent of measurements and well within the error bars of both the measurements and the calculations. The standard deviations of the measured dosimetry, based on an average of 36 readings per voltage was typically about 7% as shown in Table II. The estimated statistical uncertainties from the Monte Carlo calculation, as mentioned earlier, was 1-2%. The area of greatest discrepancy between measurement and calculation may be indicative of either additional losses outside of the beam path(and we have observed this intermittently in this electron beam) or additional error involved in the treatment of the physics within the calculation. Acknowledging it is difficult to make general error estimates for any electron code, Martin Berger, in Chapter 8 of Reference *1*, suggested that the physics package of ETRAN can "commonly" result in error estimates of 5% for simple quantities and 10-20% for more complex ones. Since the physics engine of ITS is based on that of ETRAN, we can expect similar estimates for the ITS codes. However, the ACCEPT calculations are a clear improvement over the simple 1-dimensional TIGER ones which consistently overestimate the dose by about 30%. This overestimate can be mostly accounted for by the 3D hibachi structure which effectively blocks 22% of the open path between the electron source and the dosimeter in this particular electron beam but must also include other factors as well. When TIGER output is reduced by 71%(k'), the numbers are in good agreement with dosimetry as shown in Figure 3. The ability to apply a single correction factor is indicative that the loss factor is consistent over the voltage range studied.

Depth/dose plots of the TIGER simulations in Figure 4 show the effects of voltage and attenuation in the modeled cross-sections. The differential dose across the dosimeter subzones is a very steep descent at low voltages(<150 kV) and penetration of the dosimeter is incomplete at 100 kV. The plot of the depth/dose distribution at 200 kV shows even and complete penetration of the dosimeter.

Dose Distribution in the Target. Figure 5 shows a downweb distribution of dose as measured(data supplied by ESI) and as calculated by ACCEPT. The appearance of the distribution and spread is similar and shows that there is an envelope of dose rates centered about the window and that it is concentrated primarily in the +/- 5 cm region from window center. The source of the spread(scattering of electrons) is a result of this particular combination of window gap, foil thickness, hibachi and source width. The results of two ACCEPT simulations are shown; one assuming the source to approximate the 2.54 cm wide cathode grid opening observed and the other assuming a more narrow source (2.0 cm wide). The assumption that the source was more narrow was justified by the fact that the extraction grid, located behind the cathode grid was measured and found to be 2.1 cm wide. The present calculations were based on the 2.54 cm width. The 2.0 cm width assumption appears to better fit the experimental data observed by ESI, indicating that some care needs to be exercised here in describing the source.

Energy Deposition in the Web Chamber Area. Figure 6 shows ACCEPT calculations of where the energy is deposited in the various components of the equipment as a function of voltage. The curves show that the window captures less radiation as the voltage increases and this is reflected in the rising K values up to 175 kV. The curves also show that the ribs stop electrons and the captured energy increases in a linear fashion in proportion to the accelerating voltage. This is reflected, experimentally, in temperature studies we have done to determine the mechanisms of window heating(*16*). As the voltage reaches 250 kV and beyond, the amount of energy that reaches the beam stop rises rapidly as more electrons are able to penetrate the layers of web materials with higher residual amounts of energy. The aluminum window clamp captures a small but fairly consistent amount of energy suggesting that it may be the additional 3D loss factor that is included in k' along with the hibachi to correct TIGER output.

Voltage Calibration. A stack of six dosimeters was passed through this same electron beam in order to get a depth/dose profile through a thickness of about 270 microns. Similarly, a stack of six dosimeters(each is 45 microns thick) was modeled using the TIGER code in the same manner as described in the Monte Carlo simulation section. The dose in Mrads was calculated using equation 1, k' = 0.78 for center only. The results between simulation and dosimetry were compared at 125, 175 and 300 kV. The results were in very good agreement at each voltage and the comparison at 175 kV is shown in

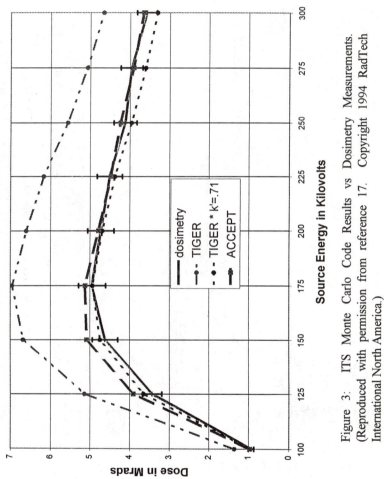

Figure 3: ITS Monte Carlo Code Results vs Dosimetry Measurements. (Reproduced with permission from reference 17. Copyright 1994 RadTech International North America.)

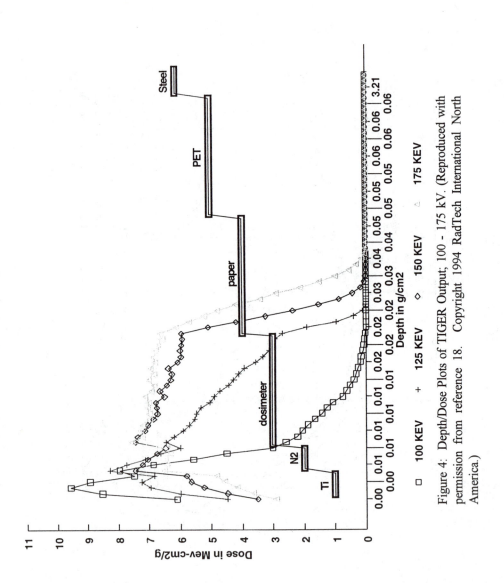

Figure 4: Depth/Dose Plots of TIGER Output; 100 - 175 kV. (Reproduced with permission from reference 18. Copyright 1994 RadTech International North America.)

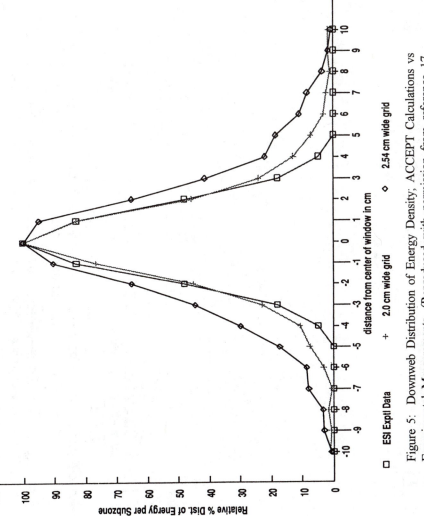

Figure 5: Downweb Distribution of Energy Density; ACCEPT Calculations vs Experimental Measurements. (Reproduced with permission from reference 17. Copyright RadTech International North America.)

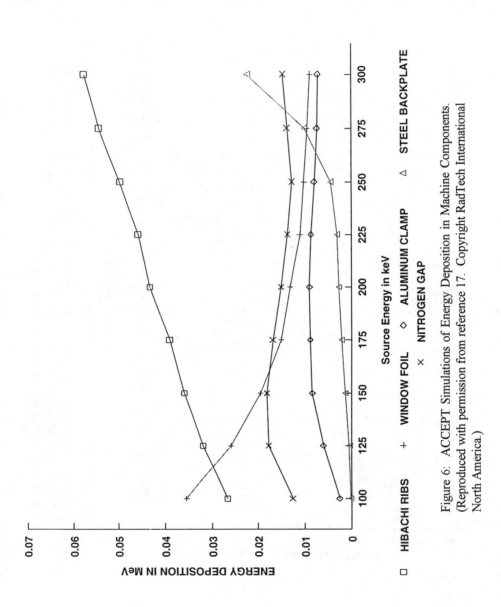

Figure 6: ACCEPT Simulations of Energy Deposition in Machine Components. (Reproduced with permission from reference 17. Copyright RadTech International North America.)

Figure 7. The ability to identify an operating voltage by fitting the 1-D simulation through the use of k' and equation 1 to the observed depth/dose distribution over a range of voltages is very useful as a method to test voltage calibration given that the nitrogen gap is accurately specified in the code.

Conclusion

An experimental/computational correlation of an ESI "Electrocurtain" electron beam has been established through dosimetry. Calibration of an electron beam by transfer dosimetry should establish a single correction factor for that machine that can be used in conjunction with TIGER to predict actual dose output from that machine through any combination of materials. In this case, the average k' value of 0.71 would suffice.

ACCEPT calculations do not require any correction factors if all of the essential 3D elements have been accurately specified. Electron beams may have other losses of current than those directly in the beam path. The code can be used in these cases to help determine how much loss is occurring as a result of these other loss mechanisms.

A calibrated Monte Carlo simulation model can be used to calculate a reliable dose

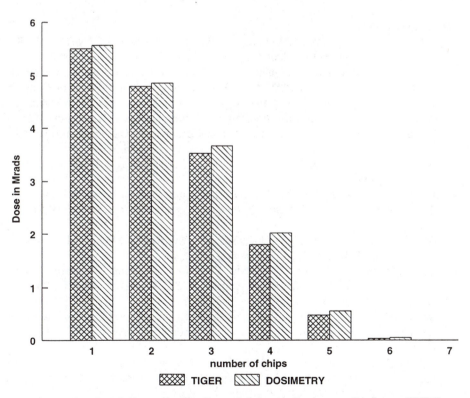

Figure 7:   Depth/Dose Profile Recorded by a Dosimeter Stack vs TIGER Simulations; 175 kV, 10 mA/50 fpm.

anywhere in a construction of materials at any voltage and under any operating conditions - real or theoretical. With such a model, one could test the effect of alternative materials, different air gaps, different backscatter elements and higher or lower voltages than available.

Others have reported good correlation between machine dosimetry and slab codes similar to TIGER(5) but this may be the first report of a correlation between a 3D Monte Carlo code and a low voltage curtain beam. The 3D-functionality of the ACCEPT simulation allows modeling the effects of machine modifications on dose and mapping the dose distribution in non-web articles.

Acknowledgments

The authors express their gratitude to Ms Amanda C. Meeks for doing the dosimetry on our electron beam during her participation in the Engineering Scholars Program at 3M Company.

The work performed at Sandia National Laboratories was supported by the U.S. Department of Energy under Contract No. DE-AC04-94AL 85000.

Literature Cited

1. *Monte Carlo Transport of Electrons and Photons*, Jenkins,T. M.; Nelson, W. R.; Rindi, A., Eds.; Plenum Publishing Corporation, New York and London, **1988**.
2. Halbleib,J. A.; Kensek,R. P.; Mehlhorn,T. A.; Valdez,G. D.; Seltzer,S. M.; Berger,M. J. *IEEE Trans. Nucl. Sci.* **1992**, *39*, 1025. "Integrated TIGER Series of Coupled Electron/Photon Monte Carlo Code System" ITS 3.0(CCC-467), is available for purchase with a users manual from the Radiation Shielding Information Center Computer Code Collection, Oak Ridge National Laboratory. Oak Ridge also puts out a monthly newsletter which notifies users of any updates or revisions.
3. Nelson,W. R. ; Hirayama,H. ; Rogers,D. W. O. *Stanford Linear Accelerator Report* SLAC-265, **1985**.
4. *MCNP-A General Monte Carlo Code for Neutron and Photon Transport, Version 3A*, Briesmeister, J. F., Ed, Los Alamos National Laboratory Report LA-7396-M, Rev. 2, **1986**.
5. McLaughlin,W. L.; Kahn,H. M.; Farahani, M.; Walker,M. L.; Puhl,J. M.; Seltzer,S. M.; Soares,C. G.; Dick,C. E.; *beta-gamma*, **1991**, *4*, 20.
6. Colbert,H. M. "SANDYL: A Computer Code for Calculating Combined Photon-Electron Transport in Complex Systems," SLL-74-0012, Sandia National Laboratories, **1974**.
7. SABRINA(PSR-242) from the Radiation Shielding Information Center at Oak Ridge National Laboratories), Los Alamos National Laboratory Report LA-UR-93-3696, **1993**.
8. Colborn,B. L.; Potter,D. W.; Armstrong,T. W. "A Space Station Freedom CAD Mass Model for Ionizing Radiation Analyses,"Science Applications International Corporation Report SAIC-TN-9301, **1993**.
9. McLaughlin,W. L. *RadTech '90 North America Conf Proceedings*, **1990**, *2*, 91.

10. GNUPLOT: GNU Emacs is free, copyright 1988. Richard M. Stallman. Accessible from Internet via host 'prep.au.mit.edu' using anonymous login.   See file '/u2/emacs/GETTING GNU SOFTWARE' or write Free Software Foundation, 675 Mass Ave, Cambridge, MA 02139. GNU Emacs Manual, 6th ED, Emacs Version 18 for Unix Users, Feb 1988.
11. Humphreys,K. C.; Kantz,A. D. *Radiat. Phys. Chem.*, **1977**, *9*, 737.
12. Humphreys,K. C.; Rickey,J. D.; Wilcox,R. L. *Radiat. Phys. Chem.*, **1990**, *35*, 713.
13. Ningno,T.; Ehlermann,D. A. E. *Radiat Phys Chem*, **1994**, *43*, 569.
14. McLaughlin,W. L.; Humphreys,J. C.; Ba, W.-Z.; Khan,H. M.; Al-Sheikhly,M.; Chappas,W. J. *Proceeding of International Symposium on High Dose Dosimetry for Radiation Processing*, Vienna, **1991**, 305.
15. Postma, N. B.; Weiss, D. E.; Kalweit, H. W.; Janus,L. R. *RadTech '90 North America Conf Proceedings*, **1990**, *1*.
16. Kalweit,H. W.; Carter,C. G.; Postma,N. B.; Weiss,D. E. *Radtech '92 North America Conf Proceedings*, **1992**, *1*, 381.
17. Weiss, D. E.; Kensek,R. P. *Radtech '94 North America Conf Proceedings*, **1994**, *1*, 130.
18. Kalweit, H. W.; Weiss, D. E. *Radtech '94 North America Conf Proceedings*, **1994**, *1*, 104.

RECEIVED October 10, 1995

Chapter 9

A Recent Advance in the Determination of Scission and Cross-Linking Yields of Gamma-Ray Irradiated Polymers

David J. T. Hill, K. A. Milne[1], James H. O'Donnell[†], and Peter J. Pomery

Polymer Materials and Radiation Group, Department of Chemistry, University of Queensland, Brisbane, Queensland 4072, Australia

Scission of main chain bonds and crosslinking between polymer molecules, the two major effects of ionizing radiation on polymers, are discussed. The historical development of the theoretical treatments for radiation induced scission and crosslinking is reviewed. In particular, scission and crosslinking yields, G(S) and G(X) respectively, may be determined by observing the variation of the number average, weight average and z-average molecular weights with gamma radiation dose. The interpretation of the experimental results depends on the initial polymer molecular weight distribution. The molecular weight versus dose relationship is considered for the simple case of an initial most probable (random) distribution. For the more general case of a Schulz-Zimm (or Poisson) distribution, the equation proposed by Inokuti and Dole(1), is now able to be solved exactly by numerical methods(2). Precisely determined values of G(S) and G(X) for various experimental systems are presented and compared with the values determined by approximate methods.

During the early studies of the physical and chemical effects of ionizing radiation on materials the special case of high polymers became clear. In these polymers, not only the usual radiation chemical and physical events take place, but by virtue of the large number of monomer units composing the polymers, two events, scission of the polymer backbone and crosslinking of two or more polymer molecules, have great importance for the physical properties of these polymers. Depending on the specific structure of the polymer, other radiation-induced events include the formation of gaseous products, the reduction of existing unsaturation, and the production of new unsaturation(3). If oxygen is present, peroxy species are formed and undergo further reactions depending on the polymer composition(3).

The term scission is defined as any event which results in the breakage of one polymer molecule into two parts. It may occur as a result of ionization of the irradiated

[1]Corresponding author

[†]Deceased

polymer molecule and a direct rearrangement of the backbone into two separate entities, or the ionized polymer molecule may undergo loss of sidegroups and consequent rearrangement, which still has the final result of dividing the molecule into two separate entities, e.g. beta scission in poly(methyl methacrylate)(4). Sometimes scission is called degradation since the breakage of the long polymer molecules results in loss of structural strength and plasticity and the physical properties are degraded or diminished.

Crosslinking occurs when two polymer molecules join to form one large molecule. It may occur when hydrogen is abstracted from two neighbouring polymer molecules leaving two radicals in close proximity, which may then react to form a crosslink, such as an H-link, endlink or Y-link. The present work discusses H-links where a crosslink is formed between sites on the polymer molecules which are not at the ends of the molecules, thus forming a H shaped structure where the horizontal line in the H represents the crosslink. Crosslinking may result in improvement of the mechanical properties of a material since the molecular weight increases. For example, crosslinking causes a reduction of solubility, elimination of the melting point, and increased resistance to corrosive attack, all desirable material properties.

The present work does not distinguish between the mechanisms of scission or the mechanisms of crosslinking. The important physical fact is that one scission causes one molecule to become two, and one crosslink causes two molecules to become one.

It is generally accepted that during irradiation of a polymer, both scission and crosslinking events are occurring randomly and simultaneously. The predominance of scission over crosslinking depends on the polymer structure, temperature, crystallinity, and the presence of air. In the case of scission, the peroxy radicals prevent the geminate recombination of the radical chain ends(3). If scission predominates, then degradation of physical properties occurs, and the polymer may become unusable. If crosslinking predominates, then gelation will eventually occur at high enough doses. The gel point is defined as the point where there is at least one crosslinked monomer unit in each molecule.

Charlesby(5) reviewed the development of mathematical models for the effect of random simultaneous scission and crosslinking on polymer properties. The germinal idea was to use the average molecular weights of the polymer samples to measure the amount of scission or crosslinking. The effects of radiation dose depend critically on the initial molecular weight distribution of the polymer i.e. the molecular weight distribution before irradiation. Charlesby assumed a random or initial most probable distribution, thus simplifying the problem. His equations for this case are used at the present time although his symbolism of p and q for the representation of scissions and crosslinked units has been replaced with the symbols G(S) and G(X) where G(S) represents the number of scissions formed and G(X) is the number of crosslinks formed per 100eV of energy absorbed by the polymer sample. Note that Charlesby defined q as the proportion of crosslinked monomer units. Since there are two crosslinked monomer units for every crosslink formed, it follows that the proportion of crosslinks is $q/2$. Thus G(X) must be directly compared with $q/2$. This point has caused some confusion in the past. For the general case of an arbitrary initial molecular weight distribution, no general treatment was proposed by Charlesby, but he indicated that for large radiation doses, where the number of scissions per average initial molecule is greater than about three, the initial distribution no longer affects the final result and may be replaced by a random distribution of the same number average molecular weight. His contribution was noteworthy as it formed the basis of the ensuing determinations of p and q. Later workers were, however, to develop more precise theories which did not depend on the assumption of an initial random distribution.

Inokuti(6) considered the problem of simultaneous crosslinking and scission where the crosslinking predominates over the scission and gel formation occurs. His work included the case of the generalized Poisson-type initial distribution. As an extension of this work, Inokuti and Dole(1) derived equations for the dependence of the weight- and z-average molecular weights on radiation dose for a generalized Poisson initial distribution (also called a Schulz-Zimm distribution). The assumptions of Inokuti and Dole(1) are (i) random crosslinking and scission occur simultaneously (ii) G(S) and G(X) are constant with radiation dose (iii)scission and crosslinking may be treated separately and their effects added (iv) not more than one crosslink connects any two polymer molecules (v) no cyclic structures are produced (vi) molecular weight changes due to other effects (such as decay of vinyl unsaturation) are absent and (vii) a generalized Poisson or Schulz-Zimm initial molecular weight distribution. Their equations are given later in the present work, and are the basis of later attempts(7,8) to utilize the theory to interpret experimental data. Due to the mathematical complexity of the equations, the task of solving the equations to give values of G(S) and G(X) has been formidable, and various approaches have been used. O'Donnell, Smith and Winzor(7) use a binomial expansion and discard the cubic and higher terms. O'Donnell, Winzor and Winzor(8) also use a binomial expansion discarding the squared and higher terms. Both of these approaches, while correct in their context, do not give a universal approach to solving the equations and in particular to making the solutions accessible for efficient interpretation of the experimental data.

The work of Saito(9) has parallelled that of Charlesby(5) and Inokuti and Dole(1,6). Indeed, some of Saito's equations have been the starting point for the deductions of Inokuti and Dole(1). Saito(9) has summarized his contribution to the field, but since he also has been concerned mainly with gelation, he has not provided an efficient way of interpreting experimental molecular weights to give values of G(S) and G(X).

The aim of the present work is to present the essential points of a recent advance in the interpretation of experimental molecular weight/dose relationships, to show that computer methods may now be utilized to give quick reliable numerically accurate determinations of G(S) and G(X), and to apply these new methods to some experimental results which have been published elsewhere to show that the new methods give agreement with results calculated by other means.

THEORY

The Equations
The full set of general equations(1) for the dose dependence of each of the molecular weights, $\overline{M}_n$(number average), $\overline{M}_w$(weight average), and $\overline{M}_z$(z average) are given below:

$$\overline{M}_n(D) = \frac{\overline{M}_n(0)}{(1 + (\dot{\tau}/\dot{\chi} - 1)u\dot{\chi}D)} \quad ....(1)$$

$$\overline{M}_w(D) = \frac{2\,\overline{M}_n(0)\,\phi_1(u\dot{\tau}D,\sigma)}{(u\dot{\tau}D)^2[1 - (4\dot{\chi}/u\dot{\tau}^2D)\phi_1(u\dot{\tau}D,\sigma)]} \quad ....(2)$$

$$\overline{M}_z(D) = \frac{3\ \overline{M}_n(0)\,[\,\phi_2(u\dot{\tau}D,\sigma)/\phi_1(u\dot{\tau}D,\sigma)\,]}{[\,1\ -\ (4\dot{\chi}/u\dot{\tau}{}^2D)\,\phi_1(u\dot{\tau}D,\sigma)\,]^{\,2}} \qquad\qquad ....(3)$$

$$\phi_1(u\dot{\tau}D,\sigma) = u\dot{\tau}D - 1 + [1+(u\dot{\tau}D/\sigma)]^{-\sigma} \qquad\qquad ....(4a)$$

$$\phi_2(u\dot{\tau}D,\sigma) = 1 + [1+(u\dot{\tau}D/\sigma)]^{-(\sigma+1)} - (2/u\dot{\tau}D)\{1 - [1+(u\dot{\tau}D/\sigma)]^{-\sigma}\} ....(4b)$$

D denotes the radiation dose in grays, u is the number–average degree of polymerization ($u = \overline{M}_n(0)/M$ where M is the molecular weight of the monomer unit), $\dot{\tau}$ and $\dot{\chi}$ are the respective probabilities per gray of scission and crosslinking of a single monomer unit, and $\sigma = 1/[(\overline{M}_w(0)/\overline{M}_n(0)) - 1]$ and is a measure of the width of the initial molecular weight distribution.

Strategies for solution
(a) Initial most probable distribution ($\sigma = 1$)
For the case of the initial most probable distribution ($\sigma = 1$), the general equations become simplified and are given below:

$$\overline{M}_n(0)/\overline{M}_n(D) = 1 + (\dot{\tau}/\dot{\chi} - 1)u\dot{\chi}D \qquad\qquad ....(1a)$$

$$\overline{M}_w(0)/\overline{M}_w(D) = 1 + (\dot{\tau}/\dot{\chi} - 4)u\dot{\chi}D \qquad\qquad ....(5)$$

$$\overline{M}_z(0)/\overline{M}_z(D) = (1 + u\dot{\tau}D - 4u\dot{\chi}D)^2/(1 + u\dot{\tau}D) \qquad\qquad ....(6)$$

These equations are then amenable to solution as simultaneous equations in pairs by some of the more popular symbolic mathematics computer programs e.g. MATHEMATICA(*10*) and MACSYMA(*11*). The solutions are given below:

Considering $\overline{M}_n$ and $\overline{M}_w$

$$\dot{\tau} = \frac{4C\ -\ A\ -\ 3}{3\,u\,D} \qquad\qquad ....(7a)$$

$$\dot{\chi} = \frac{C\ -\ A}{3\,u\,D} \qquad\qquad ....(7b)$$

Considering $\overline{M}_w$ and $\overline{M}_z$

$$\dot{\tau} = \frac{A^{\,2}\ -\ B}{B\,u\,D} \qquad\qquad ....(8a)$$

$$\dot{\chi} = \frac{A^{\,2}\ -\ AB}{4\,B\,u\,D} \qquad\qquad ....(8b)$$

Considering $\overline{M}_n$ and $\overline{M}_z$

$$\dot{\tau} = \frac{-(48BC\ +\ B^{\,2})^{\,1/2}\ +\ 24C\ +\ B\ -\ 18}{18\,u\,D} \qquad\qquad ....(9a)$$

$$\dot{\chi} = \frac{-(48BC\ +\ B^{\,2})^{\,1/2}\ +\ 6C\ +\ B}{18\,u\,D} \qquad\qquad ....(9b)$$

Where

$$A = \frac{\overline{M}_w(0)}{\overline{M}_w(D)}, \quad B = \frac{\overline{M}_z(0)}{\overline{M}_z(D)}, \quad \text{and } C = \frac{\overline{M}_n(0)}{\overline{M}_n(D)} \quad \text{....(resp. 10a,10b,10c)}$$

There are two sets of solutions for $\overline{M}_n$ and $\overline{M}_z$. The negative roots of the function $(48BC + B^2)^{1/2}$ are accepted since they provide physically significant solutions and the positive ones do not. $\dot{\tau}$ and $\dot{\chi}$ are related to $G(S)$ and $G(X)$ by the equations:

$$G(S) = 9.65 \times 10^9 \, u\dot{\tau}/\overline{M}_n(0) \, ; \quad G(X) = 9.65 \times 10^9 \, u\dot{\chi}/\overline{M}_n(0) \qquad \text{....(resp. 11a,11b)}$$

(b) The general case $(\sigma \neq 1)$
In this case, equations (1),(2) and (3) were taken in pairs and solved simultaneously:

Considering $\overline{M}_n$ and $\overline{M}_w$
From equation (2), by simple rearrangement,

$$(u\dot{\tau}D)^2 \, \overline{M}_w(D) - 4u\dot{\chi}D\phi_1\overline{M}_w(D) - 2\phi_1\overline{M}_n(0) = 0 \qquad\qquad \text{....(12)}$$

From equation (1) by rearrangement and multiplication of both sides by $\phi_1 \overline{M}_w(D)$,

$$4\phi_1\overline{M}_w(D) + 4u\dot{\tau}D\phi_1\overline{M}_w(D) - 4\phi_1\overline{M}_w(D) \, \overline{M}_n(0)/\overline{M}_n(D) = 4u\dot{\chi}D\phi_1\overline{M}_w(D) \text{ ....(13)}$$

Equation (13) may then be used to eliminate the term in $\dot{\chi}$ in equation (12) to give,

$$(u\dot{\tau}D)^2\overline{M}_w(D) - 2\phi_1\overline{M}_n(0) - 4\phi_1\overline{M}_w(D) - 4u\dot{\tau}D\phi_1\overline{M}_w(D)$$

$$+ 4\phi_1\overline{M}_w(D)\overline{M}_n(0)/\overline{M}_n(D) = 0 \qquad \text{....(14)}$$

Considering $\overline{M}_w$ and $\overline{M}_z$
Squaring both sides of equation (2) and dividing by equation (3) and rearranging eliminates $\dot{\chi}$ and gives,

$$3\phi_2 (u\dot{\tau}D)^4 [\overline{M}_n(0)/\overline{M}_z(D)] [\overline{M}_w(D)/\overline{M}_n(0)]^2 - 4\phi_1^3 = 0 \qquad \text{....(15)}$$

Considering $\overline{M}_n$ and $\overline{M}_z$
Equation (3) may be written as

$$\overline{M}_z(D) = \frac{3 \, \overline{M}_n(0) \, \phi_2/\phi_1}{X^2} \qquad\qquad \text{....(16)}$$

where $X = 1 - 4\dot{\chi}\phi_1/u\dot{\tau}^2D$ \qquad \text{....(16a)}

From rearrangement of equation (1),

$$\frac{4\dot{\chi}}{u\dot{\tau}^2D} = \frac{4}{(u\dot{\tau}D)^2} + \frac{4}{u\dot{\tau}D} - \frac{4 \, \overline{M}_n(0)}{(u\dot{\tau}D)^2 \, \overline{M}_n(D)} \qquad \text{....(17)}$$

Substituting equation (17) into equation (16a) gives

$$X = 1 - \frac{4\phi_1}{(u\dot{\tau}D)^2} - \frac{4\phi_1}{u\dot{\tau}D} + \frac{4\phi_1}{(u\dot{\tau}D)^2}\frac{\overline{M}_n(0)}{\overline{M}_n(D)} \quad ....(18)$$

and equation (16) becomes

$$X^2\overline{M}_z(D) - 3(\phi_2/\phi_1)\overline{M}_n(0) = 0 \quad ....(16b)$$

where X is defined by equation (18) and ϕ_1 and ϕ_2 are defined previously.

Thus, taking pairs of the three molecular weights in turn, the problem reduced to equations in one unknown. In the general case, all attempts to find analytical solutions to equations (14), (15) and (16b) failed despite the use of the best symbolic computation programs available e.g. MATHEMATICA(*10*) and MACSYMA(*11*). Computer programs were written using minimization(*12*) and numerical methods(*13*) to solve the equations exactly and numerically. The methods and computer programs generated are discussed elsewhere(*2,14*).

EXPERIMENTAL DETERMINATION OF MOLECULAR WEIGHTS

Molecular weights of polymers are important because they determine a large number of physical properties of polymer materials (see Table 1). Molecular weights may be experimentally determined by a number of methods (see Table 2).

Table 1
Materials properties affected by molecular weights

Yield Strength	Fracture Properties	Abrasion Resistance
Tensile Strength	Melt Viscosity	Solution Viscosity
Glass Transition Temperature	Orientation	Refractive Index
Solution Osmotic Pressure	Solubility	Surface Tension
Processing Properties	Biodegradation	Fibre Formation
Adsorption on Substrates	Diffusivity Through Polymer	
Polymer Pair Compatibility	Kinetics of Polymer Reactions	

Table 2
Techniques for Measuring Molecular Weights

Gel Permeation Chromatography (GPC) or Size Exclusion Chromatography (SEC)	$\overline{M}_n$, $\overline{M}_w$, $\overline{M}_z$
Sedimentation Equilibrium	$\overline{M}_w$, $\overline{M}_z$
Laser Light Scattering	$\overline{M}_w$
Osmometry	$\overline{M}_n$
Viscosity	$\overline{M}_v$
Cryoscopy, Ebulliometry	$\overline{M}_n$
Vapour Pressure Lowering	$\overline{M}_n$

In order to determine the two unknowns G(S) and G(X), two sets of molecular weight versus dose data are required. If all three are known, as is possible with the use of GPC, then it is possible to calculate three sets of G(S) and G(X) values, thus giving an opportunity to crosscheck the calculations and/or discard results which are not reliable.

APPLICATION TO EXPERIMENTAL RESULTS

The purpose of the present work is to clarify and enable the determinations of G(S) and G(X) from experimental molecular weight versus dose relationships. Errors may be determined by successive determination of the G(S) and G(X) values for the higher and lower limits of the input parameters. For instance, where the molecular weights are determined as given as   M ± a   then   M + a   and   M - a   are the upper and lower limits of the value of M and if respectively substituted in the computer program, then the obtained values of G(S) and G(X) for the upper and lower limits of M will give the range of values expected for G(S) and G(X). From these values, a result of   G(S) ± h   should be able to be determined.

Hill, O'Donnell, Winzor and Winzor(15) determined scission and crosslinking yields of poly(acrylic acid) (PAA) and poly(methacrylic acid)(PMAA) using sedimentation equilibrium to determine the weight average and z-average molecular weights. Their analysis was not exact since they used a binomial expansion in their simplification of the equations, and truncated at squared terms and higher. Table 3 summarizes their results and gives the result of an exact determination using their experimental results and the exact solutions demonstrated in the present work. In most cases there is agreement between the approximate and exact values, the discrepancy in the values of G(S) for PAA being caused mainly by a large experimental error in the measurements as indicated by the range of possible values shown in the Table 3. The exact values were determined by taking the numerical results of Hill et al.(15) at each dose and determining an exact value of G(S) and G(X) for each dose. The error ranges arise from the experimental scatter, rather than the calculation or the form of the equations. Table 3 shows that the approximations of Hill et al.(15) did not introduce any further significant error into their work.

Table 3
Comparison of the present work with the results of Hill et al.(15)
(errors are 2 x standard deviation)

PAA
Approximate   Exact
(Average of five doses)
           G(S) = 0.0   ± 0.6   G(S) = 0.62 ± 0.4
           G(X) = 0.44 ± 0.09   G(X) = 0.47 ± 0.1
PMAA
Approximate   Exact
(Average of six doses)
           G(S) = 6.0 ± 0.5   G(S) = 6.0   ± 1.2
           G(X) = 0.0 ± 0.2   G(X) = 0.04 ± 0.12

Carswell(*4*)determined G(S) and G(X) for isotactic poly(methyl methacrylate) (PMMA). In the present work, G(S) and G(X) have been determined for each dose point (see Table 4). G(X) was found to be 0 within experimental error in agreement with Carswell(*4*). In the more precise determination of the present work, there can be seen a large variation of G(S) with dose. It is not clear at present whether this is due to experimental error, or whether it is real. Certainly the value of Carswell(*4*) of 1.5 ± 0.2 does not cover the range of values observed here. Much more careful work needs to be performed before the presence of a variation of G(S) with dose is clear.

Mitomo et al.(*16*) studied the radiation-induced degradation of poly(3-hydroxybutyrate)(PHB), a biodegradable polyester produced by bacteria. They have calculated G(S) to be 3.0 for PHB irradiated in air. Their experimental values of number average and weight average molecular weight have been extracted and analysed by the present methods to give G(S) = 1.65 ± 0.4 for $\sigma = 1.3$. If $\sigma = 1.0$ is assumed then the exact calculation gives G(S) = 1.8 ± 0.5. These values of G(S) do not agree with the value of Mitomo et al.(*16*) and their graphical method may have incorporated other minor errors. Certainly, the value of G(S) for $\sigma = 1.0$ is closer to Mitomo's value than for $\sigma = 1.3$.

Table 4
G(S) values for isotactic PMMA (Carswell(*4*))

DOSE/kGy	G(S) (present work)	G(S) (Carswell(*4*))
44	0.806	
94	2.43	
140	2.01	
180	1.66	
180	1.39	
AVERAGE	1.66	
Minimization over		
5 doses	1.71	1.5 ± 0.2

Thus, the discrepancy between their values and the present work may be partly due to the fact that they used the equation for an initial most probable distribution ($\sigma = 1$) when in fact for their initial distribution $\sigma = 1.3$, and partly due to other unexplained factors. G(X) was found to be negligible. This is an example where close attention to the exact solution of the equations produces a different result to the approximate case. The polymer appears to become monodisperse with increasing radiation dose, which is not expected if random scission were occurring alone. It is possible that some depolymerization is occurring which may further complicate the interpretation of the experimental molecular weights. The present section has shown that reliable precise determinations of G(S) and G(X) may reveal information and previously unobserved effects in the polymer systems being studied. This work has provided an exact method for determination of G(S) and G(X) which should be used in the future for the analysis of new data.

CONCLUSION

A recent advance in the analytical techniques used in calculation of G(S) and G(X) values makes possible the computerized determination of numerically exact values G(S) and G(X) from molecular weight versus dose relationships. This is the first time such a method has become available. The calculations are performed by a computer program/programs thus reducing the time needed for calculations and removing the need for the use of approximate methods. The assumption of an initial most probable distribution of molecular weights has previously been used to expedite the calculation of G(S) and G(X) even when the assumption was not justified. At low radiation doses, this assumption may introduce significant errors particularly when the molecular weight distribution is very different from an initial most probable distribution. The present work has shown that G(S) and G(X) may be determined numerically for any Schulz-Zimm initial molecular weight distribution for all doses.

Prof. Jim O' Donnell passed away on 29th April 1995 while this work was being reviewed for publication.

LITERATURE CITED

1. Inokuti, M.; Dole, M. *J. Chem. Phys.* **1963**, *38*, 3006 - 3009
2. Bremner, T.; Milne, K. A.; O'Donnell, J. H. *Polymer Bulletin*, 1995
3. O'Donnell, J. H.; Sangster, D. F. *Principles of Radiation Chemistry* Edward Arnold, London UK, 1970; p 119
4. Carswell, T.G. *The Polymerization and Radiation Degradation of Polyesters* Ph. D Thesis University of Queensland, Queensland, Australia July 1991
5. Charlesby, A. *Atomic Radiation and Polymers;* International Series of Monographs on Radiation Effects in Materials; Pergamon Press: Oxford UK ,1960; pp 171 - 174
6. Inokuti, M. *J. Chem. Phys.* **1963**, *38*, 2999 - 3005
7. O'Donnell, J. H.; Smith, C. A.; Winzor, D. J. *J. Polym. Sci., Polym. Phys. Ed.* **1978,** *16*, 1515 - 1518
8. O'Donnell, J. H.; Winzor, C. L.; and Winzor, D. J. *Macromolecules* **1990,** *23*, 167 - 172
9. Saito, O. In *The Radiation Chemistry of Macromolecules;* M. Dole, Ed.; Academic Press: New York, NY, 1972, Vol.1, Chapter 11; p.223
10. MATHEMATICA, Wolfram Research Inc., P.O. Box 6059, Champaign, Illinois 61821 (1989)
11. MACSYMA, Macsyma Inc., 20 Academy St., Arlington, MA 02174
12. Chandler, J. P. *STEPT: A Family of Routines for Optimization and the Fitting of Data;* Department of Computing and Information Sciences, Oklahoma State University: Stillwater, Oklahoma, 1975
13. Press, W. H.; Flannery, B. P.; Teukolsky, S. A.; Vetterling, W. T. *Numerical Recipes: The Art of Scientific Computing;* Cambridge University Press: Cambridge, UK, 1986
14. Milne, K. A.; O'Donnell, J. H. *Proceedings of the Australian Institute of Nuclear Science and Engineering (AINSE) Conference on Radiation Biology and Chemistry;* University of Melbourne, Melbourne, Victoria, Australia, 16 - 18th November, 1994; AINSE: Sydney, NSW, Australia, 1994; pp 90-91
15. Hill, D. J. T.; O'Donnell, J. H.; Winzor, C. L.; Winzor, D. J. *Polymer* **1990**, *31*, 538 - 542
16. Mitomo, H.;Watanabe, Y.;Ishigaki, I.;Saito,T. *Polym. Degrad.Stab.* **1994**, *45*, 11 - 17

RECEIVED October 11, 1995

Chapter 10

Determination of New Chain-End Groups in Irradiated Polyisobutylene by NMR Spectroscopy

David J. T. Hill, James H. O'Donnell[†], M. C. Senake Perera, and Peter J. Pomery

Polymer Materials and Radiation Group, Department of Chemistry, University of Queensland, Brisbane, Queensland 4072, Australia

High Resolution NMR spectra of polyisobutylene after γ irradiation in vacuum show a large number of extremely sharp resonances. DEPT spectra and C-H COSY spectra have enabled the identification of methyl, methylene, methine, quaternary, vinyl protons and carbons which have been assigned to a variety of new end group structures resulting from main chain scission. These assignments support some previous proposals for the mechanism of radiation degradation of polyisobutylene and exclude others.

Polyisobutylene is known to undergo chain scission during exposure to high energy radiation. The degradation mechanism has been studied by many workers. An intra molecular disproportionation reaction leading to the same end products (S1 and S2) via an activated polymer molecule reaction (1a) or primary main chain scission reaction (1b) into free radicals were suggested by Alexander et al(1) and Chapiro(2) respectively.

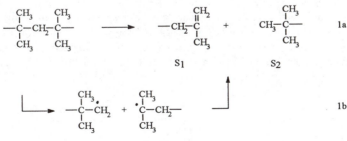

[†]Deceased

0097–6156/96/0620–0139$12.00/0
© 1996 American Chemical Society

Miller et al.(3) assumed that the primary radiation-chemical event resulted in the scission of a methyl C-H bond, leading to radical R3 which spontaneously undergoes β-cleavage according the reaction 2 to give S1

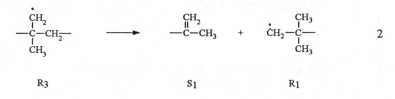

A similar mechanism was suggested by Wall(4). Slovokhotova and Karpov(5) suggested that the free radical which was expected to form with the highest probability is radical R4, due to the lower energy of C-H bond in a methylene group as compared to methyl. This free radical was assumed to undergo ß-cleavage according to reaction 3 to give structure S3.

Slovokhotova(6) reported that methyl radicals formed in the system may react to form ethyl groups according to the reaction scheme 4.

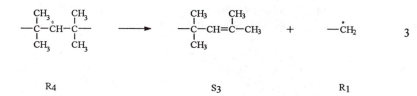

The ESR spectra of polyisobutylene, irradiated with high energy radiation and measured at low temperatures show a broad doublet with a hyperfine splitting constant of 20G, and was attributed by Ranby and Cartensen(7) to the radical R4 formed by the cleavage of C-H bond of the methylene group. Hori and Kashiwabara(8), interpreted this broad doublet as a mixture of R3 and R4 radicals formed by the loss of hydrogen from methyl and methylene groups respectively. The only radical remaining at temperatures above 213 K is R4. Although it is well established that polyisobutylene undergoes scission during irradiation, the main chain scission radicals (R1 and R2) were not observed in either of these ESR studies. Bartos(9) studied the decay of the radical R4 in the region 223-243K and found that it followed second order kinetics with an activation energy of 77.5 kJ mol^{-1}. Therefore, the reaction 5 was suggested(8,9) for the decay of free radicals in the whole temperature range of 77-303K.

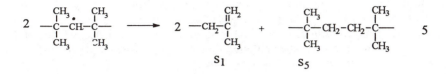

However, the radical R4, should decay(4) to the structure S3 rather than the structure S1, since S3 is sterically more stable. The transformation of S3 to S1 has not been explained.

The above reaction mechanisms appear to account for most of the structures observed by Turner and Higgins (10) in infrared spectra of irradiated polyisobutylene, such as ethyl groups, vinylidene double bonds, tetrasubstituted double bonds and T-T linkages (S5) of isobutylene units etc.. These various reactions would lead to the formation of one double bond per main chain scission. The experimental value of 1.35 of the reaction of double bonds to scission obtained by Turner(11), therefore cannot be explained by considering only the above reaction paths. Also the detection of ethyl groups, tetrasubstituted double bonds and T-T linkages based on IR spectra are ambigous.

The aim of the present study was to identify the changes in the molecular structure of polyisobutylene produced by irradiation using NMR spectroscopy and hence to critically evaluate previous proposed mechanisms of radiation degradation.

Experimental

Materials and Methods- Polyisobutylene (Aldrich) was found to have Mn=1.99*10^5, Mw=4.33*10^5,Mz=7.22*10^5 and Mw/Mn=2.2 (GPC). Polymer samples were precipitated twice from chloroform with methanol and were dried in a vacuum oven at room temperature (until no trace of chloroform could be detected in the ^{13}C NMR spectrum). Samples were evacuated at 10^{-5} Torr at room temperature for 24 hours in glass tubes, sealed under vacuum and irradiated at 303K using a ^{60}Co source with dose rate of 3.0 kGy/h. Samples were left at room temperature for one week before opening the glass tubes to make sure that all the radicals had reacted before exposure to oxygen. We (12) have shown, by ESR spectroscopy, that the radicals are not stable at room temperature.

Nuclear Magnetic Resonance Spectroscopy- ^{1}H and ^{13}C NMR spectra were obtained using a Joel GX 400 spectrometer operating at 100 MHz for carbon. For ^{1}H NMR, free induction decays were accumulated in 8K data points, spectral width of 4400 Hz, 7.0 μs(90^0) pulse and a recycle time of 4 seconds. For ^{13}C NMR spectra, free induction decay were accumulated in 32K data points, spectral width of 22000Hz, 9.1 μs(90^0) pulse width, and 10 seconds repetition time. Spectra were determined at 298 K in $CDCl_3$ (10% W/V) with TMS as an internal standard. Spectral intensities were measured using integration. The DEPT pulse sequence (flip angle of 135 ^{0}) was used to identify methyl, methylene, methine and quaternary carbon resonances.

The proton decoupled C-H cosy experiment was performed by using the Bax-Ruter sequence(13). The spectrum was obtained for 25% irradiated polyisobutylene in chloroform. A total of 512 scans were accumulated over 64 T1 increments with a relaxation delay of 1.9 seconds. The initial matrix size was 4K and 128 w(2000 Hz) in F2 and F1 respectively. A sine-bell apodization function without phase shift was applied in both dimensions prior to Fourier Transformation.

Results and Discussion

The ^{1}H NMR spectrum of polyisobutylene irradiated to 9 MGy is shown in Figure 1. The peaks are designated as H-1 to H-16. H-1 and H-2 are the methyl and methylene resonances of unirradiated PIB(14). H-14 and H-15 are the proton resonances assigned to exo-methylene group ($-C=CH_2$) and H-16 is a olefinic proton of the backbone unsaturated group ($-C=CH-$) (15). The assignments of the other proton resonances are discussed later.

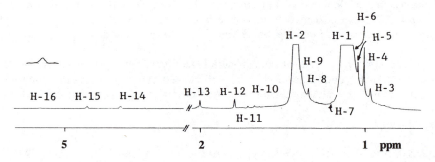

Figure 1: 1H NMR spectrum of polyisobutylene irradiated to 9 MGy at 303 K

The ^{13}C NMR spectra of unirradiated PIB and PIB irradiated to 9 MGy are shown in Figure 2. The spectrum of unirradiated PIB has been assigned(16). The spectrum of the unirradiated polymer was free from any small peaks, confirming that the isobutylene units are linked only in a head to tail pattern and no other structures are present. The ^{13}C NMR spectrum of irradiated PIB (Figure 2b) showed a number of new peaks in both the aliphatic and olefinic regions.

The expansion of the region at 10-60 ppm is given in figure 3a. The DEPT spectrum (Θ=135) shown in the Figure 3b was used to distinguish between the

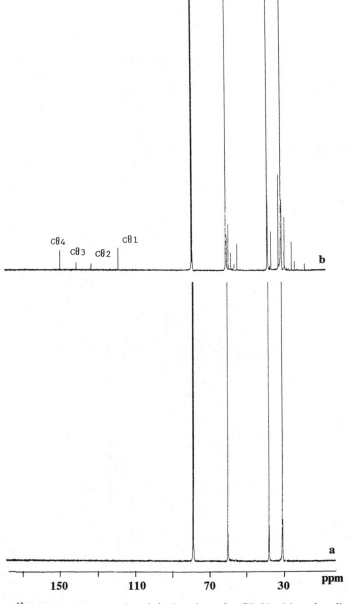

Figure 2: ^{13}C NMR spectra of polyisobutylene in CDCl$_3$ (a) unirradiated (b) irradiated to 9 MGy in vacuum at 303K.

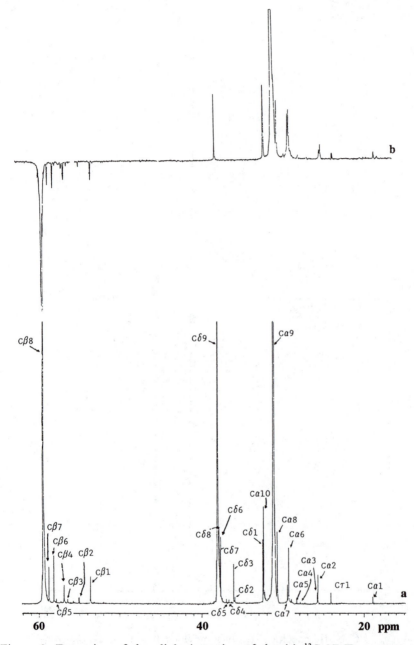

Figure 3: Expansion of the aliphatic region of the (a) ^{13}C NMR spectrum of irradiated polyisobutylene (b) DEPT spectrum of the same sample

different types of carbon resonances. For convenience the peaks are designated as Cα-methyl,Cβ-methylene,Cτ-methine,Cδ-quaternary carbons and CΘ-olefinic carbons. The assignments of the peaks were made by calculation of chemical shifts of all the possible chemical structures resulting from irradiation and comparing these with the observed chemical shifts. The calculations were made using the methods Lindeman and Adams(17) for alkanes, and the Dorman et al.(18) and the Beebe(19) for alkenes. The constants provided by these authors were used in the calculations. The shielding factors proposed by Chu and Vukov(20) to account for the gem dimethyl effect in isobutylene were also taken into account in the calculations. The calculated and observed chemical shifts are given in the Table 1 and 2.

Peaks CΘ1 and CΘ4 were assigned to the carbons 1 and 2 of the structure S1. The only methine peak, Cτ1, in the spectrum was assigned to the carbon 1 in the structure S6 (Table 1). The most upfield quaternary carbon peak, Cδ1, was assigned to the carbon 2 of structure S2. The olefinic peaks CΘ2 and CΘ3 were also assigned to the carbons 2 and 3 of structure S3. The assignment of the remaining peaks given in Table 1 were made by comparing the peak intensities of the spectra of samples irradiated to different radiation doses. For example, the peaks Cβ1, Cβ3, Cβ4, Cα2 and Cα6 were assigned on the basis that they had similar intensities to the peaks CΘ1 and CΘ6, which were assigned to structure S1. The calculation of the chemical shift of the carbon 4 of the S3 structure was not possible because it is a quaternary carbon next to a double bond. However, a similar carbon in 2,4,4 trimethyl-2-pentene gives a peak at 32.2 ppm, therefore, the peak Cδ2 may be assigned to this carbon.

Since there is no peak at 8 ppm (the expected chemical shift of a methyl carbon of an ethyl end group) the structure S4 is evidently not formed in significant yield. Reaction 5 therefore cannot be an important process in radiation induced scission. No quantitative estimation of ethyl groups was carried out by Turner(11). ^{13}C NMR obtained in the present work confirms previous reports of the presence of structures S1,S2 and S3, and identifies the formation of the structure S6, which has not been observed before. Since $G_{C=C}$ was found by Turner(11) to be higher than G_s, ($G_{C=C}$ = 1.35 G_s), a radiolysis mechanism yielding unsaturation without causing scission was considered by Dole(20). A mechanism involving a loss of a methyl group and hydrogen to form structure S7 or S8 shown in Table 2 was proposed and S7 was preferred since IR studies indicated the presence of exomethylene groups. The internal double bonds were estimated by Turner(11), using ozonolysis, to have a G value less than 0.3. IR spectra do not allow differentiation between vinyledene groups in S1 and S7, but the ^{13}C NMR data in Table 2 indicate that neither S7 nor S8 is a major product formed in a significant amount to account for a radiation yield of 0.3. The methylene group of the structure S6 should give a resonance at 36 ppm, but in the DEPT spectra there is no evidence of methylene carbons in this region. This structure, which was proposed to be formed during the bimolecular radical reaction 5, should be present with a radiation yield of about 1.3 (= $G_s/2$). Obviously the data in Table 2 indicate that this is not the case. Previous identification of this group by IR is not convincing since it was based on a slight increase in intensity of a major peak at 1450 cm[-1].

Table 1: Peak assignments of end groups

Sym	Structure	no	Chemical shift ppm		Peaks	
			cal	obs	^{13}C	^{1}H
S1	$-CH_2\overset{5}{-}\underset{CH_3}{\overset{\overset{6}{CH_3}}{\overset{4}{C}}}\overset{3}{-}CH_2\overset{2}{-}\underset{}{\overset{\overset{7}{CH_3}}{C}}\overset{1}{=}CH_2$	1	114.8	114.4	Cθ1	H14/H15
		2	143.7	143.8	Cθ4	
		3	54.7	53.6	Cβ1	H13
		4	36.8	36.0	Cδ3	
		5	56.6	56.8	Cβ4	H9
		6	25.4	25.7	Cα2	H3
		7	28.8	29.3	Cα6	H6
S2	$-CH_2\overset{5}{-}\underset{CH_3}{\overset{\overset{6}{CH_3}}{\overset{4}{C}}}\overset{3}{-}CH_2\overset{2}{-}\underset{}{\overset{\overset{7}{CH_3}}{C}}\overset{1}{=}CH_2$	1	30.6	30.7	Cα8	H1
		2	32.1	32.5	Cδ1	
		3	58.8	58.7	Cβ7	H2
		4	37.5	37.6	Cδ6	
S3	$-CH_2\overset{6}{-}\underset{CH_3}{\overset{\overset{5}{CH_3}}{\overset{4}{C}}}\overset{3}{-}CH=\overset{2}{\underset{}{\overset{\overset{1}{CH_3}}{C}}}-CH_3$	1	17.1	19.0	Cα1	H11
			25.3	25.7	Cα3	H12
		2	129.8	127.9	Cθ2	
		3	135.1	135.0	Cθ3	H16
		4	28.8	28.3	Cα5	H10
		5	56.6	57.7	Cβ5	H2
		6	32.2	34.2	Cδ2	
S4	$-\underset{CH_3}{\overset{CH_3}{C}}-CH_2\overset{1}{-}CH_3$	1	7.8	nm	nm	nm
S6	$-CH_2\overset{4}{-}\underset{CH_3}{\overset{\overset{6}{CH_3}}{\overset{3}{C}}}\overset{2}{-}CH_2\overset{1}{-}\underset{CH_3}{\overset{\overset{5}{CH_3}}{C}}-H$	1	23.9	24.2	Cτ1	H11
		2	54.2	55.0	Cβ2	H7
		3	36.2	35.8	Cδ2	
			36.9		Cδ4	
		4	55.9	56.4	Cβ3	H8
		5	24.6	25.7	Cα3	H12
		6	30.3	29.4	Cα7	H4

The proton decoupled 2D C-H COSY spectrum of PIB irradiated to 9 MGy is shown in Figure 4. The crosspeaks observed in this spectrum have been used to assign the proton NMR peaks which are shown in the Figure 1. The correlation of the methylenecarbons Cβ1, Cβ2, Cβ3, Cβ4 and Cβ6 with H-13, H-7, H-8, H-9 and H-8 respectively are clearly observed in crosspeaks 1-5 in the figure. The Cβ7 (crosspeak 6) and the main methylene peak Cβ8 both correlate with H-2. The proton resonance that correlates with Cβ5 (crosspeak 7) is also overlapped by the main H-2 peak. Similarly, the methyl carbons Cα1, Cα2, Cα3, Cα5 and Cα6 correlate with H-11, H-3, H-12, H-10 and H-6 respectively, as observed in crosspeaks 8-12. Cα7 and Cα10 both correlate with H-4 (crosspeak 14) and Cα8 correlates with the main methyl peak H-1. The correlation of the methine carbon Cτ1 to H-11 Crosspeak (15) could also be

Table 2: Peak assignments of the main chain modifications

Symbol	Structure	Carbon	Chemical Shift ppm	
			cal	obs
S5	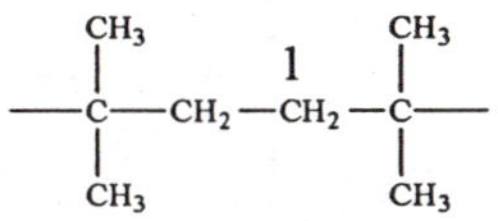 ![structure: CH₃–C(CH₃)(1)–CH₂–CH₂–C(CH₃)–CH₃]	1	37.3	nm
S7	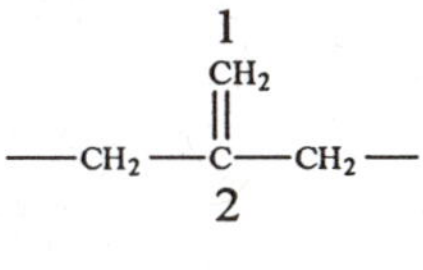 ![structure: –CH₂–C(=CH₂(1))(2)–CH₂–]	1 2	108.8 149.7	nm nm
S8	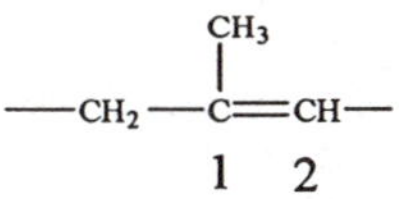 ![structure: –CH₂–C(CH₃)(1)=CH(2)–]	1 2	139.6 128.2	nm nm

clearly observed. These correlations therefore facilitate the assignment of the ^{1}H NMR spectrum to the different structures shown in Table 1. Another interesting observation in the C-H COSY spectrum is the absence of any peaks corresponding to 36 ppm (^{13}C NMR), confirming that there are no methylene peaks (structure S6) in this region.

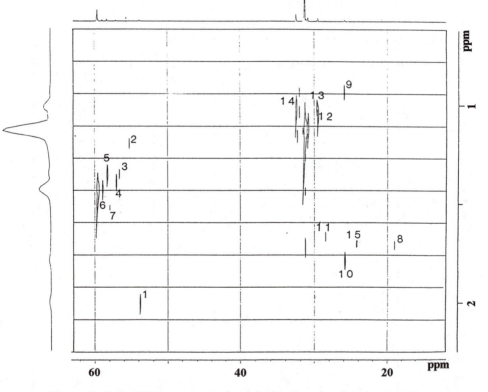

Figure 4: C-H COSY spectrum of polyisobutylene irradiated to 9 MGy

The ^{13}C NMR spectra therefore indicate the presence of only the structures S1,S2,S3 and S5. The intensities of the peaks CΘ1,Cδ1,CΘ2 and Cτ1 were used to calculate the number of these structural units per 10,000 carbon atoms. These peaks were selected for this calculation because they do not overlap with the main peaks in the spectrum. For these quantitative measurements the spectra were accumulated with a 10 seconds pulse delay and without NOE. The pulse delay of 10 seconds was selected since the spin lattice relaxation time of the quaternary carbons of the polyisobutylene has been found(16) to be about 1.5 seconds. The radiation yields of different structural units, G(st), were then calculated using the equation 1.

$$G(st)= (n*N)/(6.24*10^{13} *R* 2500*M)   \qquad 1$$
where n = units per 10,000 carbons (calculated from NMR)
N = Avogadro number
M = molecular weight of the monomer
R = dose in Gray

and the values are given in the Table 3. Assuming that each scission gives rise to two terminal groups, G(s) was calculated by (G(S1)+G(S2)+G(S3)+G(S6))/2.

Table 3: Radiation yields of different structural units for polyisobutylene

	13 C NMR	1 H NMR
G(S1)	2.4	2.2
G(S2)	3.6	
G(S3)	0.5	0.7
G(S6)	0.7	1.0
Total	7.2	
G(s)	3.6	
G(-C=C-)	2.9	2.7
G(s)/G(-C=C-)	1.2	

The proton NMR spectra may also be used to calculate the yields of some of the new structures proposed above. Only some of the proton peaks could be used in such calculations because of the overlap of other peaks. The amount of S1 may be calculated from either the H-14, H-15 or H-13 peaks and S3 may be calculated from either the H-16 or H-10 peaks. Unfortunately all of the proton resonances in the structure S2 overlap with the backbone isobutylene peaks. The amount of S5 could be calculated using the resonance H-11 after correcting for the S3 structure. H-7 and

H-8 were difficult to integrate accurately due to the presence of large peaks close by. A similar equation to that of 1 was used to convert the number of units/ 10,000 protons to radiation yields, which are given in Table 3.

Conclusions

^{13}C NMR spectra of the irradiated polyisobutylene indicated that reactions 4 and 5 are not the major pathways in the radiation scission of polyisobutylene while supporting reactions 1,2 and 3.

Acknowledgements

The authors wish to thank Australian Institute of Nuclear Science and Energy and the Australian Research Council for providing funds to carry out this research.

References

1. Alexander P., Black R.M. and Charlesby A., Proc.Roy.Soc. (London), (1955), A232,31
2. Chapiro A., J.Chim.Phys., (1956),53,295
3. Miller A.A., Lawton E.J. and Balwit J.S., J.Polym.Sci.,(1954), 14,503
4. Wall L.A.,J.Polym. Sci., (1955), 17,141
5. Slovokhotova N.A. and Karpov .L., Trudy I Vsesoyuz. Soveshchaniya po Radiatsionni Khim., Academy of Sciences of the USSR, Moscow, (1955), 1966.
6.   Slovokotova N.A., ibid, (1958), 263
7.   Ranby B. and Carstensen P., Advances in chemical series, (1967),256
8.   Hori Y. and Kashiwabara H., J Polym Sci; Polym Phys ed,(1981), 19,1141
9.   Bartos J., Collection Czechoslovak Chem Commun, (1985),50,1699
10. Turner D.J. and Higgins G.M.C., J.Polym.Sci.,A2, (1964),1713
11. Turner D.J., J.Polym.Sci.,A-2,4 (1964),1699
12. Bremner T.,Hill D.J.T., O'Donnell J.H., Perera M.C.S. and P.J.Pomery, submitted J.Polym.Sci.,Polym.Chem.ed.,1995
13. (a) Bax A., J.Mag.Reson.,(1983),53,517
       (b) Ruter V., J.Magn.Reson.,(1984),58,306.
14.   Pham Q., and Petiaud R., Proton and Carbon NMR of polymers, John Wiley and Sons,(1983),167.
15.   Simons W.W., The Sadtler Handbook of Proton NMR Spectra Sadtler Research Laboratory Inc., Pennsylvania, (1978),29
16.   Corno C., Proni A., Priola A. and Cesca S., Macromolecules, (1980),13,1092
17.   Linderman L.P. and Adams J.Q.,Analytical Chemistry, (1971),43,1245
18.   Dorman .E., Jautelat M. and Roberts J.D., J.Org.Chem., (1971),36,2757
19.   Beebe D.H.,Polymer, (1978),19,231
20.   Chu C.Y. and Vukov R., Macromolecules, (1985),18 1423
21.   Dole M., The Radiation Chemistry of Macromolecules, Vol 2, M. Dole ed., Academic Press, New York, 1972

RECEIVED August 16, 1995

RADIATION PROCESSING: POLYMERIZATION, CURING, AND STERILIZATION

Chapter 11

Radiochromic Solid-State Polymerization Reaction

William L. McLaughlin[1], Mohamad Al-Sheikhly[1], D. F. Lewis[2], A. Kovács[3], and L. Wojnárovits[3]

[1]Ionizing Radiation Division, Physics Laboratory, National Institute of Standards and Technology, Technology Administration, U.S. Department of Commerce, Gaithersburg, MD 20899–0001
[2]Advanced Materials Group, ISP Technologies, Inc., Wayne, NY 07470
[3]Radiation Chemistry Department, Institute of Isotopes, Hungarian Academy of Sciences, H–1525 Budapest, Hungary

Radiochromic films (GafChromic DM 1260 and MD 55), consisting of thin, colorless transparent coatings of a polycrystalline, substituted-diacetylene sensor layer on a clear polyester base, are studied by pulse radiolysis and flash photolysis, in terms of the kinetics of their response to ionizing radiation and ultraviolet light, respectively. These films have recently been established for broad applications in radiographic imaging, nuclear medicine, and dosimetry for radiotherapy, blood irradiation, insect population control, food irradiation, and industrial radiation processing. The radiochromic reaction is a solid-state polymerization, whereby the films turn deep blue proportionately to radiation dose, due to progressive 1,4-trans additions as polyconjugations along the ladder-like polymer chains. The pulsed-electron-induced propagation of polymerization has an observed first-order rate constant of the order of 10^3 s^{-1}, depending on the irradiation temperature (activation energy $\approx$ 50 kJ mol^{-1}). The UV-induced polymerization is faster by about one order of magnitude ($k_{obs}=1.5\times10^4$ s^{-1}). In the case of the electron beam effect, the radiation-induced absorption spectrum exhibits a much slower blue-shift of the primary absorption band ($\lambda_{max}=675$ nm $\rightarrow$ 660 nm) on the $10^{-3} - 10^{+1}$ second time scale. This effect is attributed to crystalline strain rearrangements of the stacked polymer strand units.

Unlike the attempts to find polymers resistant to radiation-induced coloration, we report here on the intentional darkening of organic conjugated polymers by ionizing radiation and on the kinetic behavior of the radiochromic process as analyzed by pulse radiolysis and flash photolysis. We have investigated a sensitive radiochromic medium based on solid-state polymerization of two forms of proprietary sensor coatings of a thin-layer, fine polycrystalline dispersion on a transparent plastic film base. The radiochromic film forms a very high-resolution image (>1200 lines/mm) that can be analyzed quantitatively with very small pixel size, at wavelengths within the characteristic visible absorption spectrum, using scanning microdensitometry or spectrophotometry (*1,2*).

Progressive coloration and image formation can arise from the partial polymerization of monomers that yield main-chain polyconjugated polymers as characterized when diacetylenes undergo such polymerization. The color is determined by the planarity of the linear polymer backbone (effective conjugation length), with a typical blue color appearing in a highly planar conformation that is characteristic of the trans-addition partial polymerization of disubstituted diacetylenes (*3,4*). In such reactions, the propagation chain length increases homogeneously (*4*). Subsequently, mechanical stresses arise between the stacked monomer units and the polymer crystal repeat units, where the degree of polymerization increases without induction period (*5-7*).

A simple first-order kinetics model has been suggested for such monophase monomer-to-polymer conversion in solid-state linear systems (*6,8,9*). The step-wise behavior is characteristic of 1,4-trans additions occurring in radiation-induced solid-state topochemical polymerization (*7*).

Radiochromic Films

New proprietary radiochromic films (GafChromic DM 1260 and GafChromic MD 55, ISP Technologies, Inc.)[*] consist of a thin radiochromic sensor dispersion coated on a polyester base (*1,10*). The colorless, transparent film responds to ultraviolet light and to ionizing radiation by turning blue, with two distinct absorption bands (λ_{max} =675 nm and 610 nm). The radiation-induced color change is formed directly without thermal, optical, or chemical development and the original blue image is stable at temperatures up to about 60 °C, above which the color of the image changes abruptly from blue to red. Very little dependence of the film darkening on relative humidity occurs during irradiation, but there is a marked temperature dependence, the degree of which varies with radiation dose (*1,10,11*). The γ-ray and electron response shows negligible dose-rate dependence (*1*). The approximate usable absorbed dose ranges for these films for dosimetry are 2×10^1-10^4 Gy for DM 1260 and 5 to 10^3 Gy for MD 55 (*1*). Response curves are shown in Fig. 1 at the low dose levels. (All spectrophotometric measurements given in this paper, except where otherwise noted, were taken at 22 °C; estimated combined uncertainties for these measurements as $U = k u_c = \pm 5$ percent, where u_c is estimated standard deviation and the coverage factor k=2) (*1*).) Higher dose levels can be measured by judicious selection of spectrophotometric wavelength. It has been demonstrated that the DM 1260 film provides very high spatial resolution images. The sensor-layer on both films has atomic constituents in the following proportions by mass (*1*): H 0.099; C 0.606; N 0.112; O 0.193. For ionizing photons

---

[*]The mention of commercial products throughout this paper does not imply recommendation or endorsement by the National Institute of Standards and Technology, nor does it imply that the products identified are necessarily the best available for the purpose.

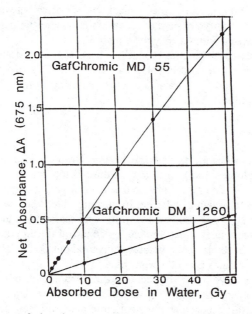

Fig. 1. Increase of absorbance at 675 nm wavelength as a function of ^{60}Co γ-radiation dose, for GafChromic MD 55 and GafChromic DM 1260 radiochromic films (2).

and electrons in the energy region 0.1 to 10 MeV, the sensor simulates the radiation absorption properties of water and muscle, in terms of ratios of photon mass energy-absorption coefficients and electron mass collision stopping powers, within ± 2 percent (*12*).

Pulse Radiolysis Study

Two electron accelerators were used to provide chemical kinetic evaluations by spectrophotometric analysis immediately following short single pulses of electrons incident on 20 mm × 10 mm pieces of radiochromic films. The 4 MeV LPR-4 Linac at the Institute of Isotopes, Budapest provides single 2.5 μs pulses (13 Gy/pulse). The kinetics of the photopolymerization reactions were determined by following the changes in optical absorption at different wavelengths. The measurements were made using a xenon lamp and shutter with the light transmitted to the photomultiplier through an Applied Photophysics monochromator and 2.5 meter Suprasil quartz fiber optics bundle (*13*). In the experiments two filters were applied between the Xe lamp and the films in order to eliminate the light above about 720 nm and below about 500 nm. The second accelerator used was at NIST and it is a 2.5 MeV Febetron Model 705 field emission machine with single 20 Gy, 0.05 μs pulses. Spectrophotometric absorbance values were followed digitally over time scales from microseconds to 10 seconds.

Figure 2 shows, for DM 1260, pulse radiolysis kinetics data for the transient absorbance increase (at 675 and 660 nm) as a function of time after 0.05 μs single-pulsed electron beam exposure at 20 Gy/pulse. The insets on right side of the figure show the semilog kinetics data for the time period between the vertical lines of the exponential kinetics curves at the left. These results indicate a first-order process for color formation at 675 nm (top curves), but at 660 nm, a slight hypsochromic shift and broadening of the absorption band apparently distort the first-order appearance.

The spectral band broadening becomes pronounced at longer times after the pulse, as seen in Fig. 3 for the time period up to 9 seconds. The inset shows the long-term transient behavior at three wavelengths near the absorption peak, where the relative absorbance is normalized at t=0. While the absorbance is relatively stable at 670 nm, the shift to shorter wavelengths causes a decrease with time at 675 nm and an increase at 665 nm. On a longer time scale, this blue-shift goes slightly farther, slowing down exponentially for up to several weeks.

The time profiles of the build-up of absorbance on the μs-ms time scale (Fig. 4) and the absorption spectra of the irradiated MD 55 film (Fig. 5) are similar to those shown in Fig. 2 for the DM 1260 film. As shown in Fig. 4, the first-order rate of color formation and the yield of color-forming species increase with rising temperature during the radiation pulse. The rate constant of the build-up here also shows a wavelength dependence, being slightly faster at the higher wavelengths of analysis. The absorption spectra in Fig. 5 reveal a hypsochromic shift with time after the pulse, similar to that shown in Fig. 3 for GafChromic DM 1260. The position of maximum absorbance shifts from about 675 nm to about 655 nm in the time interval 8 ms to 9 s after the 2.5 μs, 13 Gy pulse.

The temperature dependence of the build-up of the absorbance in the 100 μs to 3 ms time window was investigated on the example of the MD 55 film in the 22 to 50 °C temperature range. The upper temperature was limited by the melting

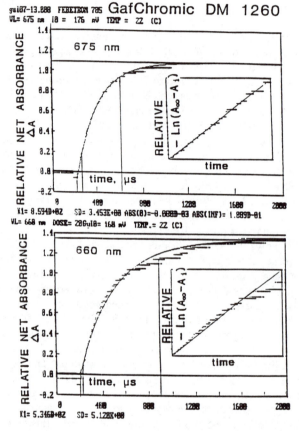

Fig. 2. Kinetics curves from pulse radiolysis of GafChromic DM 1260 radiochromic film, measured at 675 nm (top) and at 660 nm (bottom), following a 0.05 μs pulse of radiation with a dose of 20 Gy. The right-hand insets are semilog plots of the absorbances as a function of time for the period between the two vertical lines on the kinetics curves at the left. The linear semilog function indicates first-order kinetics for solid-state polymerization. At 675 nm, a rate constant of 0.9×10^3 s^{-1} is observed and propagation is terminated within 2000 μs.

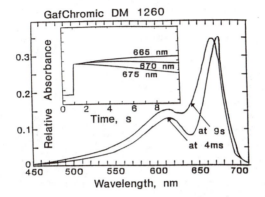

Fig. 3. Radiation-induced absorption spectra of GafChromic DM 1260 film measured at 4 ms and 9 s after the 2.5 μs electron pulse with a dose of 13 Gy. The inset shows the long-term relative absorbance change due to a hypsochromic shift (where the relative absorbances have been normalized at zero time). The faster kinetics shown in Fig. 2 appears here as vertical increase in absorbance.

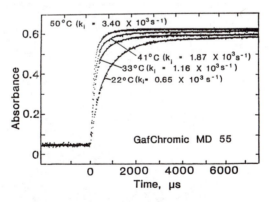

Fig. 4. Kinetic curves from pulse radiolysis of GafChromic MD 55 radiochromic film, measured at 660 nm, with the electron pulse and kinetic spectrophotometric measurement made at the indicated temperatures; the first-order rate constants of each are also indicated, showing an increase with temperature rise (see Fig. 6).

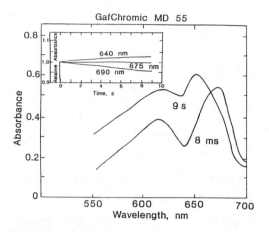

Fig. 5. Radiation-induced absorption spectra of GafChromic MD 55 film measured at 8 ms and 9 s after the 2.5 μs electron pulse with a dose of 13 Gy. The inset shows the long-term relative absorbance change due to a hypsochromic shift.

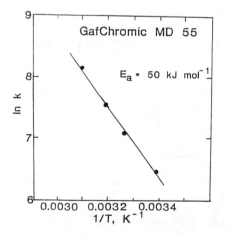

Fig. 6. Arrhenius-type plot of the rate constant of absorbance build-up of the MD 55 film in the 100 to 3000 μs time range as a function of temperature in the 22 to 50 °C range. The slope of the straight line indicates an activation energy of about 50 kJ mol^{-1}.

point of the active component. The monitoring wavelength was 660 nm. In this temperature range the rate constant increased from 6.3×10^2 s^{-1} to 3.4×10^3 s^{-1} indicating an activation energy of about 50 kJ mol^{-1} (12 kcal mol^{-1}), according to the Arrhenius plot of Fig. 6. This value falls between the values of 2 kcal mol^{-1} for γ ray initiated and 22 kcal mol^{-1} for thermally initiated partial polymerization of para-toluene sulfonate-substituted diacetylene monomer single crystals, reported by Chance and Patel (6).

Flash Photolysis Study

In order to compare the polymerization kinetics of UV-induced color formation with those of pulsed electron-beam radiolysis results, laser flash photolysis experiments were carried out on DM 1260 film, using single pulses from a Nd YAG laser (λ =266 nm, pulse length 16 ns, energy/pulse 30 mJ) and digital analysis of the kinetic spectrophotometric data.

The resulting absorption spectra taken at two times after the pulse (1 and 90 μs) (see Fig. 7) show essentially no blue color formation at the shorter time, but at 90 μs, a spectrum matching that induced by ionizing radiation. The features are a major peak at ≈ 660 nm and a minor one at ≈ 610. As indicated by the two small, kinetic UV-induced absorbance build-ups of the film with or without the sensor coating (see Figs. 8a and 8b) being completed in about 1 μs, the small background absorbance (λ =430 nm) at the short time (1 μs in Fig. 7) is due primarily to the typical yellow coloration of the polyester material when irradiated by large fluences of deep UV or ionizing radiation.

Evaluation of the flash photolysis data using absorbance build-up of the sensor-coated film, at around the major absorption-band peak (660 nm), as shown in Figs. 9a and 9b for the two time scales, indicates a complicated kinetics behavior and a progressive slowing down of color build-up during the first 30 μs after the UV pulse. At this wavelength the build-up between 30 and 100 μs appears to occur with first-order kinetics at an observed rate constant of 1.5×10^4 s^{-1}. In Fig. 9b the lowest curve represents the difference between the measured kinetics curve and the linear extrapolation fit in this time scale.

Thus, the UV-induced coloration appears to show a similar absorption spectrum. Although the product of polymerization in the sensor is essentially the same, the kinetics of polymerization are different for photolysis and radiolysis. Since the UV-induced build-up is at least an order of magnitude faster than that of the kinetics of the ionizing radiation-induced process and is more complicated at very short times after the pulse (<30 μs), one would expect the UV photolysis to lead to a different polymerization mechanism from radiolysis with electrons.

The photo-induced polymerization kinetics of the present system are similar in reaction rate to results previously published by Niederwald *et al.* (*14*), for flash UV photolysis of single para-toluene sulfonate-substituted diacetylene monomer single crystals at room temperature (T=293 K).

Discussion

Radiation-induced products of the solid-state array of disubstituted diacetylene monomer units represent the first reported conjugated polymers exemplifying ideal

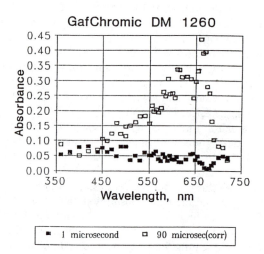

Fig. 7. Absorption spectra of GafChromic DM 1260 film at 1 μs and 90 μs after 0.016 μs, λ =266 nm laser light pulse. The spectrum at 1 μs is actually due to the ultraviolet-induced yellow coloration of the polyester backing material.

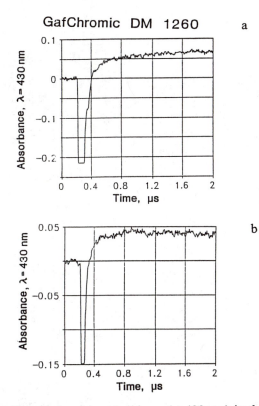

Fig. 8. Time-profiles of absorbance build-ups (at 430 nm) in the 0 to 2 μs time window, after 0.016 μs, λ =266 nm laser light pulse irradiation of the GafChromic DM 1260 film with (Fig. 8a) or without (Fig. 8b) the sensor coating.

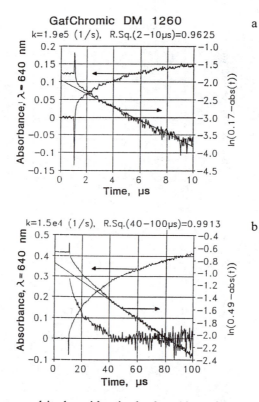

Fig. 9. Absorbance and its logarithm in the 0 to 10 μs time window (Fig. 9a) and in the 0 to 100 μs time range (Fig. 9b), after 0.016 μs, λ =266 nm laser light pulse, of the GafChromic DM 1260 film. In Fig. 9b the lowest curve represents the difference between the measured and extrapolated curve.

linear polymer chains with propagated extended forms (*15*). Chance and Patel (*6*) have shown differences in the characteristics of thermal-, UV- and γ-ray initiated polymerization, the latter two being less dependent on crystal strain and hence less autocatalytic than thermal polymerization.

In the present work with UV photons and high-energy electrons, the fast kinetic analyses of the two radiochromic blue-forming films were performed using single 2.5 μs and 0.05 μs electron or 0.016 μs light pulses. The results obtained with the light and electron irradiations show similarities, however, the kinetics are different. The coloration occurs with essentially the same absorption spectrum in electron and UV irradiation. The build-up of the absorbance occurs with first-order kinetics, but the rate is slightly higher at the longer wavelengths than at the shorter wavelengths. The rate constants in pulse radiolysis measurements vary from 7×10^2 s^{-1} at 550 nm to 1.5×10^3 s^{-1} at the main absorption peak region at ≈ 670 nm. The UV light-induced color-forming reaction on the μs time scale is found to be more than an order of magnitude faster than the reaction observed after electron irradiation. The rate followed at 660 nm gradually slows down with the time after the light pulse in the 1 to 30 μs range. The rate constant decreases from an apparent initial value of about 2×10^5 s^{-1} to 1.5×10^4 s^{-1}.

The films thus behave as a classic partial polymerization of concentrated dispersions of disubstituted diacetylenes coated on a polymer substrate (*6,7*) as a micro-crystalline thin layer. The colorless solid monomeric system is known to have overlapping lamellar-like array units and to contain certain intrinsic unspecified impurities also from the host matrix. When the polymerization is initiated and the chains then propagate in essentially one-dimensional alignments, the diacetylene units undergo 1,4-trans-polymerization, either by rotation or translation or both (*8*):

$$R_1 - C \equiv C - C \equiv C - R_2 \rightarrow \left[\begin{array}{c} R_1 \\ \diagdown \\ \diagup \\ \end{array} C - C \equiv C - C \begin{array}{c} \diagup \\ \diagdown \\ R_2 \end{array} \right]_n \tag{1}$$

where R_1 and R_2 represent bulky organic molecular end fragments with molecular weights over 150. The product can eventually become a fully polymerized coordinated chain connected by crosslinks when taken to very large radiation doses (> 100 kGy) or high temperatures (> 60 °C).

Heat-induced polymerization of some diacetylene monomers occurs as a sudden onset to form an opaque, metallic-like reddish appearance, which can be identified as being due to the autocatalytic process (a sigmoid response curve shape) and this full polymerization takes place over a relatively small range of temperature increase (*6,8,16*). An effect was observed in the present work during electron pulse radiolysis, where the long-term post-irradiation interrogation (10 to 20 seconds) by the bright xenon-light beam in the long-time measurements caused a sudden red color to appear in the area of the light-beam cross-section resulting from heat build-up. This could be due to an autocatalytic effect or the blue-to-red transition occurring with some polymerizing diacetylenes.

Models of reaction kinetics for γ-ray irradiated diacetylene systems have been given by Baughman (*8*), to explain monophase solid-state polymerization, including the possible roles played by impurities and mechanical stresses.

Indications are that, counter to the autocatalytic behavior of thermal polymerization, γ-ray-induced chain initiation is relatively insensitive to crystalline strain. Moreover, the greater the concentration of initial monomer (Z_o) units and of impurities (W_o), the less autocatalytic (with sigmoid characteristic response curve) is the polymerization-versus-radiation dose behavior. The present results, involving a relatively concentrated crystalline monomer complex dispersed in a host material, supplying controlled impurity concentrations, are representative of a high value of ($Z_o W_o$). This is indicated by the non-sigmoid γ-ray response curves shown in Fig. 1. The radiation-induced partial polymerization giving the steadily increasing optical absorption with increasing dose can then be construed in terms of an orderly increase of double-bond concentration in the main-chain, accompanied by excitonic absorption due to a quasi-one-dimensional guest-polymer strand matrix under some lateral strain, but still isolated in the monomer host crystal (*17*).

The subsequent long-term blue-wavelength-shifts in the absorption spectrum, shown by the relatively slow process over seconds time scales after the pulse (see Figs. 3 and 5), are apparently due to mechanical strain rearrangements in the close ladder-like packing at aligned tilts to the chain axes (*4,18*).

The initial stage of radiation-induced topochemical polymerization is then conjectured according to the following simplified model into a continuing close ladder-like packing at a certain tilt to the main-chain axis (*9*).

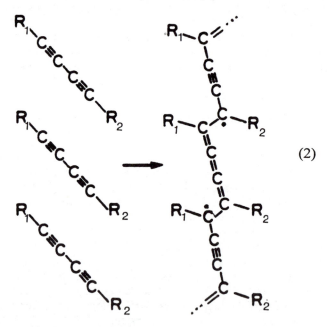

(2)

The product of this model reaction indicates where transitory radicals are likely to be located at the polymerization initiation stage. ESR spectrometry has shown evidence of a modest increase in the concentration of stable paramagnetic centers at relatively low radiation doses administered and measured under cryogenic conditions (*9,19-21*).

Photo-initiated polymerization (λ =310 nm) of para-toluene sulfonate-substituted single-crystal diacetylene monomer at low temperature (77 K) has been explained as being due to reaction to the diradical states and further multiplicates through carbenes and dicarbenes, due to direct and indirect singlet or triplet photoexcitations (*21,22*). The termination products are stable oligomer states.

It has been suggested that in γ-ray radiolytic diacetylene polymerizations, the diradicals, carbenes, and dicarbenes are readily convertible, by energy absorption, to stable oligomers with ten or more repeat units. The diradicals visualized in reaction (2) are stable for short chain lengths at cryogenic temperatures. At room temperature they quickly convert to the carbene species (*23*). Carbene and dicarbene radicals and the associated triplet carbenes localized at non-propagating polymer chain ends are conjectured to be the sites of initiation of radiation-induced polymerization (*4,9*). Bloor (*9*) has pointed out that the kinetics of polymerization depends on the way the energy is transferred to or trapped at the sites of initiation (diradicals). It is also pointed out that, at the short times of initial partial polymerization, the guest polymer strands and the surrounding monomer hosts are under approximately equal strain. This is substantiated by the present finding that absorption spectra due to UV and ionizing radiation represent the same blue color, thus giving evidence of a high degree of planarity of polymer chains (relatively uniform conjugation lengths). The long-term hypsochromic shifts in the absorption bands (see Figs. 3 and 5) over time scales of seconds and longer after the radiation pulse, represent slight slow rearrangement (small blue shifts) due to mechanical stresses between the adjacent polymer arrays. This causes departures from polyconjugated planarity to slight skewness of the chains. In the case of heat-induced polymerization, planarity is fully disrupted by sudden autocatalytic chain reaction to a fully polymerized system (a metallic red color).

Conclusion

The GafChromic radiochromic film is a colorless microcrystalline dispersion of an active monomer in a gel binder. It has been developed for direct radiographic imaging and dosimetry. This relatively sensitive sensor may be used as a thin-coating on a transparent or reflecting substrate for broad-range dosimetry ($5-10^4$ Gy) and for registering permanent blue images.

Pulse radiolysis studies show that the primary radiation-induced image formation (λ_{max} =675 nm) and propagation of the polymerization with a first-order rate constant of $\approx 10^3$ s^{-1} are terminated within 2000 μs. With ionizing radiation, there follows a much slower exponential hypsochromic shift of the absorption maximum from about 675 nm to 650 nm. The radiation-induced polymerization effect gives approximately the same spectral characteristics as the ultraviolet effect, but the fast UV-induced kinetics are more complicated, with rate constants greater by about one-order of magnitude.

The radiochromic films are currently finding application in clinical radiography, stereotactic radiosurgery, food irradiation, blood irradiation, insect population control, and industrial radiation processing (including medical sterilization and quality control).

Acknowledgement: This work was supported by U.S.–Hungary Joint Science and Technology Program JF-139.

Literature Cited

1. McLaughlin, W.L.; Chen, Y.-D.; Soares, C.G.; Miller, A.; VanDyke, G.; Lewis, D.F. *Nucl. Instr. Methods in Phys. Res.* **1991**, *A302*, 165-176.
2. McLaughlin, W.L. "Future Dosimetry Systems" in *International Standardization and Quality Assurance of High-Dose Dosimetry, Tech-Doc Report Co453*, International Atomic Energy Agency, Vienna **1993**, pp 42-63.
3. Chance, R.R. *Macromol.* **1980**, *13*, 396-398.
4. Wegner, G. In *Molecular Metals*; Hatfield, W.E., Ed.; NATO ASI Series E **1978**, pp. 209-242.
5. Schmidt, G.M.J. In *Reactivity of the Photoexcited Organic Molecule*; J. Wiley, New York, NY, **1967**, p. 227.
6. Chance, R.R.; Patel, G.N. *J. Polym. Sci. Polym. Phys. Ed.*, **1978**, *16*, 859-881.
7. Eckhart, H.; Prusik, T.; Chance, R.R. In *Polydiacetylenes*; Bloor, D.; Chance, R.R.; Eds.; NATO ASI Series E; Martinus Nijhoff: Dordrecht, **1985**, Vol. 102; pp 25-42.
8. Baughman, R.H. *J. Chem. Phys.* **1978**, *68*, 3110-3121.
9. Bloor, D. In *Quantum Chemistry of Polymers-Solid State Aspects*; Ladik, J., Ed.; **1984**, pp. 191-220.
10. Saylor, M.C.; Tamargo, T.T.; McLaughlin, W.L.; Khan, H.M.; Lewis, D.F.; Schenfele, R.D. *Radiat. Phys. Chem.* **1988**, *31*, 529-539.
11. Janovsky, I.; Mehta, K. *Radiat. Phys. Chem.* **1994**, *43*, 407-409.
12. McLaughlin, W.L.; Soares, C.G.; Sayeg, J.A.; McCullough, E.C.; Kline, R.W. Wu, A.; Maitz, A.H. *Med. Phys.* **1994**, *21*, 379-388.
13. Földiák, G.; Hargittai, P.; Kaszanyiczki, L.; Wojnárovits, L. *J. Radioanal. Nucl. Chem., Articles*, **1989**, *125* 19-28.
14. Niederwald, H.; Eichele, H.; Schwoerer, M. *Chem. Phys. Letters*, **1980**, *72*, 242-246.
15. Wegner, G. *Z. Naturforsch.* **1969**, *24B*, 824-832; *Makromol. Chem.* **1971**, *145*, 85-94; *Makromol. Chem.* **1972**, *154*, 35-48.
16. Patel, G.N.; Chance, R.R.; Witt, J.D. *J. Chem. Phys.* **1979**, *70*, 4387-4392.
17. Sixl, H.; Warta, R. In *Electronic Properties of Polymers and Related Compounds*, Kuzmany, H.; Mehring, M.; Roth, S., Eds.; Springer Verlag: Heidelberg, Berlin, **1985**, pp. 246-249.
18. Gross, H.; Sixl, H. *Chem. Phys. Lett.* **1982**, *91*, 262-267.
19. Eichele, H.; Schwoerer, M.; Huber, R.; Bloor, D. *Chem. Phys. Lett.* **1976**, *42*, 343-349.
20. Bubeck, C.; Sixl, H.; Wolf, H.C. *Chem. Phys.* **1978**, *32*, 231-237.
21. Bubeck, C.; Sixl, H.; Neumann, W. *Chem. Phys.* **1980**, *48*, 269-275.
22. Patel, G.N.; Chance, R.R.; Turi, E.A.; Khanna, Y.P. *J. Am. Chem. Soc.* **1978**, *100(21)*, 6644-6649.
23. Niederwald, H. Thesis, University of Bayreuth **1982**.

RECEIVED December 1, 1995

Chapter 12

Deposition of Plasma-Polymerized Styrene under Ion Bombardment

Jung H. Lee, Dong S. Kim, and Young H. Lee

Department of Chemical Engineering, Drexel University, Philadelphia, PA 19104

Ion bombardment on a growing film is an effective way to modify film properties. In this study, plasma polymerized styrene thin films are deposited on silicon wafers, glass plates, and polyimide films under varying degrees of ion bombardment in a 13.56 MHz, asymmetric plasma reactor. The degree of ion bombardment is varied by adjusting the power and pressure. FTIR spectra of the deposited films show that sp^2 fraction and hydrogen concentration decrease with increasing ion bombardment on the substrate. RBS data show that the atomic fraction of carbon is uniform throughout the film. Film adhesion is very poor above -325 V of substrate bias voltage at pressure of 70 mTorr. At lower substrate bias voltages, NH_4OH cleaning enhances the adhesion of the film on silicon substrates, whereas higher substrate temperature gives better adhesion on glass plates.

The depositions and characterizations of plasma polymerized thin films have been the subject of intensive research in recent years. Plasma polymerization is simpler than the conventional methods for polymer preparation since fewer fabrication steps are involved. However, the resulting polymer films are very different from the corresponding conventional polymer films. Properties of plasma polymerized films depend strongly on the deposition parameters such as the power input, chamber pressure, substrate temperature , and the reactor configuration (1,2). The main applications of plasma polymerization are sensors (3,4), protective coatings (5,6), microelectronics (7,8), and biomedical area (9,10).

Conventional polystyrene has been used as the membrane layer in constructing micro sensors. One of the problems with these sensors is the weak adherence of the membrane to the sensor surface. The dip-coated 'conventional' polystyrene membrane peels from the surface easily and this occurrence causes excessive drift and noise in the output signal (11,12).

0097–6156/96/0620–0167$12.00/0

Ion bombardment of the substrate during film deposition has been found to be effective in optimizing many film properties (*13-15*). In this method, the surface of the growing film is bombarded by positive ions attracted by the negative bias developed on the substrate. As a result, the film properties such as density, stoichiometry, and morphology are altered depending on the degree of ion bombardment. Also, the ion bombardment can improve film adhesion (*16*). Plasma polymerization of styrene has been studied previously (*17-20*). However, there have been few studies on the effect of ion bombardment on film properties.

The purpose of this paper is to analyze the plasma phase and the film properties for the plasma polymerized styrene under varying degree of ion bombardment. Also, film adhesion on silicon, glass, and polyimide sheet (Kapton) is investigated in terms of substrate cleaning, degree of ion bombardment, and substrate temperature.

Experimental

Monomeric styrene was purchased from Aldrich (Milwaukee, WI). It was purified by passing the styrene solution through a disposable column for inhibitor removal. Film depositions were carried out in a capacitively coupled, asymmetric plasma reactor (Plasma-therm, NJ) driven by a 13.56 MHz rf power supply (Figure 1). The internal dimension of the reactor was 26.5 cm square and 34 cm height. The distance between the powered electrode and the top plate was 9.5 cm. A self-bias voltage was developed at the powered electrode due to the asymmetry of electrodes (*21*). A Penning gauge and a Baratron pressure gage were used for measuring reactor pressure. The flow rates of argon and styrene were adjusted by a mass flow controller and a metering valve, respectively.

The substrates used were (100) oriented p-type silicon wafers, glass plates (microscope slides), and polyimide films (Kapton). After 10 min. of cleaning in an ultrasonic cleaner with acetone, the substrates were immersed in a ammonium hydroxide (NH_4OH, Aldrich) and deionized water (D.I. H_2O) solution of the volume ratio of NH_4OH:D.I. H_2O=100:75 at 75 °C. The cleaning times were 30 sec. for silicon wafers and 10 min. for glass plates. The temperature of the substrate holder of the plasma reactor was controlled by a recirculating water bath. An oil diffusion pump was used to achieve a base pressure below 4×10^{-4} Torr. After the base vacuum was established, argon gas was introduced into the reactor. The desired partial pressure of the reactor was achieved by adjusting the argon flow rate and the throttle valve. Before film deposition, the substrates were cleaned in-situ by an argon plasma at 100 mTorr, and 40 W for 10 min. The metering valve on the styrene line was controlled to establish the desired pressure. The total pressure was varied from 50 to 100 mTorr. After the gas flow rates and the pressure were stabilized, rf power was applied and adjusted to the desired level. When the plasma was extinguished, the deposited films were exposed to pure styrene vapor at 70 mTorr for 3 min. in an effort to consume the remaining radical sites. Following this post reaction, argon gas was introduced into the reactor for 5 min.

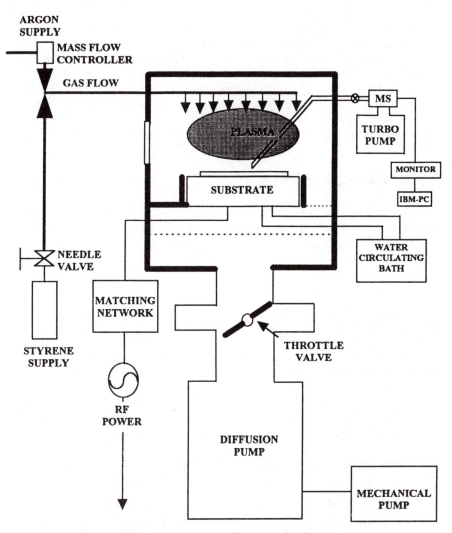

Figure 1. Schematic diagram of asymmetric plasma reactor.

For measuring the bias voltage on the powered electrode, an oscilloscope (Hitachi V-680) equipped with a 100X probe and a current sensor was used. The on-line mass spectrometer (MS) used is a differentially pumped Ametek Dycor M200, working in the range 1-200 amu. The background pressure of the MS was less than 1×10^{-6} Torr. The FTIR spectra of the films were obtained with a Perkin-Elmer FTIR spectrometer (Model 1800). NaCl was used as the support in FTIR measurement and the sample was scanned 100 times. In RBS (Rutherford Backscattering Spectrometer) measurement, the specimen was irradiated with a collimated beam of monoenergetic He ions (1.5 MeV). The index of refraction and the thickness of the films were measured using a Gaertner Scientific Dual Mode Automatic Ellipsometer L116A. The wave length used for the ellipsometer was 632.8 nm and the incident angle was 70 degree. The thickness of the films was also measured with a profilometer (Dectak IIA). Scotch tape test was used to evaluate the adhesion of the film on the substrates after deposition.

Results and Discussion

Self-bias Voltage. The negative bias voltage (V_{dc}) on the substrate depends on both applied power (W) and reactor pressure (P). Figure 2 shows that it approximately follows

$$V_{dc} \propto (W)^{\frac{3}{4}} (P)^{-\frac{1}{4}}$$

for pure argon. V_{dc} is slightly higher in argon-styrene mixture than in pure argon (Figure 3). It should be noted that in certain ranges of the applied power, the discharge does not fill the whole space between two electrodes uniformly. Because V_{dc} depends not on the geometrical area of the electrodes, but on the actual area which comes in contact with the discharge, the result of Figure 3 suggests that the active cathode area is slightly reduced in the argon-styrene discharge. The negative bias at the substrate attracts the positive ions from the plasma region, causing ion bombardment during film growth. The degree of ion bombardment affects the composition, density, and the sp^2/sp^3 ratio of carbon in the film (*13*).

Gas Phase Kinetics. The mass spectrum of styrene before initiating the discharge shows the following major peaks: $C_2H_2^+$ (M=26), $C_2H_3^+$ (M=27), $C_3H_2^+$ (M=38), $C_4H_3^+$ (M=51), $C_4H_4^+$ (M=52), $C_5H_3^+$ (M=63), $C_6H_5^+$ (M=77), $C_6H_6^+$ (M=78), $C_8H_7^+$ (M=103), and $C_8H_8^+$ (M=104) (Figure 4). Figure 5 shows the relative intensities of $C_2H_2^+$, $C_4H_3^+$ and $C_8H_8^+$ peaks at different applied power. The peak intensity of $C_2H_2^+$ (M=26) is shown to increase linearly up to 40 W, then levels off. However, the peak intensities of $C_4H_3^+$ and $C_8H_8^+$ decrease linearly as the power is increased. These results indicate that the decomposition rate of styrene increases with an increase in power. Therefore, it is quite possible that the structure of the deposited films depends upon the power input.

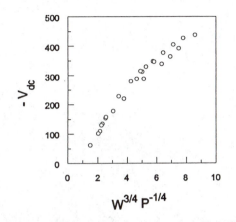

Figure 2. Substrate self-bias voltage as a function of applied power and reactor pressure for argon. Substrate temperature is kept at 25 °C. W in watt and P in mTorr.

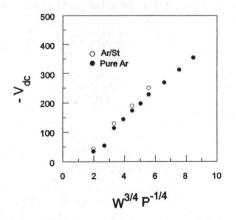

Figure 3. Effect of styrene on the self-bias voltage. Reactor pressure is 67 mTorr and substrate temperature is 25 °C. W in watt and P in mTorr.

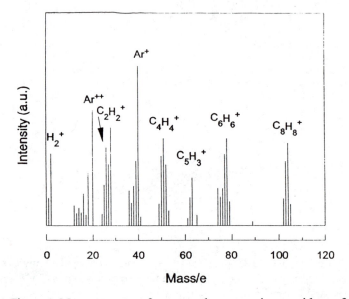

Figure 4. Mass spectrum of argon and styrene mixture without rf power. Reactor pressure is 100 mTorr and substrate is at 25 °C.

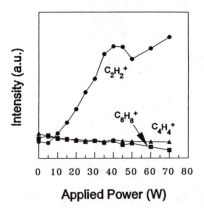

Figure 5. Relative peak intensities in mass spectra as a function of applied power at 100 mTorr, and 25 °C.

Film Structure and Hydrogen Concentration. Figure 6 shows the FTIR spectra in the C-H stretching region for films prepared at different V_{dc}. The chamber pressure is maintained at 70 mTorr. At low V_{dc} (-65 V), the film shows sp^3 C-H at 2916, sp^2 C-H (olefin) at 3027, sp^2 C-H (aromatic) at 3051, and sp C-H stretching at 3297 cm^{-1}, respectively. However, beyond -195 V, the peak intensities of sp C-H , sp^2 C-H (olefin), and sp^2 C-H (aromatic) stretching reduce significantly. This suggests that the fractions of sp, sp^2, and sp^3 bonded carbon species vary with increasing degree of ion bombardment. The MS data indicate that the decomposition of styrene to smaller molecules is increased with applied power. Therefore, it appears that the aromatic ring of styrene is not conserved at higher applied power, and this results in less sp^2 C-H stretching in the deposited film.

The hydrogen concentration in the film affects film density and mechanical properties. It is reported that the hydrogen concentration of amorphous hydrogenated carbon films decreases with increasing V_{dc} at a fixed chamber pressure (*22,23*). The C-H content of the film can be obtained from the integrated intensities of the C-H stretching modes in the FTIR spectra using the following formula (*24*)

$$N_{C-H} = \frac{A_s}{\upsilon_o} \int \alpha(\upsilon)\, d\upsilon$$

where N_{C-H} is the concentration of C-H bonds per cm^3, $\alpha(\upsilon)$ is the absorption coefficient at a given wave number, and υ_o is the wave number at the maximum absorption. A_s is a proportionality constant, and its value is 1.7×10^{21}/cm^2 (*25*). The dependence of N_{C-H} on V_{dc} is shown on Figure 7: the hydrogen concentration of the film decreases with increasing V_{dc}.

Figure 8 shows a typical RBS spectrum of the deposited film on a silicon wafer prepared at 30 W, 70 mTorr, and 45 °C. The depth profile of the atoms was obtained by simulating the RBS channeling data with the software ("Rump") that came with the instrument. Note that the atomic fraction of carbon is relatively constant, indicating a uniform elemental composition in the film. Also, the figure shows that very small amounts of oxygen and argon are detected in the film. Their atomic fractions are below 0.005. Note that the oxygen peak is mainly due to silicon oxide on the surface of the substrate. Although the silicon oxide is removed during NH$_4$OH cleaning, the RBS data show that it can grow on the surface during the initial stage of deposition.

Deposition Rate and Refractive Index. Figures 9 and 10 show the effect of process parameters on the deposition rate and the refractive index. The deposition rate increases initially but decreases as the substrate bias voltage is increased. The initial increase may be due to an increase in the deposition precursor in the plasma phase as a result of power increase. Subsequent decrease appears to be due to an increase in etching rate as the bias voltage is increased. Note that etching is expected to be significant at higher ion bombardment. The effect of temperature on the deposition rate is shown in Figure 10. The deposition rate is not changed significantly between 25 °C and 65 °C, but a significant increase is observed at temperature above 75 °C.

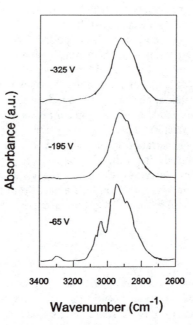

Figure 6. FTIR spectra at different self-bias voltages in the region of C-H stretching. Substrate is NaCl and film thickness is about 600 nm. Reactor pressure is 70 mTorr and substrate is at 25 °C.

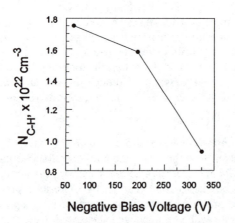

Figure 7. N_{C-H} concentration as a function of self-bias voltage. Deposition conditions are the same as in Figure 6.

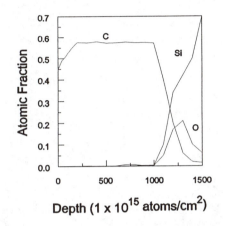

Figure 8. RBS depth profile of C, O, and Si. Substrate is silicon wafer. Deposition conditions are 30 W, 70 mTorr, and 45 °C.

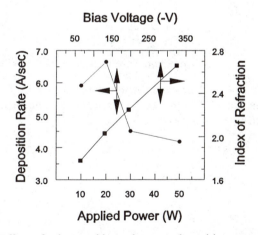

Figure 9. Effect of substrate bias voltage on deposition rate and the index of refraction at 70 mTorr, and 25 °C. Filled circle: film deposition rate. Filled rectangle: index of refraction.

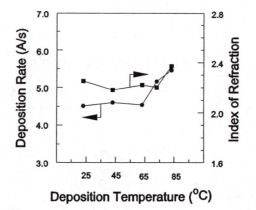

Figure 10. Effect of substrate temperature on deposition rate and index of refraction at 30 W, and 70 mTorr. Filled circle: film deposition rate. Filled rectangle: index of refraction.

With increasing power input, which corresponds to an increase in the bias voltage, a linear increase in the refractive index is observed (Figure 9). Similar trends have been found for plasma polymerized methyl methacrylate (*26*), hexamethyl disiloxane (*27*), and tetramethyl silane (*28*). The refractive index of plasma polymerized styrene thin film is 1.78 at -130 V and 2.27 at -195 V at 70 mTorr. It is known (13) that the refractive index of the amorphous carbon films prepared by rf PECVD depends on the V_{dc} used. At low V_{dc}, the film is soft and has low refractive index (usually less than 1.8), thus polymer-like. In the medium range of V_{dc}, the deposited film is hard ("diamond-like") and has refractive index between 2.0 and 2.3. We found that, at -130 V of the V_{dc}, the film was soft and polymer-like. However, the film was much harder at the -195 V of V_{dc} primarily due to greater degree of ion bombardment. As shown in Figure 10, the refractive index of the films depends also on the substrate temperature. However, the dependence is less pronounced compared with that of the ion bombardment.

Adhesion. It has been known that the amorphous carbon film prepared by plasma polymerization is nonpolar and hydrophobic (*29*). The film adheres well on polymer substrates, but the adhesion is generally poor on substrates that do not contain carbon atoms. In this study, the deposited films show very good adhesion on Kapton regardless of the deposition conditions and the pre-cleaning conditions.

The adhesion on silicon and glass depends on both deposition and pre-cleaning conditions. Below -325 V of V_{dc} at pressure of 70 mTorr, the deposited films are stable. Beyond that, the films peel after they are exposed to air. Considering that the film adhesion depends on the stress in the film and the bonding strength at the substrate-film interface, it seems that the weak bonding at the interface is deteriorated by absorbing moisture in the air.

The stable films prepared at relatively low V_{dc} are tested by the Scotch tape method to determine the effects of pre-cleaning and substrate temperature. Table I shows the results. For this comparison, V_{dc} is fixed at -195 V. The results of sample A and B show clearly that cleaning with NH_4OH solution, which can be used to remove surface particles in substrates, improves adhesion only on silicon. While glass remains unaffected in NH_4OH solution, silicon is etched in this solution. The chemical etching of the silicon wafer for 30 sec. makes the surface rough and this surface roughness may have resulted in a better adhesion.

It is reported that, in certain cases, in-situ argon plasma treatment enhances film adhesion (30). In this study, in-situ Ar ion bombardment prior to film deposition is tried but it turns out to be less effective than the other cleaning methods. When the glass and the silicon wafer are pretreated with Ar plasma without any cleaning, the deposited films fail to pass the adhesive tape test. However, plasma pre-treatment is necessary to prevent contamination after the chemical cleaning.

The substrate temperature plays an important role in promoting the adhesion of the deposited films on glass plates. Films deposited at higher substrate temperature (75 °C) pass the Scotch tape test (sample C), while those deposited at lower temperature (25 °C; sample B) fail. Glass surfaces are notorious for adsorbing moisture. At high temperature, moisture adsorption tends to be less and this appears to lead to a better adhesion.

Table I: Adhesion characteristic of films deposited at 30 W, 70 mTorr (-195 V_{dc})

Sample	Pre-cleaning	Substrate temperature	Results
A	x, z	20 °C	Fail on both silicon and glass
B	x, y, z	20 °C	Fail on glass, but pass on silicon
C	x, y, z	75 °C	Pass on both silicon and glass

Pre-cleaning; x: 10 min. ultrasonic cleaner with acetone.
               y: 30 sec. for silicon and 10 min. for glass in NH_4OH solution.
               z: Ar plasma for 10 min. at 40 W, and 100 mTorr.

Conclusion

Plasma polymerized styrene thin films are deposited on glass, silicon wafer, and polyimide sheet under varying degrees of ion bombardment. The bias voltage on the substrate is increased by increasing applied power and decreasing chamber pressure, with the former being more effective. On-line analysis of the plasma phase shows that decomposition reaction is prevalent at higher power input. FTIR spectra of the deposited films show that the sp^2 fraction of C-H stretching decreases with increasing degree of ion bombardment. Also, hydrogen concentration in the film decreases with increasing bias voltage. As the substrate bias voltage is increased, the deposition rate increases initially, but the rate decreases upon further increase in the bias voltage. The refractive index increases with increasing bias voltage on the substrate. The substrate temperature also affects the refractive index, but it is less effective than the bias voltage. The results of adhesion test show that chemical pretreatment is more effective than Ar ion bombardment with silicon wafer, while substrate temperature is more critical for the adhesion on glass. Higher substrate temperature gives better adhesion on glass.

Literature Cited

1. Biderman, H.; Osada, Y. *Plasma Polymerization Processes*; Elsevier: NY, 1992.
2. d'Agostino, R *Deposition, Treatment and Etching of Polymers*; Academic: San Diego, 1990.
3. Yamaha, N.; Hatai, T.; Kusanagi, S. *JP*; 04363652; Dec. 16, 1992.
4. Doi, K.; Fujioka, T.; Watabe, Y.; Kusanagi, S.; Yamaga, N. JP; 05273176; Oct. 22, 1993.
5. Leiber, J.; Michaeli, W. *Annu. Tech. Conf.-Soc. Plast. Eng.* **1992,** *50th(2),* 2583.

6. Zeik, D. B.; Clarson, S. J.; Taylor, C. E.; Boerio, F. J.; van Ooij, W. J.; Sabata, A. *Polym. Prepr.* **1993,** *34(1),* 693.
7. Tanaka, K.; Nishio, S.; Matsuura, Y.; Yamabe, T. *J. Appl. Phys.* **1993,** *73(10),* 5017.
8. Weidman, T. W.; Joshi, A. M. *Proc. SPIE-Int. Soc. Opt. Eng.* **1993,** *1925,* 145.
9. Cao, W.; Zhou, K. *Hecheng Huaxue* **1994,** *2(1),* 57.
10. Marchant, R. E.; Danilich, M. J. *Polym. Prepr.* **1993,** *34(1),* 655.
11. Kim, J. Y.; Lee, Y. H. *Biotech. and Bioeng.* **1990,** *35,* 850.
12. Luk, S.; Lee, Y. H. *AICHE J.* **1986,** *32,* 1546.
13. Robertson, J. *Pure & Appl. Chem.* **1994,** *66(9),* 1789.
14. Greene, J. E.; Barnett, S. A.; Sundgen, J. E.; Rockett, A. *Ion Beam Assisted Film Growth*; Elsevier: Amsterdam, 1988.
15. Hey, H. P. W.; Sluijk, B. G.; Hemmes, D. G. *Solid State Technol.* **1990,** *33(4),* 139.
16. Takagi, T. *Thin Solid Films* **1982,** *92,* 1.
17. Prohaska, G. W.; Johnson, E. D.; Evans, J. F. *J. Polym. Sci. Polym. Chem.* **1984,** *22,* 2953.
18. Ozden, B.; Hacaloglu, J.; Akovali, G. *Eur. Polym. J.* **1991,** *27(12),* 1405.
19. Thomson, L. F.; Mayhan, K. G. *J. Appl. Polym. Sci.* **1972,** *16,* 2317.
20. Ugolini, D.; Tuilier, M. H.; Eitle, J.; Schelz, S.; Wang, J. Q.; Oelhafen, P. *Appl. Phys.* **1990,** *A51,* 526.
21. Kasper, W.; Böhm, H.; Hirschauer, B. *J. Appl. Phys.* **1992,** *71(9),* 4168.
22. Robertson, J. *Prog. Solid St. Chem.* **1991,** *21,* 199.
23. Fourches, N.; Turban, G. *Thin Solid Films* **1994,** *240,* 28.
24. Gat, E.; El Khakani, M. A.; Chaker, M.; Jean, A.; Boily, S.; Pepin, H., Kieffer, J. C.; Durand, J.; Cross, B.; Rousseaux, F.; Gujrathi, S. *J. Mater. Res.* **1992,** *7,* 2478.
25. Nakazawa, K.; Ueda, S.; Kumeda, M.; Morimoto, A.; Shimizu, T. *Jpn. J. Appl. Phys.* **1982,** *21,* L176.
26. Tobin, J. A.; Denton, D. D. *Appl. Phys. Lett.* **1992,** *60(21),* 2595.
27. Rochotzki, R.; Arzt, M.; Blaschta, F.; Kreyβig, E.; Poll, H. U. *Thin Solid Films* **1993,** *234,* 463.
28. Catherine, Y.; Zamouche, A. *Plasma Chem. Plasma Process* **1987,** *5,* 353.
29. Sharma, A. K.; Yasuda, H. *J. Vac. Sci. Technol.* **1982,** *21(4),* 994.
30. Kondo, I.; Yoneyama, T.; Kondo, K.; Takenaka, O. *J. Vac. Sci. Technol.,* **1992,** *A10(5),* 3166.

RECEIVED July 29, 1995

Chapter 13

Synthesis of Polysaccharide Chemical Gels by Gamma-Ray Irradiation

Annamaria Paparella[1] and Kinam Park[2,3]

[1]Fidia Advanced Biopolymers, Via Ponte della Fabbrica 3/a, 35031 Abano Terme (PD), Italy
[2]School of Pharmacy, Purdue University, West Lafayette, IN 47907

The aim of this study was to compare the ability of different polysaccharides to form chemical gels by gamma-irradiation. Dextran, alginic acid, hyaluronic acid, benzyl esters of hyaluronic acid, and gellan were functionalized to introduce double bonds through reaction with glycidyl acrylate. All the polysaccharides used in our study formed chemical gels by gamma-irradiation, although the extent of gel formation was different. The effects of the polymer concentration, the amount of glycidyl acrylate, and the gamma-irradiation dose on the hydrogel formation were studied by determining the degree of swelling of gels in water. The acidification of polyelectrolytes to pH 3 allowed them to form chemical gels more easily. Alginic acid, hyaluronic acid, and benzyl ester of hyaluronic acid at 25% of esterification degree formed chemical gels at lower polymer concentrations and at lower gamma-irradiation dose at pH 3 than at pH 6. The ability to form chemical gels using various natural polymers may be useful in the development of biodegradable hydrogels for various applications.

Hydrogels have been widely used in various applications ranging from controlled drug delivery to agriculture. Recently many biodegradable hydrogel systems have been developed in the area of controlled drug delivery. Hydrogels made from natural polymers present advantages in controlling the drug release profiles and avoiding the necessity of removing the delivery systems after their use (1). Generally, the preparation of polymeric networks is carried out by direct crosslinking of homopolymers or copolymers in solution using a small amount of a crosslinking agent or by simultaneous copolymerization and crosslinking reactions of monofunctional and multifunctional monomers (2). Alternatively, modified biopolymers can be crosslinked by gamma-irradiation (3). Since the modified biopolymers can be purified before exposure to gamma-irradiation, the formed gels do not contain undesirable components, such as unreacted monomers or crosslinking agents (3).

We have previously prepared chemical gels from dextran and gelatin by gamma-irradiation (3-4). The polymers were functionalized with glycidyl acrylate to introduce double bonds and then crosslinked by gamma-irradiation. This is a simple and efficient method of hydrogel formation which can be applied to a variety of water

[3]Corresponding author

soluble polymers. The resulting gels do not need any further purification, since the functionalized polymers are purified prior to gel formation. In addition, this approach allows the preparation of hydrogels without the use of any externally added crosslinker. Drug molecules can be mixed with the functionalized polymers prior to the gel formation (3). This allows incorporation of a large amount of high molecular weight drugs such as proteins which are difficult to load by other approaches, e.g., equilibrium swelling of hydrogels in a drug solution.

This paper describes the synthesis of gels by gamma-irradiation from various polysaccharides such as dextran, gellan, alginic acid, hyaluronic acid and ester derivatives of hyaluronic acid. All the polysaccharides, except for dextran, are polyelectrolytes characterized by peculiar solution properties and ion sensitivity (5-7). Although gamma-irradiation provides a clean and efficient method for gel formation, it can also lead to degradation of polymer chains and loss of viscosity (2,8). We explored the pH-dependent ionization of polyelectrolytes to make chemical gels by gamma-irradiation under mild conditions which do not degrade polysaccharides. The formation of chemical gels at lower gamma-irradiation dose (i.e., less than 0.5 Mrad) will also minimize the loss of bioactivity of the incorporated drugs such as bioactive proteins (3,9,10).

Materials and Methods

Preparation of Functionalized Polymer Solutions

Functionalization of proteins (11-13) and polysaccharides (14-16) has been reported in the literature. Polysaccharides were functionalized following the procedure established in our laboratory (3,4). The ability to form gels was studied for samples of dextran (Sigma, mol. wt. 2,000,000), gellan (Gelrite, Kelco), alginic acid (Sigma, Medium viscosity), hyaluronic acid (HA, mol. wt. of 200,000 and 2,000,000), and benzyl esters of hyaluronic acid at 25% and 50% of esterification degree (HYAFF11-p25 and HYAFF11-p50, respectively). The benzyl esters of hyaluronic acid were prepared from HA of mol. wt. 200,000. Both HA and HAYFF11 samples were supplied by Fidia Advanced Biopolymers (Abano Terme, Italy).

The polymers were dissolved in distilled deionized water to obtain the final concentrations of 5% (w/v) for dextran, alginic acid, HA 200,000, and HYAFF11-p25, 1.5% for HA 2,000,000, and 3.5% for HYAFF11-p50 and gellan. The initial concentrations of polysaccharide solutions were chosen based on the viscosity of the solution. Glycidyl acrylate (Aldrich) was added directly to polysaccharide solutions while stirring. The amount of glycidyl acrylate was varied from 0.8 ml to 1.3 ml/g of polymer. After 1 day (except for 2 days for dextran samples) glycine solution (20%) was added to stop the functionalization process. The solutions were then dialyzed against distilled deionized water. The final volumes of functionalized solutions of alginic acid, HA 200,000, HYAFF11-p25, and HYAF11-p50 increased significantly after dialysis. The solutions were then centrifuged and the precipitates removed. Finally the supernatants were concentrated using a rotoevaporator to obtain a concentration range between 1 and 3% (w/v). The final pH of all solutions was 6. Portions of the functionalized solutions of gellan, alginic acid, hyaluronic acid, and benzyl esters of hyaluronic acid at different concentrations were acidified to pH 3 with citric acid.

The acrylic group content in the functionalized dextran was determined spectrophotometrically after bromination of the double bonds as well as spectroscopically by FT-^{1}H-NMR analysis. The former method consisted in measuring the decrease in absorbance at 480 nm in methanol-water (80/20) solution of bromine (0.2% v/V) (17). Acrylamide was used as a standard for this assay. The bromination of polysaccharides did not work very well and often resulted in a slow and incomplete reaction (18). The proton NMR analysis of dextran solutions at the concentration of 10% (w/v) in deuterium oxide produced more precise results than the bromine analysis. The degree of substitution of dextran was obtained from the peak ratio between anomeric proton and the acrylic proton (19). The results indicated that one acrylic group was introduced per 20 glucose residues. For the other polysaccharides it was not possible to determine the degree of substitution by NMR spectroscopy because of high viscosity of the solutions.

Preparation of Hydrogels

The functionalized polymer solutions were gamma-irradiated from 1 h to 8 h using a dose rate of 0.0606 Mrad/h. The irradiation was performed with a ^{60}Co source. The samples were not purged with nitrogen prior to irradiation because of the known damaging effect of nitrogen on hyaluronic acid solutions (20). The gels were air dried for 24 h and oven-dried at 60°C for 12 h.

The effect of gamma-irradiation dose on the gel formation was studied by examining the equilibrium swelling of the dried hydrogels. The gels were allowed to swell in distilled deionized water at room temperature. The time required to reach the equilibrium was determined by monitoring the change in the weight of hydrogels. The swelling ratio, Q, was calculated from the following equation:

$$Q = W*/W$$

where $W*$ and W are the weights of the swollen and dried gels, respectively (21).

Results and Discussion

The functionalized polysaccharides showed different abilities to form chemical gels. Of the polysaccharides tried in our study, dextran formed the gel most readily and did not degrade easily. Functionalized dextran solutions resulted in gel formation just after 0.12 Mrad of gamma-irradiation. As the gamma-irradiation time increased from 0.24 Mrad to 0.48 Mrad, the swelling ratio decreased due to the higher crosslinking density by higher gamma-irradiation dose.

The gel formation of functionalized alginic acid was dependent on the polymer concentration, gamma-irradiation dose, degree of functionalization, and the pH of the solution. Figure 1 shows the gel formation by alginate at two different pH values as a function of alginate concentration and gamma-irradiation dose. The lines indicate the minimum concentrations required to form a chemical gel at a given gamma-irradiation dose. At pH 6, at least 2% concentration of alginate was necessary to form gels. When the concentration of glycidyl acrylate used for alginate modification was 0.8 ml/g polymer instead of 1.3 ml/g polymer, the alginate concentrations required to form gels at pH 6 were at least 0.5% higher than those shown in Figure 1. When the functionalized solutions were acidified to pH 3, gels were formed more readily. At pH 3, the alginate concentrations required to form gels were lower than those at pH 6. For example, a solution of 1% alginate at pH 3 was able to form a gel even by 0.12 Mrad of gamma-irradiation, while a higher concentration (>2%) was necessary at pH 6. This behavior occurs probably because the acidification deionizes the carboxylic groups and thus decreases the repulsion between the molecules. The polymer chains can come closer to each other and this favors the crosslinking reaction (8). Furthermore, the acidification can cause the formation of hydrogen bonds between the polymer chains. The presence of physical interactions accelerates the chemical gel formation. As the gamma-irradiation time increased from 0.12 Mrad to 0.48 Mrad at pH 3, a higher alginate concentration was required to form a gel. This is probably due to degradation of the polysaccharide at higher gamma-irradiation dose. In fact, the formed gels degraded and turned into very viscous solutions after long exposure to gamma-irradiation.

The trend of the gel formation by HA (Mol. Wt. of 200,000) was similar to that by alginates as shown in Figure 2. Gels were formed at lower concentrations if pH

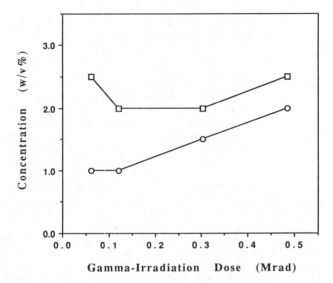

Figure 1. Gel formation by alginate as a function of alginate concentration and gamma-irradiation dose at pH 3 (O) and pH 6 (□). Glycidyl acrylate used to functionalize alginate was 1.3 ml/g polymer. The lines indicate the minimum concentration of alginate necessary to form a three-dimensional network.

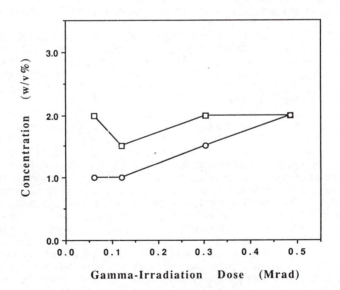

Figure 2. Gel formation by hyaluronic acid (HA, Mol. Wt. 200,000) as a function of HA concentration and gamma-irradiation dose at pH 3 (O) and pH 6 (□). Glycidyl acrylate used to functionalize alginate was 0.8 ml/g polymer. The lines indicate the minimum concentration of alginate necessary to form a three-dimensional network.

was reduced to 3. It is interesting to notice that the trend of gel formation at pH 3 is the same as that of alginate. HA with higher molecular weight, however, showed different behavior. Solutions of functionalized HA (Mol. Wt. of 2,000,000) at pH 6 were very viscous and formed gels at 1.5% upon 0.06 Mrad of gamma-irradiation. As the gamma-irradiation dose increased more than 0.48 Mrad, the formed gels were found to degrade. The acidification of HA solutions before gamma-irradiation did not result in any improvement in the gel formation. This behavior also occurred at higher polymer concentrations. Comparison of the HA gels at two different molecular weights showed that HA with higher molecular weight degraded to a greater extent.

Figure 3 shows the concentrations of HYAFF11-p25 necessary to form chemical gels at different gamma-irradiation doses. Solutions of HYAFF11-p50 at pH 3, however, formed microgels which are dispersed in solution. The gamma-irradiation did not result in a macroscopic three dimensional network formation. Functionalized gellan also did not form macroscopic three-dimensional networks at pH 6. The acidification of gellan resulted in formation of microgels. Further study is required to find out the reasons for the formation of microgels, instead of macroscopic hydrogels, by HYAFF11-p50 and gellan under acidic conditions.

The characterization of polyelectrolytes gels was carried out by swelling studies. Figure 4 compares the equilibrium swelling ratios of alginate gels at two different concentrations (2.5% and 3% w/v) obtained by reactions with the same amounts of glycidyl acrylate (1.3 ml/g polymer). The equilibrium swelling ratio increased with decreasing polymer concentration. Moreover, as the gamma-irradiation dose increased from 0.12 Mrad to 0.48 Mrad, the equilibrium swelling ratio increased. This means that the gel crosslinking density is reduced by long exposure to gamma-irradiation. The only explanation for this is that crosslinking is increased by long exposure to gamma-irradiation but at the same time alginate molecules are degraded. Thus, the overall effective crosslinking density is reduced. The results of the swelling study with HA (shown in Figure 5) shows that the degradation of HA by gamma-irradiation is more pronounced than that of alginate.

Figure 6 shows the equilibrium swelling ratios of hyaluronic acid gels obtained at two different pH values. The gels prepared from solutions at pH 6 did not maintain their shape during the swelling and therefore it was difficult to recover them for the measurement. The higher mechanical strength and the higher equilibrium swelling ratios of gels prepared at pH 3 probably means that the degradation at pH 3 was reduced compared to that at pH 6.

Our preliminary study on the formation of gels from various functionalized polysaccharides suggests that the gel formation is strongly dependent on the type of biopolymer. The strength of dextran gels did not change significantly as the gamma-irradiation dose increased up to 0.48 Mrad. On the other hand, gels formed from alginic acid, hyaluronic acid, and ester derivatives of hyaluronic acid degraded at high gamma-irradiation doses. These three polysaccharides formed gels more readily if the solutions were acidified to pH 3 before gamma-irradiation. The acidified solutions formed gels at lower polysaccharide concentrations and lower gamma-irradiation doses. The effect of gamma-irradiation on the degradation of polysaccharides varies depending on the type of polysaccharide. While more study is necessary to understand the mechanisms of degradation and formation of microgels instead of hydrogels, polysaccharide hydrogels with different properties can be prepared by adjusting the experimental conditions, such as the gamma-irradiation dose, the polymer concentration, the degree of functionalization, and the pH of the solutions.

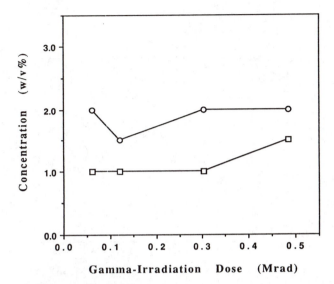

Figure 3. Gel formation by benzyl ester of hyaluronic acid at 25% of esterification degree (HYAFF11-p25) as a function of HYAFF11-p25 concentration and gamma-irradiation dose at pH 3 (O) and pH 6 (□). Glycidyl acrylate used to functionalize alginate was 0.8 ml/g polymer. The lines indicate the minimum concentration of alginate necessary to form a three-dimensional network.

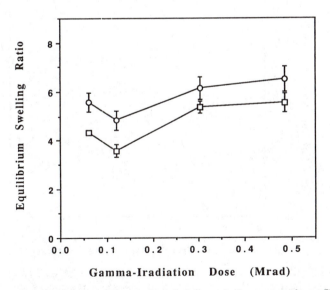

Figure 4. Equilibrium swelling ratio of alginate gels prepared at pH 3 as a function of the gamma-irradiation dose. The concentrations of alginate were 2.5% (O) and 3% (□). Glycidyl acrylate used to functionalize alginate was 1.3 ml/g polymer.

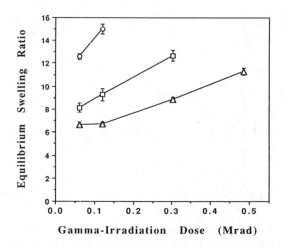

Figure 5. Equilibrium swelling ratio of HA gels prepared at pH 3 as a function of the gamma-irradiation dose. The concentrations of HA were 1% (O), 1.5% (□) and 2% (Δ). Glycidyl acrylate used to functionalize alginate was 0.8 ml/g polymer.

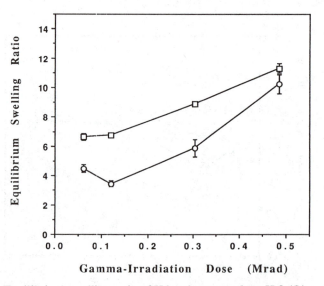

Figure 6. Equilibrium swelling ratio of HA gels prepared at pH 3 (O) and at pH 6 (□) as a function of the gamma-irradiation dose. The concentrations of HA was 2%. Glycidyl acrylate used to functionalize alginate was 0.8 ml/g polymer.

Literature Cited

1. Kamath, K. R.; and Park, K. *Adv. Drug Delivery Rev.*, **1993**, *11*, 59-84.
2. Peppas, N. A., Ed., *Hydrogels in Medicine and Pharmacy*, CRC Press, Boca Raton, FL., 1987, vol. 1, pp 1-25.
3. Kamath, K. R.; and Park, K. *ACS Symp. Ser.*, **1994**, *545*, 55-65.
4. Kamath, K. R.; and Park, K. *Proceed. Intern. Symp. Control. Rel. Bioact. Mater.*, **1992**, *19*, 42-43.
5. Sanderson G. R. In *Food Gels*; Harris, P., Ed.; Elsevier Science Publishers Ltd., New York, 1990, pp 201-232.
6. Shah, C.B.; and Barnett, S. M. *ACS Symp. Ser.*, **1992**, *480*, 116-130.
7. Sime W. J., In *Food Gels*; Harris, P., Ed.; Elsevier Science Publishers Ltd., New York, 1990, pp 53-78.
8. Alexander P.; Charlesby A. *J. Polym. Sci.*, **1957**, *23*, 355-375.
9. Maeda, H.; Suzuki, H.; Yamauchi, A.; Sakimae, A. *Biotechnol. Bioeng.*, **1974**, *16*, 1517-1528.
10. Maeda, H.; Suzuki, H.; Yamauchi, A. *Biotechnol. Bioeng.*, **1973**, *15*, 607-610.
11. Torchilin, V. P.; Maksimenko, A. V.; Smirnov, V. N.; Berezin, I. V.; Klibanov, A. M.; Martinek, K. *Biochim. Biophys. Acta*, 1979, 567, 1-11.
12. Plate, N.A.; Postnikov, V.A.; Lukin, N. Y.; Eismont, M. Y.; Grudkova, G.; *Polymer Sci. U. S. S. R.*, **1982**, *24*, 2668-2671.
13. Plate, N.A.; Malykh, A. V.; Uzhinova, L. D.; Mozhayev, V.V. *Polymer Sci. U. S. S. R.*, **1989**, *31*, 216-219.
14. Plate, N.A.; Malykh, A. V.; Uzhinova, A.D.; Panov, V.P.; Rozenfel'd, M. A. *Polymer. Sci. U.S.S.R*, **1989**, *31*, 220-226.
15. Edman, P.; Ekman, B.; Sjoholm, I. *J. Pharm. Sci.*, **1980**, *69*, 838-842.
16. Guiseley, K. B. In *Industrial polysaccharides: Genetic, Engineering, structure/property relations and applications*, M. Yalpani, Ed.; Elsevier Science Publishers B. V., Amsterdam, 1987, pp 139-147.
17. Hoppe, H.; Koppe, J.; Winkler, F. *Plaste Kautsch,* **1977**, *24*, 105.
18. Lepisto, M.; Artursson, P.; Edman, P.; Laakso, T.; Sjoholm, I. *Anal. Biochem.*, **1983**, *133*, 132-135.
19. Katsura, S.; Isogai, A.; Onabe, F.; Usuda, M. *Carbohydr. Polym.*, **1992**, *18*, 283-288.
20. Lal, M. *J. Radioanal. Nucl. Chem. Art.*, **1985**, *9*, 105-111.
21. Shalaby, W. S. W.; Park, K. *Pharm. Res.*, **1990**, *7*, 816-823.
22. Laurent, T. C. *Acta Chem. Scand.*, **1964**, *18*, 274-275.

RECEIVED June 20, 1995

Chapter 14

Interpenetrating Polymer Network Formation by Electron-Beam Curing of Acrylated Epoxy Resin Blends

Mohamad Al-Sheikhly[1] and William L. McLaughlin[2]

[1]Department of Materials and Nuclear Engineering, University
of Maryland, College Park, MD 20742–2115
[2]Ionizing Radiation Division, Physics Laboratory, National Institute
of Standards and Technology, Technology Administration,
U.S Department of Commerce, Gaithersburg, MD 20899

Continuous manufacturing processes of complicated structures of
epoxy resin-graphite fiber composites require higher control of the
level and dispersity of polymerization. This goal was achieved by
using a two-step curing process involving different blends of
unmodified epoxy resins with modified acrylated epoxy resins. The
first step was designed to achieve partial curing of the blend by
using high-energy electron beam (e-beam) irradiation, followed by
full thermal curing. Higher glass transition temperature (T_g) values
were obtained by using modified fully unsaturated resins with
unmodified epoxy resins, as compared to blends of modified mono-
saturated resins with unmodified epoxy resins. No partial curing
was accomplished by e-beam irradiation of the unmodified epoxy
resin alone. The present results demonstrate that the introduction of
the fully-unsaturated resin blend increases the efficiency of the
e-beam radiation curing through free-radical polymerization and then
free-radical crosslinking, which leads to IPN formation.

Over the past decade, high-energy electron radiation has been used to initiate the
polymerization of acrylated epoxides. Unlike radiation-induced cationic
polymerization of epoxides, polymerization of acrylated epoxides by ionizing
radiation propagates by carbon-centered free radicals through the unsaturated end
groups (*1,2*). In the present work, the two-part epoxy-based resin matrix is cured
by a two-step process: (1) In the first step, an electron beam pre-cure supplies
form stability; (2) a thermal cycle then completes the cure. The ultimate objective
is to provide a controlled epoxy polymerization reaction through the blending of
fully unsaturated epoxy with an unmodified epoxy. Such a blended resin contains
a component which, when exposed to ionizing radiation, forms a three-dimensional

structure within the unmodified epoxy resin. Thus, upon thermal curing of the unmodified epoxy resin component within the radiation cured matrix, an interpenetrating polymer network (IPN) is produced. This two-step approach, however, is not new. For example, it has been shown that radiation curing of a mixture of epoxy oligomer, acrylate, and styrene gives a higher gel content than chemically initiated curing of the same mixture and that post-irradiation heat treatment further increases the gel component to 92 to 98 percent (*3-5*). The required dose for achieving complete curing of this system was reported to be 100 to 120 kGy. It has also been demonstrated that, prior to irradiation, production of unsaturated ester by reaction of epoxy resins with acrylic acid enhances the radiation chemical yield of C-C reaction to 400 times that by irradiation of epoxide alone. In addition, crosslinking is enhanced by post-irradiation heating or by irradiation at elevated temperatures (*6*). In fact, in the presence of acrylates thermally initiated polymerization alone can occur at 160 °C (*7*). Another example of the post-irradiation thermal treatment of an epoxy oligomer-acrylate-polyamide-plasticizer mixture has given better gel properties than with thermal curing of the same mixture (*8*). In addition, a combination of irradiation and heat treatment can be used to produce a resin network with high strength and thermal stability from a mixture of a bisphenol-A epoxy, polyamide and acrylic acid or acrylamide (*9,10*).

It has been reported that high-energy electrons are also effective in curing carbon fiber prepregs (*11*). These prepregs (the type used in the aircraft industry for hand lay-up applications in polymer-carbon fiber composites) have been shown to contribute as integral components of high-strength, low-weight construction materials. Unfortunately, the radiation polymerization of most epoxy compounds proceeds by a cationic mechanism, a process that is inhibited by trace amounts of water. This obstacle, however, can be overcome by acrylating the terminal epoxy groups of epoxy oligomers which changes the system to a free-radical process, not inhibited by traces of water. Although not included in the present study, it should be noted that, by introducing certain onium salts to the system, gamma and e-beam-induced cationic polymerizations of epoxy resins can be achieved.

Recent studies have shown that the initiation mechanism involves the production of radiation-induced radical cations of the epoxy monomer (or oligomer), followed by a deprotonation process to produce open-ring carbon-centered radicals. The carbon-center is then oxidized by an onium salt through electron-transfer mechanisms to produce the monomer cation. The polymerization reaction then propagates through the addition of the cation to the monomer (*12*).

Gamma rays have also been used to cure carbon fiber composites (*13*). Acrylate epoxy compounds were selected to achieve an aircraft-quality composite. The general formula that meets all the mechanical specifications imposed by a major U.S. aerospace company is: 20 percent dipentaerythritol monohydroxypenta-acrylate; 30 percent polybutadiene diacrylate; and 50 percent epoxy diacrylate. A plain-weave carbon fabric was selected for the prepreg material. A solvent process using methylethylketone was used to impregnate the fabric with about 35 percent (by mass) resin. The prepared material was then irradiated at a dose-rate of 17 kGy/h to a total absorbed dose of 50 kGy.

The ultimate objective of the work presented in this paper is to provide a controlled epoxy polymerization reaction by blending fully unsaturated epoxy with unmodified epoxy. Such a blended resin contains a component which reacts completely by radiation exposure to create a fully polymerized structure within the unreacted thermally-polymerizable unmodified epoxy resin. Thus an IPN can be formed by thermally crosslinking the unmodified epoxy-resin component within the radiation-cured unsaturated resin component. It is suggested here that by reaching these objectives, continuous radiation/thermal curing of epoxy-based resin matrices can be achieved on a production scale.

EXPERIMENTAL APPROACH

In the present work, the experimental approach can be outlined as follows:

1. to devise a sequence of controlled epoxy polymerization using a blend of a fully unsaturated acrylated epoxy with an unmodified epoxy (without acrylated groups);
2. with irradiation, the unsaturated component should be capable of providing a polymerized structure within the thermally-polymerized unmodified epoxy resin;
3. subsequently, an IPN can be formed by crosslinking the unmodified epoxy within the radiation-cured unsaturated resin network.

Thus, for controlling the polymerization process, the experimental approach is based on the following two-step curing approach:

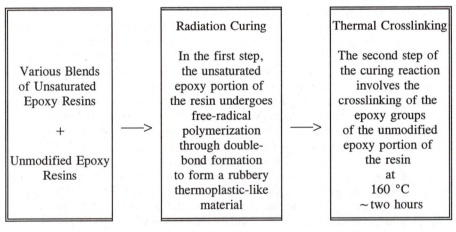

	Radiation Curing	Thermal Crosslinking
Various Blends of Unsaturated Epoxy Resins + Unmodified Epoxy Resins	In the first step, the unsaturated epoxy portion of the resin undergoes free-radical polymerization through double-bond formation to form a rubbery thermoplastic-like material	The second step of the curing reaction involves the crosslinking of the epoxy groups of the unmodified epoxy portion of the resin at 160 °C ~two hours
	Step 1	**Step 2**

Step 1: Radiation-induced polymerization: The unsaturated epoxy portion of the resin undergoes free-radical polymerization through the double bonds forming a rubbery thermoplastic material with high frictional properties. In order to enhance the radiation-induced free radical polymerization, partially and fully unsaturated resins are prepared by esterification of the epoxide resin with unsaturated acid.

Step 2: Thermally-induced polymerization: Crosslinking of the unmodified epoxy portion of the resin is initiated by high-temperature treatment (160 °C for 2 h).

A 2 to 9 MeV Varian V7715J Electron LINAC[*] was utilized to irradiate the neat epoxy resin and the 6-ply epoxy fiber prepared composite. An accelerating potential of 8 MeV was chosen to ensure full penetration of the electrons through the samples. The surface dose rate for the irradiation was 0.05 kGy per 3 μs pulse at a repetition rate of 20 pulses per second. FWT-60-00 radiochromic film dosimeters, traceable to the National Institute of Standards and Technology (NIST), were used to measure the total dose and dose per pulse (*14*). The combined estimated uncertainty of these measurements is documented as $U = k u_c = \pm 3$ percent, with the estimated standard deviation $u_c = \pm 3$ percent and the coverage factor k = 1) (*15,16*).

In order to minimize the effect of the temperature during irradiation, the total dose was fractionated into increments of 100 kGy or less, and a constant flow of cool air was aimed at the samples during and after irradiation. Several minutes cooling between fractionations was adequate to keep the irradiation temperature below 30 °C.

The first stage of this study was the electron-beam curing of neat modified epoxies (acrylated epoxies), while the second stage was the electron-beam irradiation of the modified epoxy resin graphite composite. Graphite composite samples were produced by impregnating a plain-weave AS4 graphite fiber fabric using a typical wet layup method. Six-ply laminates were produced after preparing resin-impregnated single-ply elements.

A Perkin Elmer DSC7 was used to measure the glass transition temperature (T_g) values for the cured samples as an indication of degree of curing. For each T_g measurement, five specimens from the 6-ply composite were used. The thermograms were measured at a constant heat-up rate of 20 °C per minute. All measurements were carried out by employing ASTM standard practice D3418-82 for DSC analysis (*17*). Due to water contamination in the sample holder, the sample pans, and/or the laminate specimen, a few samples revealed slight transitions around 0 °C and 100 °C.

RESULTS AND DISCUSSION

The objective of the first part of this two-step process for curing the epoxy-based resin matrices by electron-beam irradiation was to achieve a rubbery thermoplastic-like (partially cured) composite. Thus, various blends of mono-unsaturated and fully-unsaturated resins with unmodified epoxy resin were irradiated by the electron beam at different dose levels in the range 100 to 300 kGy. At this stage no dose rate effect on yield was observed. Immediately after irradiation, T_g values were measured. As expected, no significant polymerization was found for the irradiated unmodified epoxy resin, even at comparatively high doses (~ 300 kGy).

---

[*]The mention of commercial products throughout this paper does not imply recommendation or endorsement by the National Institute of Standards and Technology, nor does it imply that the products identified are necessarily the best available for the purpose.

This can be explained by the fact that the radiolytically produced cations transfer a proton to the H_2O molecules converting the protonated epoxy species to stable epoxy oligomer. On the other hand, the blended samples show some degree of polymerization.

The second curing step was the post-irradiation thermal treatment of the samples. T_g measurements were also carried out after the thermal treatment. The results of the electron irradiation of neat epoxy and different blend ratios of neat mono-unsaturated epoxy (modified) resins mixed with unmodified epoxy resins and their effects on the T_g values at different absorbed levels and after thermal treatment are summarized in Table I. The combined uncertainty of the T_g measurements is estimated to be $U = k u_c = \pm 8$ percent, with a standard deviation of $u_c = \pm 8$ percent and the coverage factor k = 1 (16).

Table I. T_g For Neat and Modified Epoxy Resins

Blend Formulation	Absorbed Dose (kGy)	T_g after Irradiation (°C)	T_g after Post Cure (°C)
	0	N/M[*]	152.2
un-modified	92	N/M	155.0
epoxy resin	183	N/M	153.7
	300	N/M	156.5
un-modified			
epoxy resin	0	N/M	113.4
(100 parts)	92	−8.9	129.4
+	183	−7.5	128.9
mono-unsaturated	300	−9.5	134.6
epoxy			
(50 Parts)			
un-modified			
epoxy resin	0	N/M	78.4
(100 parts)	92	−5.6	130.8
+	183	−4.3	135.2
mono-unsaturated	300	−4.0	132.7
epoxy resin			
(200 parts)			

[*]not measurable, the blend still in a liquid phase after irradiation

The results in Table I show that increasing the percentage of the mono-unsaturated resin component leads to an increase in the glass transition temperature due to free-radical polymerization (see scheme 1). These data also show that the unmodified epoxy resin reaches full curing by thermal treatment despite the fact that no appreciable polymerization was achieved by irradiation alone.

The same experimental two-step (irradiation + thermal) approach was used for curing different blends of modified epoxy resins in matrix graphite-fiber

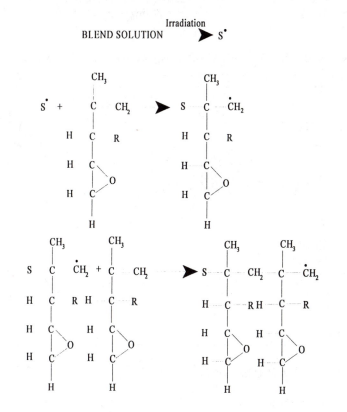

Scheme 1.

composite samples. Electron beam-induced partial curing of two-resin compositions was carried out: (1) a blend of unmodified epoxy resin (100 parts) with a mono-unsaturated epoxy resin (50 parts); and (2) a blend of unmodified epoxy resin (100 parts) with fully unsaturated epoxy (23) parts. T_g values resulting from two-step curing of different blends of modified epoxy resins in matrix graphite-fiber composites are summarized in Table II. Higher T_g values were found for fully-unsaturated blend samples compared to those of the mono-unsaturated samples when treated by electron-beam irradiation. However, at very high dose (300 kGy), the results indicate that degradation starts predominating (T_g lowered by ~5 percent).

Table II. T_g For Modified Epoxy Graphite Composites

Blend Formulation	Absorbed Dose (kGy)	T_g after Irradiation (°C)	T_g after Post Cure (°C)
un-modified epoxy resin (100 parts) + mono-unsaturated epoxy (50 parts)	100	7.4	116.9
	200	5.6	125.9
	300	7.6	127.2
un-modified epoxy resin (100 parts) + fully-unsaturated (the diester) (23 parts)	100	6.7	137.4
	200	9.9	136.2
	300	11.0	129.9

SUMMARY AND CONCLUSION

The present work shows that epoxy resin polymerization solely by electron irradiation initiation cannot be achieved. This is likely due to the presence of traces of water. Such proton acceptors have been shown to inhibit cationic polymerization (*18*). On the other hand, electron irradiation-induced partial curing occurs in the presence of unsaturation, since polymerization proceeds through a free-radical mechanism (scheme 1). While polymerization through cation propagation is inhibited by the presence of proton donors, free radical-induced polymerization can be impeded by the presence of dissolved oxygen and at higher dose rates. Oxygen reacts very fast with carbon centered radicals to produce the corresponding peroxy radicals. In this way the dissolved oxygen will react with modified epoxide [mono- and/or fully unsaturated epoxy] free radicals, E˙, thus impeding the additional reaction of E˙ with another monomer molecule (propagation reaction). Hence the reaction

$$E^{\cdot} + O_2 \rightarrow EO_2^{\cdot} \tag{1}$$

interferes with

$$E^{\cdot} + E \rightarrow E\,E^{\cdot} \quad \text{(propagation)} \tag{2}$$

Additionally, the radical-radical reaction, which can be enhanced at high dose rates, also competes with the propagation reaction

$$E^{\cdot} + E^{\cdot} \rightarrow E - E \quad \text{(dimerization)} \tag{3}$$

Moreover, dissolved oxygen reacts extremely fast with the electron in competition with the monomer. As a result, the efficiency of radiation-induced polymerization will be decreased.

Despite the negative effects of oxygen, the results demonstrate that by implementing a two-step curing method (irradiation and post-irradiation thermal curing) of compositions blended with mono-unsaturated epoxy and fully unsaturated epoxy, the sequence of controlled epoxy polymerization can be optimized, thus providing manufacturing advantages in the production of stable cured resins. Equally important is the possible formation of IPN's by the introduction of the fully-unsaturated resin blend, which increases the efficiency of electron-beam curing by free-radical polymerization and subsequent free-radical crosslinking, giving controlled T_g values. Since the main objective of this work has been to introduce a novel method of continuous mass production of composite curing for complex geometrical structures, by using electron beams for partial curing followed by complete thermal curing, no attempts were made here to irradiate under anaerobic conditions to minimize the oxygen inhibition effects.

LITERATURE CITED

1.   Crivello, J. V.; Fan, M.; Bi, D. *J. Appl. Polymer Sci.* **1992**, *44*, 9.
2.   Gotoda, M.; Miyashita, Y.; Takeyama, K. In Fundamental Studies in Osaka Laboratory for Radiation Chemistry. **1975**, Annual Report: April 1 1973-March 31, 1974, *JAERI-5030*, 125-135.
3.   Kuznetsova, V. M.; Meleshevich, A. P.; Vishev, Yu. V.; Gorodnichenko, A. I. Plast. *Massy* **1978a**, *2*, 46.
4.   Kuznetsova, V. M.; Meleshevich, A. P.; Vishev, Yu. V.; Zhikharev, V. S.; Derevyanko, A. G. Plast. *Massy* **1978b**, *2*, 44.
5.   Kuznetsova, V. M.; Meleshevich, A. P.; Fedina, L. P.; Vishev, Yu. V.; Gerasimenko, E.V. *Plast. Massy* **1978c**, *2*, 49.
6.   Persinen, A. A. Proc. 5th Tihany Symp. *Radiat. Chem.* **1982**, *2*, 1005.
7.   Takács, E. Radiation Chemistry Group, Institute of Isotopes of the Hungarian Academy of Sciences, Budapest, private communication.
8.   Vishev, Yu. V.; Furman, Ye. G.; Melesgevich, A. P.; Kuznetsova, V. M.; Valatina, O. A. *Vysokomol. Soyed.* **1983**, *A25*, 970.

9.   Barunin, A. A.; Plachenov, B. T.; Krasovskaya, I. A.; Vesnebolotskii, K.
      I.; Masloboev, D. S. *Mekh. Kompoz. Mater.* (Zinatne) **1983**, *5*, 935.
10.   Ivanov, G. M.; Neschastnova, L. M.; Vinogradova, V. V.; Shiryaeva, G.
      V.; Prishchepa, N. D.; *Plast. Massy* **1985**, *3*, 24.
11.   Dikson, L. B.; Singh, A. *Radiat. Phys. Chem.* **1988**, *31*, 587.
12.   Crivello, J. V; Mingxin, F.; Daoshen, B.; *J. Appl. Polymer Sci.* **1992**, *44*,
      9.
13.   Saunders, C. B.; Dickson, L. W.; Singh, A. A.; *Polymer Composites* **1988**,
      9, No. 6, 389.
14.   McLaughlin, W. L.; Humphreys, J. C.; Wei-Zhen, BA; Khan, H. M.; Al-
      Sheikhly, M.; Chappas, W. J. *High Dose Dosimetry for Radiation
      Processing*, Proceedings of International Symposium, IAEA STI/PUB/846,
      held in Vienna, 5-9 November 1990. International Atomic Energy Agency:
      Vienna, **1991**, pp 305-316.
15.   McLaughlin, W. L.; Boyd, A. W.; Chadwick, K. H.; McDonald, J. C.;
      Miller, A. *Dosimetry for Radiation Processing* (Taylor and Francis,
      London) **1989**
16.   Taylor, B. N.; Kuyatt, C. E. *NIST Technical Note 1297*, National Institute
      of Standards and Technology, Gaithersburg, MD.
17.   ASTM Standard D 3418-82, *ASTM Standard-Composite Materials*, Second
      Edition, American Society for Testing and Materials, Philadelphia PA,
      1990.
18.   Al-Sheikhly, M. and McLaughlin, W. L. Mechanism of Radiation-Induced
      Curing of Epoxy-Fiber Composites. In *Radiat. Phys. Chem.* Editor
      W. Chappas (1996) in press.

RECEIVED November 22, 1995

Chapter 15

Electron Processing of Carbon-Fiber-Reinforced Advanced Composites: A Status Report

Ajit Singh, Chris B. Saunders, Vince J. Lopata, Walter Kremers, Tom E. McDougall, Miyoko Tateishi, and Minda Chung

Research Chemistry Branch, AECL Research, Whiteshell Laboratories, Pinawa, Manitoba R0E 1L0, Canada

An emerging application of industrial electron accelerators is the production of advanced composites for the aerospace and other industries. Our work on the production of these advanced composites, using electron-curable matrices, is briefly reviewed. The dose required for curing the matrix can vary with the dose rate, and, as our results show, this dose rate dependence can vary from resin to resin. Our results on the reduction of the void content, the use of fiber-matrix coupling agent, and fabrication and curing of thick composites are also discussed.

Industrial radiation processing involves the use of natural and man-made sources of high-energy radiation (γ-radiation, electron beams), to manufacture a wide range of products (1-3). Radiation processing is a growing industry; in 1990, the value added to products by radiation processing was estimated to be in the billions of dollars (4). Both cobalt-60 sources and electron accelerators are used for radiation processing. Gamma radiation and X-rays are much more penetrating than electrons of the same energy (5,6). Therefore, thick products are typically irradiated with gamma radiation (and some with X-rays), and thin products are typically irradiated with electrons, e.g., 10-MeV electrons allow uniform irradiation of up to 9 cm of unit density material by two sided-irradiation. Gamma irradiation is almost exclusively used for sterilization of medical disposables and for food irradiation (7), though other applications, e.g., sewage irradiation, are under investigation (8,9). Electron accelerators are primarily being used for processing plastics or products derived from them (3); their other present or potential uses (8,10) include irradiation of sewage sludge (11), wastewater treatment (12,13), sterilization of medical devices and disposables (14-18), and food irradiation (19-22).

An emerging application of industrial electron accelerators is the production of advanced composites for the aerospace and other industries. Traditionally, the carbon-, aramid- and glass-fiber-reinforced composites with epoxy matrices are produced by thermal curing (23). However, equivalent composites with acrylated-

0097–6156/96/0620–0197$12.00/0
Published 1996 American Chemical Society

epoxy matrices can be produced by electron processing *(24-26)*. In this paper we present a status report on our work on electron processing of carbon-fiber-reinforced advanced composites, which has been in progress for almost a decade.

Advanced Composites

Advanced composites, specifically carbon-fiber-reinforced epoxies, are being used for many applications, primarily because of their high strength-to-weight and stiffness-to-weight ratios *(23)*. Applications for these thermosetting composites are found in the aircraft, aerospace, sporting goods, transportation and automotive industries *(23)*. Electron processing or curing of composites involves using electrons as ionizing radiation to initiate polymerization or cross-linking reactions in suitable matrix resins (e.g., acrylated epoxy oligomers) in place of the traditional thermal initiation of the polymerization and cross-linking reactions in epoxy formulations *(24,25,27,28)*.

Advantages of Electron Processing. Many advantages have been identified for using electron processing *(25,29)* rather than thermal curing of advanced composites, as outlined below.

1. Curing at Ambient Temperature: The thermal curing cycle can change the dimensions of the product and create internal stresses, which can decrease its strain to failure and the fracture toughness *(29)*. Electron processing at ambient temperature can reduce these dimensional changes in both the tool and the product, thus reducing the internal stresses in the final product.

2. Reduced Curing Times: The production speed for a 50 kW accelerator to cure composites would be up to 6 times higher than thermal curing with a typical autoclave. This is so, even though the products are cured one at a time during electron processing, compared to large batches in autoclaves.

3. Improved Resin Stability: Most electron-curable matrix resins do not auto-cure at room temperature, making low-temperature storage unnecessary.

4. Reduction in the Amount of Volatiles Produced: Electron processing eliminates the production of toxic thermal degradation products, though very small amounts of gases such as hydrogen and methane may be produced.

5. Better Material Handling: Two of the factors that contribute to more efficient material handling during electron processing are: (i) the ability to handle the resins at room temperature makes it easier to prepare prepregs, and to fabricate components from them; and (ii) the ability to electron process each item as it gets fabricated, reduces the space requirements for storage of the uncured items. In the case of electron processing, components with different resins requiring different doses can be processed one after the other. However, in thermal curing, all the contents of the autoclave need to have the same thermal curing cycle.

6. Reducing the Operating Costs, Primarily Energy Costs: The energy required for electron processing could be lower by a factor of 5 or more *(30)* with overall savings of the order of 30%, compared to thermal curing.

Constraints for Electron Processing. Electron processing of composites also faces some constraints, as follows:

1. Availability of Electron-Curable Matrix Resins: The epoxy formulations currently being used in the aerospace industry are not appropriate for electron curing. Acrylated epoxy resins that can be cured by irradiation are now commercially available, but in a much smaller selection.
2. Qualification Procedures: Extensive testing is required to develop electron-processed matrices for advanced composites that are truly equivalent to the conventional, thermally cured, matrices *(31,32)*.

Our Work. Our advanced composites work was initiated with various formulations derived from five commercially available electron-curable resins as the matrix materials *(33)*, and Hercules' AS4 carbon fabric for reinforcement. Most of the work has been done with electron irradiation using the AECL I-10/1 electron accelerator at Whiteshell *(30,34)*. Some of the work has also been done using X-rays produced from the I-10/1 electron beam and gamma radiation from an AECL Gammacell 220.

The glass transition temperature and relevant mechanical properties for conventional epoxies and for acrylated epoxies fall in similar ranges *(28,35)*. The mechanical properties obtained for a 14-ply electron processed laminate compare very well with the properties required by the aerospace industry. Recent data also show that the internal stress is lower in electron processed composites, compared to thermally cured composites *(30)*. In principle, similar advanced composites using different fibers (aramid- or glass-fiber) can also be made using the electron-processing technology. Recently, we have focussed on the use of fiber-matrix coupling agents, reduction of void content, use of adhesives to join pre-cured composites, and preparation of thick samples (up to 900 plies) *(36)*. We have used seven commercially available formulations and five formulations of our own as adhesives for advanced composites *(37)*. The results show that electron-cured adhesives can give lap shear strengths similar to, and in some cases better than, the thermally-cured adhesives.

Dose-Rate Effect. We have examined the radiation curing characteristics of several matrix resins. Some resins cure at a lower total dose on electron irradiation, while others require a lower dose on gamma irradiation. This is brought out by the data summarized in Table I, which shows the doses required to obtain a 90% gel fraction for various resins on electron and gamma (or X-ray) irradiation. It is therefore important to determine the curing characteristics of the resin of interest.

Void Content. Most of the acrylated epoxies shrink by up to 12% by volume on electron curing. In the fiber-reinforced composites containing 35% matrix resins (65% fiber, by weight), the void content in the product can be up to 8%. However, the void content has been reduced to ~4%, in the composites with 45% resin content, by appropriate compaction methods.

200 IRRADIATION OF POLYMERS

Table I. Dose Required for 90% Gel Fraction

Resin[1]	e⁻	γ kGy	Reference kGy
CN-104	25[4]	40[2]	26
C3000, Epoxy diacrylate	8	60[3]	Lopata et al.[5]
S297, 1,3-Butylene glycol dimethacrylate	28	8	Lopata et al.[5]
S604, Polypropylene glycol monomethacrylate	40	6	Lopata et al.[5]
CN964, Urethane acrylate oligomer	2	1	Lopata et al.[5]

[1]All resins supplied by Sartomer.
[2]X-rays, dose rate, 4 kGy.h⁻¹.
[3]Dose rate, 10 kGy.h⁻¹.
[4]Dose rate 5400 kGy. h⁻¹.
[5]Lopata, V.J., Kremers, W., McDougall, T.E., Tateishi, M., Saunders, C.B., Singh, A., unpublished results (1993).

To reduce the void content, the sizing was removed by treating the fiber with hot chromic acid, washing with distilled water and drying in an oven at 150°C for 24 hours. The resins used were FW3 (Applied Poleramics) and CN104 (Sartomer). An additional step was added to the normal fabrication of the composite. The vacuum-bagged composite was heated to 80°C for 4 hours to reduce the viscosity of the resin. At 80°C the viscosity for FW3 resin is 600 cps compared with 900,000 cps at room temperature. The excess resin was allowed to bleed into a breather cloth. The void content for these composites was much lower, as determined by the C-scan method (38) and confirmed by mercury intrusion measurements using a Micromeretics Auto-pore II porosimeter. The void content in these samples was ~4%, as compared to the previous samples which showed void content up to 8%.

Fiber-Matrix Coupling Agents. We have used a number of different fiber-matrix bonding agents, with sized and unsized AS4 carbon fiber from Hercules. Surface treatment of the carbon fiber can play an important role in achieving acceptable bond strengths. The treatment of the fiber surface can take many forms, e.g., a coupling agent may be used to enhance chemical bonding between the fiber and the matrix resin. A proprietary coupling agent was applied by a solvent method at a 1% loading. A heating step was used to chemically bond the isocyanate coupling agent and the fibers. The acrylated matrix resin was then applied and the fabricated composite cured with a dose of 50 kGy.

Figures 1a and b show Scanning Electron Microscopy (SEM) micrographs for the untreated and treated fiber surfaces of the fractured composite. The fibers with no coupling agent are clean, showing no adhesion between the fibers and the matrix resin (Figure 1a). Figure 1b shows that the fibers treated with the coupling agent

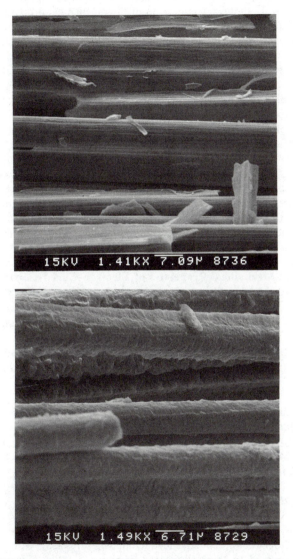

Figure 1. SEM photographs of the fiber surfaces of the fractured composite: (a) untreated fibers; (b) treated fibers. (Reprinted from ref. 37 with permission from the authors and the Society for the Advancement of Materials and Process Engineering.)

have the matrix resin adhered to them, due to bonding between the fiber and the matrix resin.

The use of a coupling agent greatly increases the mechanical properties of the composite. Table II shows a comparison of the mechanical properties of the composite samples fabricated from the sized and unsized AS4 carbon fiber with CN104, an epoxy acrylate resin, as the matrix. As can be seen from the data, the use of the coupling agent with the sized fiber improves the mechanical properties of the composite. It was expected that the composite fabricated from the unsized fiber (with the coupling agent) would also give a similar improvement in the mechanical properties, but it did not. The reason for the difference in the behaviour of the sized and unsized fiber may be attributed to the fate of active sites on the fiber with which the coupling agent can react. The sizing on the fiber most likely protects these active sites. In the case of the unsized fiber, the exposure of the fiber to the environment, following removal of the sizing, may lead to their inactivation over time, removing possible bonding sites for the coupling agent. Although the application of the coupling agent on the sized fiber gives better mechanical properties, the dose required to cure the composite is higher, by a factor of as much as 3. Our results suggest that if the coupling agent were to be applied as the sizing agent during the manufacture of the fiber, it could reduce the curing dose for the resulting composite.

Table II. Mechanical Properties of Electron-Cured Carbon-Fiber Reinforced Composites[a]: Use of a Coupling Agent

| | Sized Fiber (Hercules AS4) | |
Property	No Coupling Agent	With Coupling Agent[b]
Compression		
Modulus, GPa	56	60
Strength, MPa	156	320
Flexural		
Modulus, GPa	48	56
Strength, MPa	349	643

[a]Matrix, CN104 epoxy acrylate.
[b]An acrylated isocyanate.

CONCLUSION

Our previous work has demonstrated the feasibility of using electron curing to produce carbon fiber-reinforced advanced composites (24-28) with lower internal stresses as compared to thermally cured composites (30). The use of electron curing of adhesives for composites (37), and the production of up to 900 ply thick composites (36) has also been demonstrated. Our recent work shows that (i) the optimum curing dose for the matrix resins varies from resin to resin and also depends on the dose rate; (ii) the void content of the electron-cured composites can be reduced to ~4%; and (iii) the use of a fiber-matrix coupling agent improves the mechanical properties of the composites produced.

Literature Cited

(1) Silverman, J., Radiation Processing - The Industrial Applications of Radiation-Chemistry; *J. Chem. Educ.*, 1981, *58*, 168-173.

(2) Singh, A.; Silverman, J., Radiation Processing: An Overview. *In Radiation Processing of Polymers;* Singh, A.; Silverman, J., Eds.; Hanser Publishers: Munich, 1992, pp 1-14.

(3) Saunders, C.B., *Radiation Processing in the Plastics Industry: Current Commercial Applications;* Atomic Energy of Canada Limited Report, AECL-9569, 1988.

(4) Cook, P.M., Impact and Benefit of Radiation Technology; *Radiat. Phys. Chem.*, 1990, *35* , 7-8.

(5) Holm, W.N.; Berry, R.J., *Manual on Radiation Dosimetry;* Marcel Dekker, Inc., New York, NY, 1970.

(6) Rogers, D.W.O.; Bielajew, A.F., Monte Carlo Techniques of Electron and Photon Transport for Radiation Dosimetry; *In The Dosimetry of Ionizing Radiation;* Kase, K.R.; Bjarngard, B.E.; Attix, F.H., Eds.; Academic Press, Inc.: New York, NY, 1990, pp 427-539.

(7) Kunstadt, P.; Steeves, C.; Beaulieu, D., Economics of Food Irradiation; *Radiat. Phys. Chem.*, 1993, *42*, 259-268.

(8) Huang, Q.; Wu, J.; Takehisa, M.; Miller, A., Proc. 8th Intl Meeting Radiat. Processing, Beijing, China, 1992, *Radiat. Phys. Chem.*, 1993, *42* (1-6), 1-1053.

(9) Swinwood, J.F.; Fraser, F.M., Environmental Application of Gamma Technology: Update on the Canadian Sludge Irradiator; *Radiat. Phys. Chem.*, 1993, *42*, 683-687.

(10) Leemhorst, J.G.; Miller, A., Proc. 7th Intl. Meeting Radiat. Processing, Noordwijkerhout, The Netherlands, 1989, *Radiat. Phys. Chem.*, 1990, *35* (1-6), 1-878.

(11) Bennett, G.S.; Saunders, C.B.; Singh, A., *Radiation Disinfection of Sewage;* Atomic Energy of Canada Limited Research Company Report, RC-94, 1988. Available from Scientific Document Distribution Office (SDDO), Atomic Energy of Canada Limited, Research Company, Chalk River, Ontario K0J 1J0.

(12) Singh, A.; Sagert, N.H.; Borsa, J.; Singh, H.; Bennett, G.S., The Use of High-Energy Radiation for the Treatment of Wastewater: A Review; *In Proc. 8th Symp. on Wastewater Treatment,* Montreal, 1985, Environment Canada, Ottawa, 1985, pp 191-209.

(13) *Removal of Phenol from Aqueous Solutions Using High Energy Electron Beam Irradiation;* United States Environmental Protection Agency, Emerging Technology Bulletin EPA/540/F-93/509, 1993.

(14) Sadat, T.; Morisseau, D.; Ross, A., Electron Beam Sterilization of Heterogeneous Medical Devices; *Radiat. Phys. Chem.*, 1993, *42*, 491-494.

(15) Mehta, K., Process Qualification for Electron-Beam Sterilization; *Med. Dev. Diagnost. Ind.*, 1993, *14* (6), 122-134.

(16) Sato, Y.; Takahashi, T.; Saito, T.; Sato, T., Takehisa, M., Sterilization of Health Care Products by 5 MeV Bremsstrahlung (X-ray); *Radiat. Phys. Chem.*, 1993, *42*, 621-624.

(17) Barnard, J.B., E-Beam Processing in the Medical Device Industry, *Med. Dev. Tech.*, 1991, *1* (5), 34-41.

(18) Saunders, C.; Lucht, L.; McDougall, T., Radiation Effects on Microorganisms and Polymers for Medical Products; *Med. Dev. Diagnost. Ind.*, 1993, *2* (5), 88-82.

(19) Singh, H., Prevention of Food Spoilage by Radiation Processing; *Can. Home Econom. J.*, 1987, *37* (1), 5-10.

(20) Singh, H., *Dose Rate Effect in Food Irradiation: A Review;* Atomic Energy of Canada Limited Report , AECL-10343, 1991.

(21) Sadat, T., Progress Report on Linear Accelerators; *Radiat. Phys. Chem.*, 1990, *35*, 616-618.

(22) Borsa, J.; Iverson, S.L., The Cost and Benefits of Grain Disinfestation and Poultry and Frozen Shrimp Decontamination Using 10-MeV Electron Accelerators; *Proc. Int. Symp. on Cost-Benefit Aspects of Food Irradiation Processing*, France, 1993; IAEA-SM-328/75, pp 223-231.

(23) Margolis, J.M., Ed., *Advanced Thermoset Composites*, Van Nostrand reinhold, New York, NY, 1986.

(24) Singh, A.; Saunders, C.B., *In Radiation Processing of Polymers;* Radiation Processing of Carbon-Fiber Acrylated Epoxy Composites; Singh, A.; Silverman, J., Eds.; Hanser Publishers: Munich, 1992, pp 187-203.

(25) Saunders, C.B.; Singh, A., *The Advantages of Electron-Beam Curing of Fibre-Reinforced Composites;* Atomic Energy of Canada Limited Report, RC-264, 1989. Available from Scientific Document Distribution Office (SDDO), Atomic Energy of Canada Limited, Research Company, Chalk River, ON K0J 1J0.

(26) Singh, A.; Lopata, V.J.; Kremers, W.; McDougall, T.E.; Tateishi, M.; Saunders, C.B., Electron-Cured Fibre-Reinforced Advanced Composites; *Proc. CANCOM'93*; Wallace, W.; Gauvin, R.; Hoa, S.V., Eds.; Ottawa, 1993, pp 277-289.

(27) Dickson, L.W.; Singh, A., Radiation Curing of Epoxies; *Radiat. Phys. Chem.;* 1987, *31*, 587-593.

(28) Saunders, C.B.; Singh, A.; Czvikovszky, T., Radiation Processing of Fibre-Reinforced Composites; *Proc. 12th Annual Can. Nucl. Soc. Meeting*, 1991, pp 60-64.

(29) Weeton, J.W.; Peters, D.M.; Thomas, K.L., *Engineers' Guide to Composite Materials*; American Society of Metals, Metals Park, OH, 1987.

(30) Saunders, C.B.; Lopata, V.J.; Kremers, W.; McDougall, T.E.; Tateishi, M.; Singh, A., Electron Curing of Fiber-Reinforced Composites; Recent Developments; *Proc. 38th SAMPE Conf.*, 1993, pp 1681-1691.

(31) McCarty, J.E., *In Engineered Materials Handbook, Vol. 1;* Johnson, J.H.; Kiepura, T.; Humphries, D.A., Eds.; ASM International, Metals Park, OH, 1987, pp 346-351.

(32) Fila, J.A.; Fews, R.C., Civil Certification Methodology for Composite Materials in Primary Aircraft Structure; *Presented at CANCOM '93, 2nd Canadian Int. Conf. on Composites*, Ottawa, 1993.

(33) Saunders, C.B.; Dickson, L.W.; Singh, A.; Carmichael, A.A.; Lopata, V.J., Radiation-Curable Carbon Fiber Prepreg Composites; *Polym. Comp.*, 1988, *9*, 389-394.

(34) Barnard, J.W.; Stanley, F.W., Startup of the Whiteshell Irradiation Facility; *Nucl. Instrum. Methods Phys. Res.*, *B40/41*, 1989, 1158-1161.

(35) Beziers, D.; Capdepuy, B., Electron Beam Curing of Composites; *Proc. 35th SAMPE Conf.*, 1990, pp 1221-1232.

(36) Saunders, C.B.; Lopata, V.J.; Kremers, W.; McDougall, T.E.; Chung, M.; Barnard, J.W., Electron and X-ray Curing of Thick Composite Structures; *Proc. 39th SAMPE Conf.*, 1994, pp. 486-496.

(37) Lopata, V.J.; Chung, M.; McDougall, T.E.; Weinberg, V.A., Electron-Curable Adhesives for High-Performance Structures; *Proc. 39th SAMPE Conf.*, 1994, pp. 514-520.

(38) Henneke II, E. G. *In ASM Engineered Materials Handbook;* Dostal C.A., M; ASM International: Metals Park, OH, 1987; pp 774-778.

RECEIVED August 17, 1995

Chapter 16

Electron-Beam Manufacturing of Tank Track Pads

Byron J. Lambert[1], Ahmed A. Basfar[2], Walter J. Chappas, and Joseph Silverman

Laboratory for Radiation and Polymer Science, Department of Materials and Nuclear Engineering, University of Maryland, College Park, MD 20742–2115

The US Army's fleet of tracked vehicles operates on replaceable rubber pads which fail through wear, chunking, and chipping. The standard elastomer is styrene butadiene rubber (SBR). Evaluations in Germany of the Army's main battle tank show that the pad consumes 73% of all operating and service costs, with a replacement cost of $25 per mile. This paper describes the development of a high-wear EB cured SBR formulation and the industrial manufacture of 250 pads. The production process, including formulation, compounding, molding, and irradiation is described. Laboratory data collected at the University of Maryland and Fort Belvoir, as well as field results conducted by the Army at the Yuma Proving Grounds, demonstrate that the new pads are substantially superior to any previous sulfur-cured SBR formulation. In addition, the new formulation exhibits superior ozone resistance.

This work was funded by the US Army Tank and Automotive Command to develop a new rubber pad for the Army's fleet of tracked vehicles. The styrene butadiene SBR pads for the Army's M60 Battle Tank survive about 2,000 km (1,200 miles) on-road. Off-road, on rugged terrain, the pads last less than 900 km (500 miles). More importantly, the cost and complications of supporting the vehicle are excessive, about $15/km ($25/mile).

An SBR tank pad, while in use, experiences large cyclic deformations, both compression and extension at high temperatures. The result is cracking and loss of

[1]Current address: Guidant Corporation, 26531 Ynez Road, P.O. Box 9810, Temecula, CA 92591–4628
[2]Current address: Atomic Energy Research Institute, King Abdulaziz City for Science and Technology, Riyadh 11442, Saudi Arabia

0097–6156/96/0620–0206$12.00/0
© 1996 American Chemical Society

large sections of the pad. Elemental analysis of the failure interface (1) detected a high concentration of chemical curing agents, suggesting that a non-uniform crosslink distribution caused by inadequate mixing might be, in part, responsible for the pad's short life. Unfortunately, mixing of the additives cannot be improved without degrading the molecular weight distribution of the base rubber.

A novel formulation and curing procedure for SBR was developed in this laboratory: a sulfur precure with about 25% of that in the normal formulation followed by electron beam irradiation to full cure (2). Field tests on the new SBR pads and laboratory tests on specimens of the new SBR demonstrated the marked superiority of the elastomer.

Also, early tests demonstrated a remarkable resistance of sulfur-radiation cured SBR to ozone. This, in part, was surprising since the irradiated rubber's high unsaturation content should have led to lower ozone resistance.   In this work, the previous work is summarized and new imformation is presented on the mechanism for the unexpected large increase in ozone resistance.

Experimental

Various SBR formulations were developed, blended, pressed, and cut at the University of Maryland and at the Army's Ft. Belvoir Research Development and Engineering Center (Ft. Belvoir, VA). SBR was obtained from the Firestone Rubber Company. Laboratory test samples were irradiated at the University of Maryland and Irradiation Industries, Inc. (Gaithersburg, MD); the tank pads used for field tests were irradiated with 12 MeV electrons at the IRT Corporation (San Diego, CA). Based on our laboratory tests for mechanical properties, one of the most promising formulations was blended and molded at Firestone's Noblesville, IN, plant into standard T-142 tank pads.

The pads were molded as follows: the bolt and backing plate were degreased, cleaned, coated with an adhesive (Chemlock 205), air-dried, coated with a second adhesive (Chemlock 233), assembled, and placed in the mold. The SBR was mixed in a Banbury mixer and hot-extruded into strips, followed by hot extrusion into 15 mm x 15 mm x 20 mm blocks. The blocks were water cooled and placed in the molds. The mold top plate was held at 422 K (300 °F) and the backing plate face at 416 K (290 °F). They were bumped three times to remove air pockets, cured for 70 minutes, and removed. The specific gravity was measured to be 1.15. Trimming was the only other treatment performed. The ingredients for the standard SBR formulation and our formulation are given in Table I. The principal difference is in the sulfur content. The vulcanization and molding procedure for the conventional SBR, and the precure of our high performance SBR was the same. The precure provided dimensional stability during demolding and during the transportation to IRT for the post-molding radiation cure of 100 or 150 kGy. The chemical accelerators were provided by Firestone and were not identified. A crosslink sensitizer (3,9-divinyl-2,4,8,10-tetraoxaspiro [5,5] undecane or DTUD) was added to reduce radiation requirments but its concentration (0.1 pphr), while sufficient for a small reduction (25 kGy) of the dose for full cure, did not modify the almost conventional composition (except for sulfur content) in any meaningful way. Also, higher DTUD concentration tended to reduce the hot tear strength.

Table I. Standard and UM Formulations

Ingredients	Sulfur Cured System (pphr)	EB Cured System (pphr)
SBR-1500	100	100
Carbon Black N-110	45	45
Zinc Oxide	4	4
Stearic Acid	2	2
Sulfur	**2**	**0.5**
Sensitizer (DTUD)	-	0.1
Accelerators (Firstone proprietary compounds)	2	2
Antioxidants	2	2
Antiozonant	3	3

An adiabatic temperature increase caused by 100 kGy to raw SBR is greater than 63 K and even more to the steal backing plate to which the pad is cemented. As a result, doses were delivered to each side in 25 kGy passes beneath the radiation beam followed by cooling so as to reduce overheating of the pad and the interface between the pad and the steel backing. Attenuated Total Reflection (ATR) Fourier Transform Infrared (FTIR) spectroscopy was used to measure changes in the rubber before curing, after thermal and radiation curing, and as a function of aging as simulated by exposure to ozone. The method minimizes the effects of specimen thickness and orientation and reduces the error propagation among samples. The gripper sample clamp ensured reproducible contact between the sample and the ATR crystal. Since the sample compartment is protected, a nitrogen purge to remove atmospheric water vapor and carbon dioxide was not used. The absorption of the transvinylene (965 cm $^{-1}$) and vinyl (910 cm^{-1}) groups were the focus of interest in this study.

Dose depth profiles were theoretically calculated using EDMULT (3) and experimentally measured (Figure 1). The measurements were made by slicing a commercially produced T-142 pad, parallel to the metal backing plate. Dosimeters were then positioned between each of the slices and the pad was reassembled. Each dosimeter package contained three Far West Technology dosimeters that were calibrated at the University of Maryland's High-Dose Secondary Standards Laboratory in accordance with procedures that are fully traceable to the National Institute of Standards and Technology.

The pads were then irradiated at IRT Corporation (San Diego, CA). Since 12 MeV electrons possess insufficient energy to fully penetrate a pads thickness from one side, the pad was irradiated on each side in separate passes beneath the beam, each time receiving a surface dose of 25 kGy.

Results

The results of laboratory tests (Table II) for this elastomer are for a formulation with hot-tear (ASTM D624, Die C) properties 40-80% above those of comparable conventional rubber formulations (Figure 2), greater than 10 times the resistance to crack initiation (DeMattia Crack Initiation Test, unaged), and less than one-third the crack growth (DeMattia Crack Growth Test, unaged). The new formulation was used at the Firestone Rubber Company to manufacture 250 tank pads. In field testing by the

Table II. Comparative Properties of tank pad rubber

Mechanical Properties	Sulfur Cured SBR	EB Cured SBR
Tensile Strength	3,300 psi	3,390 psi
200 % Modulus	650 psi	655 psi
Hot Tear	130 lb/in	182 lb/in
Elongation at Break	515 %	610 %
Crack Initiation	20,000 cycles	>200,000 cycles
Crack Growth	25 mil/min	9.4 mil/min

U.S. Army on 200 pads, the pads showed a 60% wear improvement for on-road tests and a 30% improvement for the mixed on-road/off-road test (4). Even more remarkable was the rubber's resistance to aging. In ozone tests performed by the U.S. Army (according to ASTM D-1149 Bent Loop Test, using the specimen B bent loop test, at a temperature of 377 K (215 °F)), and at an ozone concentration of 50 pphm) the rubber survived 36 days "without any sign of cracks" while the standard formula failed within 5 days (5). A more severe aging test (according to ASTM D-1149 Bent Loop Test at 377 K (215 °F) and with an ozone concentration of about 400 mPa) performed at the National Institute for Standards and Technology showed no sign of cracking after 5 days (6) while the standard Army formula failed within 3 hours. These initial results demonstrate that the combination sulfur-radiation curing of rubber offers a new technology to the Army, in particular, and to industry, in general, for the manufacture of SBR materials with properties thus far unattainable by traditional chemical curing techniques.

The pads manufactured at Firestone and IRT were evaluated at the US Army's Yuma Proving Ground. The results of tests performed on a M-60 tank with a gross weight of 96,100 pounds is summarized in Table III. With the exception of three pads

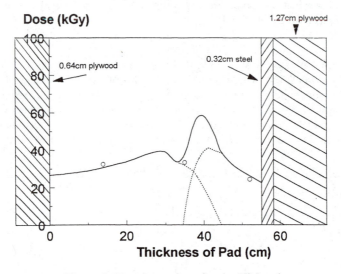

Figure 1: Hot tear v. dose for the UM pad.

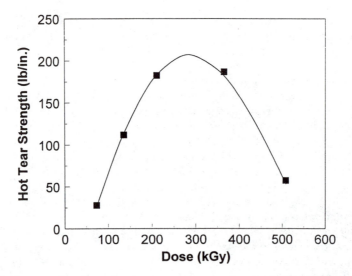

Figure 2: Dose profile through support structure and pads with 12 MeV electrons. Solid line are computer simulations and the open circles are experimental measurements.

that were lost due to adhesive failure at the steel backing plate-rubber interface, all pads exceeded the test limit of 2,000 miles. Evaluations of the weight loss indicate that the pads were approaching their life limit at 2,000 miles.

In order to characterize the effect of unsaturation on ozone resistance, sulfur-radiation cured SBR samples with a partial sulfur concentration of 0.5 pphr were irradiated at several doses. The optical densities of the transvinylene and vinyl groups decrease as the dose increases up to 1 MGy (Figures 3 and 4). Also, the IR spectra of sulfur cured samples at two sulfur concentrations (i.e. 1.5 pphr and 2.0 pphr sulfur) were obtained.. The optical density of the unsaturation for partial sulfur cured samples at zero dose are 130 % higher than that of 1.5 pphr sulfur cured samples, and 900 % higher than that 2.0 pphr sulfur cured samples. Of particular interest is the content of the vinyl group in the above mentioned systems as it relates to their ozone resistance. At a dose of 1 MGy, the vinyl content of the sulfur-radiation samples is as low as that of sulfur cured samples. Additional experiments with formulations with a partial sulfur level of 1.0 pphr were carried out in the same manner as the above. In these experiments, the same trend is observed with the exception that a lower dose is needed to reduce the vinyl content to the same low dose as that of sulfur cured systems (i.e. 200 kGy versus 1 MGy).

Discussion

Speculation concerning how radiation can improve mechanical properties for tank pad applications can be understood in light of tank pad failure modes (1). For off-road and gravel surfaces, "cutting and chunking" are reported to be the principal failures. Cutting is caused when the pad hits a "road hazard", a sharp rock or other sharp object able to produce a high point stress, with enough force to penetrate or cut the surface. Chunking can then follow by "scrubbing" the pad over rough or sharp objects. It should be noted that pad operating temperatures are high due to hysteretic heat production. Under such failure conditions, a network with high tear strength at elevated temperatures, high elongation, and high energy at break seems more to have optimum properties. The optimization of these properties is balanced by the fact that a network with very high energy at break leads to high hysteretic temperature increases and another failure mode, blowout.

A rubber with a high point tear strength might be expected from a uniform crosslink density spatial distribution of bonds of high strength at elevated temperatures. It is clear that the crosslink density distribution of sulfur crosslinked systems is dependent on the microscopic dispersion level of sulfur in the complex rubber formulation. However, dispersion is not uniform on the microscopic scale. A radiation cured system, on the other hand, is expected to have a crosslink density spatial distribution of carbon-carbon bonds that closely follows the spatial distribution of the dose. This is one potentially advantageous property of radiation which offers promise for overcoming tank-pad failures and hence improving the mechanical properties. Another is that radiation curing can be used to provide a wide range of specified spatial distributions of crosslinking density designed to overcome specific failure modes.

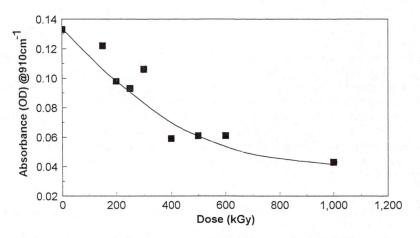

Figure 3: Transvinylene absorbance v. dose in UM formulation.

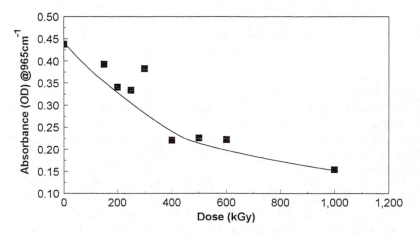

Figure 4: Vinyl absorbance v. dose in UM formulation.

Table III. Field Performance Data for T-142 Test Pads

Pad ID #	Standard Pad Miles to Failure	UM Pad Miles to Failure
1	1236	2001
2	1236	2001
3	1236	2001
4	1236	2001
5	1236	2001
6	1236	153
7	1236	2001
8	1236	2001
9	1236	2001
10	1236	2001
11	1236	153
12	1236	2001
13	1236	1613
14	1236	2001
15	1236	2001
16	1236	2001

With respect to ozone resistance, it is interesting to note that the content of unsaturation from both the transvinylene and vinyl groups is higher for partial sulfur and sulfur-radiation cured samples than for sulfur cured samples. In both of these high unsaturation formulations, ozone resistance is high.   This can be explained by examining the reactivities (7) of the various chemical bonds to ozone attack.   For example, the rate constant for ozone attack on the C=C unsaturation is 60,000 L/mol-s, whereas the rate of ozone attack on a disulfide bonds is only 47 L/mol-s and the rate of ozone attack on carbon-carbon bonds is only 0.006 L/mol-s. This demonstrates that the attack is directed overwhelmingly to the butadiene units, so that scission of crosslinks is relatively rare in comparison to the sequence of reactions said to lead to scission of the main chain. It is also reasonable to conclude from the rate constant data that attack on the carbon-carbon crosslinks is also unlikely compared to attack on main chain unsaturations.   The sensitivity to ozone is the reason for the addition of the antiozonant which has the effect of delaying the ozone interaction with the unsaturation.

The story with sulfur-radiation cured systems is totally different.   The radiation-induced crosslinks are of the C-C type which, as described, are not very reactive with the ozone molecules. Again the C=C unsaturation constitutes the major part of the SBR system. On this basis, the radiation cured formulations would suffer ozone damage in the same manner as the sulfur cured ones; but this is not the case. Radiation cured SBR formulations are associated with an amazingly high ozone resistance not observed in any unsaturated hydrocarbon elastomer in the past. During the course of this work, a paper by S. Bhagawan (8) demonstrated that 1,2-polybutadiene rubber, which contains a pendant vinyl group, has excellent ozone resistance. He stated that "Since the unsaturation of 1,2-polybutadiene occurs in the pendant vinyl group, cleavage of the double bond is not readily favored and the main chain of 1,2-polybutadiene remains intact". In a recent work by M. Golub et. al. (9), the pendant vinyl group played a significant role in protecting polybutadiene and other polymers used in lower space orbits against atomic oxygen attack. In this work, the protective effect leads to the production of crosslinks while the interaction of the other unsaturations with atomic oxygen leads to a rupture of polymer molecules. Thus, in considering the relation between vinyl unsaturation and ozone resistance, it is noteworthy that the vinyl concentration in sulfur-radiation SBR is some ten times that in standard SBR. Furthermore, the presence of the antiozonant is essential even in the latter. This high vinyl concentration near the surface provides an early delay mechanism until the antiozonant blooms to the surface.

Conclusions

This work has established the relationship between unsaturation content and ozone resistance, and that a combination sulfur-radiation cure of SBR can produce mechanically strong rubber with exceptional ozone resistance. Although the specifics of the mechanism are not fully understood, it is clear that conventional sulfur curing tends to destroy this protective agent (i.e. vinyl groups) at a higher rate than radiation curing. The mechanism appears to be as follows: In the presence of ozone both sulfur and sulfur-radiation cured SBR samples are protected against ozone by the blooming of the antiozonant to the surface where it competes with C=C unsaturation for the ozone, leaving the C=C unharmed. The antiozonant is continuously depleted to the point where it exerts no protection whatsoever on SBR samples. At this point the sulfur cured samples are left unprotected which leads to their failure (i.e. cracking), whereas sulfur-radiation cured samples remain resistant to ozone (no change in tensile properties). The key to this remarkable protection stems from two factors. The first is that the high content of the vinyl group in the bulk rubber works as an antiozonant that consumes ozone molecules at a higher rate than the transvinylene. The second (which we surmise from the necessity for the presence of antiozonant) is that the permeability into the interior of the SBR can be sharply lowered by the build up of ozonolysis products which limit further permeation.

Work continues to identify products that can be manufactured or processed to take advantage of the new technology, including solid tires for military and industrial vehicles, vibration damping pads for aircraft and marine applications, conduit, sheathing, and protective coating materials, shock absorbers for railroad cars, jounce bumpers and retread materials for trucks, and hoses and high-pressure tubing for aircraft, military and commercial vehicles.

Both current and anticipated advances in radiation cured SBR formulations and the related manufacturing technology could have a significant impact on the rubber industry which already employs radiation principally for enhancement of green strength.

The great enhancement of SBR resistance to wear and aging in an oxidative environment suggests its potential application to tank pads, shock absorbing pads and bushings, window channel, gas mask components, rubberized protective clothing, hoses, belting, etc.

To date, the focus of the research has been for the development of an SBR tank pad with the maximum possible hot tear strength for a normalized crosslink density. However, the formula can be modified in order to achieve other combinations of mechanical and physical properties for other specific applications.

Acknowledgments

The authors gratefully acknowledge the valuable technical contributions of Dr. Charles Church (Headquarters, US Army, Washington, DC), Mr. Paul Touchet and Mr. Gume Rodriguez (Ft. Belvoir R&D Center) and Mr. Jacob Patt (Tank and Automotive Command, Warren, MI) and the laboratory contributions of Debashish Mukherjee, Hou-Ching Yang, Fu-Wei Tang, and Vincent G. Adams.

We are also grateful to Irradiation Industries, Inc. for sample irradiations, the Firestone Rubber Company (now the Bridgestone Rubber Company) for blending and molding the test pads, and IRT Corporation for irradiating the test pads.

References

1. D.W. Dwight, H.R.Lawrence, and J. Patt, TACOM Technical Report Number 13062 (1985)
2. B.J. Lambert and J. Silverman, TACOM Technical Report Number 13215 (1986)
3. R. Ito and T. Tabata, "Semiemperical Code EDMULT for Depth-Dose Distributions of Electrons in Multilayer Slab Absorbers: Revisions and Applications", Radiation Center of Osaka Prefecture, ISSN 0285-8797, November 1987
4.. W.E. Brooks, Yuma Proving Ground, Report No. 607, "Report of Technical Feasibility Testing of T142 (M60) Track", April 27, 1989.
5. E.J. York, US Army Belvoir Research, Development, and Engineering Center, Letter Report to J. Silverman, May 2, 1988
6. R.D. Stiehler and M. Al-Sheikhly, National Institute of Standards and Technology, Letter Report to J. Silverman, May 2, 1988.
7. S.D. Razumorski and G.E. Zaikov, "Ozone and its Reactions with Organic Compounds", Elsevier Science (1984)
8. S.S. Bhagawan, "Thermal and Ozone Resistance of Thermoplastic Elastomers Based on 1,2-Polybutadiene", Polymer Degradation and Stability, 23, 10 (1988)
9. M. A. Golub, N.R. Lerner, and T. Wydeven, "Reactions of Atomic Oxygen with Polybutadienes and Related Polymers", ACS Symposium Series 364, 342 (1988)

RECEIVED August 31, 1995

Chapter 17

Gas-Plasma Sterilization

Paul T. Jacobs and Szu-Min Lin

Advanced Sterilization Products, Johnson & Johnson Medical, Inc., 33 Technology Drive, Irvine, CA 92718

Low temperature hydrogen peroxide gas plasma has been developed as a new method of sterilizing medical products. The process has been shown to inactivate a broad spectrum of microorganisms, including resistant bacterial spores. A Sterility Assurance Level (SAL) of 10^{-6} has been demonstrated for the process utilizing **Bacillus stearothermophilus** spores, the most resistant organism tested. Material compatibility studies have shown that the process is compatible with a wide range of metallic and non-metallic devices. When compared to r-irradiation, the low temperature hydrogen peroxide gas plasma has been found to affect the surface properties, i.e., wetting properties, of some non-metallic devices but not the bulk physical properties. Functionality studies have also shown that heat and moisture sensitive electronic, optical and mechanical devices are not adversely affected by exposure to the process.

The common methods of sterilizing packaged medical products in a hospital environment have historically involved the use of steam or dry heat for heat tolerant medical devices, and ethylene oxide gas or formaldehyde gas for heat sensitive devices. In the industrial area, radiation sterilization, mostly involving gamma irradiation (Cobalt 60), has been used for over 35 years, and more recently sterilization with electron beam has become more widespread. However, due to the large investment required to install a gamma irradiation facility or an electron beam sterilization process, the use of these technologies have not moved from the industrial to the hospital setting. Additionally, the detrimental effects of ionizing radiation on the bulk properties of some nonmetallic materials has placed some limits on the general use of these technologies on medical devices that are intended to be reused and exposed to repeated sterilization processes.

0097–6156/96/0620–0216$13.00/0

As the number of heat and moisture sensitive devices in the hospital has increased, especially in the diagnostic and less invasive surgery areas, so has the need for a rapid method of sterilizing these devices so that they can be reused many times each day. Ethylene oxide, the conventional method for sterilization heat sensitive devices, requires a long turn around time due to the extended aeration times required to remove the toxic ethylene oxide from sterilized items. In addition, ethylene oxide has been under increased regulatory pressures worldwide, due to the inherent toxicity of the ethylene oxide gas, and the detrimental effect of the chlorofluorocarbons used in the process on the earth's ozone layer. The use of formaldehyde gas has toxicological concerns similar to those expressed for ethylene oxide. A combination of all of these factors has accelerated the search for alternative methods of sterilizing heat and moisture sensitive materials. As a results of that search, Low Temperature Gas Plasma has emerged as a new technology capable of rapidly sterilizing sensitive medical devices.

Plasma Technology

Plasma is defined as a fourth state of matter, energetically distinguishable from solids, liquids, and gases. It can be produced through the action of either high temperatures, or electric or magnetic fields, and it is normally composed of a cloud of ions, electrons, and neutral species. Depending upon the environment in which it exists, i.e., field strength, interference, etc., the exact composition of the plasma will differ. For an ionized gas to be properly defined as a plasma, the number of positively and negative charged species present in the discharge need to be approximately equal. This requirement is satisfied when the dimensions of the discharge gas volume, Λ, are significantly larger than the Debye length, λ_D (*1*).

$$\lambda_D = \left| \frac{E_0 \, K \, Te}{n \, e^2} \right|^{1/2}$$

The Debye length represents the distance over which a charge imbalance can exist. In the above equation, E_0 is the permittivity of free space, K is the Boltzmann constant, Te is the electron temperature, n is the electron density, and e is the charge on the electron.

Plasmas can be conveniently divided into two categories. One category includes these plasmas that have the common characteristic of having temperatures in excess of $5000°K$ and are described as high temperature plasmas. While these plasmas exist in nature (i.e., in the sun or stars) and can be created in high pressure arcs or plasma jets, they are of no interest in plasma sterilization of medical devices due to the extreme temperatures involved.

The plasma utilized in plasma sterilization processes is in the second category known as glow discharge, or low temperature plasma. Low temperature plasmas, such as those found in neon lights or in sterilization processes, are created under vacuum conditions. These plasmas have average electron energies in the range of 1 to 10 eV and electron densities in the range of 10^9 to 10^{12} cm^{-3}. There is also a lack of thermal equilibrium in these plasmas between the electron temperature Te and the gas temperature Tg. The ratio of Te/ Tg is typically in the range of 10 to 100. For that reason, these plasmas have the unique properties of having electrons or other species that have sufficient energy to cause the rupture of molecular bonds while the temperature of atoms and molecules in the plasma are near ambient values. It is this characteristic that makes low temperature plasmas well suited to the sterilization of thermally sensitive materials.

Background of Plasma Sterilization

For over 25 years researchers have investigated the use of plasma technology to sterilize various materials and devices. One of the first reports on plasma sterilization was in a 1968 patent issued to Arthur D. Little Company (2) in which the use of a pulsed higher pressure/ high temperature plasma sterilization process was disclosed. A second patent issued in 1972 to Arthur D. Little (3) disclosed the first use of a low pressure, low temperature plasma process. This process utilized halogen gases to sterilize contaminated surfaces. Additional patents were issued in 1974 to Boeing Company (4,5) on the development of flow-through plasma sterilization systems for medical devices, in 1980 to Boucher (6) on a plasma process utilizing aldehydes in the gas phase, and in 1982 to Motorola (7,8) on sterilizing through sealed porous packaging and on the use of pressure pulsing of the plasma to enhance antimicrobial efficiency in diffusion restricted environment such as lumens.

In 1987, a patent was issued to Surgikos, Inc., a Johnson & Johnson Company (9), on the use of a low temperature gas plasma sterilization process utilizing hydrogen peroxide as the precursor chemical. One of the unique features of this process involved the use of a pretreatment stage to allow the hydrogen peroxide to diffuse through the materials and come in close proximity to the devices to be sterilized prior to the generation of the low temperature plasma. This feature allows for the efficient sterilization of packaged medical devices. The importance of this feature was discussed in a 1989 paper by Addy (10) on low temperature plasma sterilization. The first commercial plasma sterilization product, the STERRAD 100 sterilizer was developed based on this patented process.

In 1992, a patent was issued to Abtox, Inc. (11) on the use of another low temperature plasma sterilization process that also involves a pretreatment step. In this process, the pretreatment chemical involved a mixture of peracetic acid and hydrogen peroxide. The plasma is generated upstream by microwave discharge and also contains oxygen, hydrogen, and argon gases.

Although not all inclusive, these disclosures reflect the interest generated over the past 25 years on the application of low temperature gas plasma technology to the sterilization area. As the development of low temperature gas plasma sterilization has reached the commercialization stage, the interest in this area has continued to increase.

Generation of Low Temperature Plasma

Plasmas are normally referred to by the type of processor gas or vapor from which they are formed. For example, in the STERRAD Process, hydrogen peroxide and water molecules are initially introduced into a vacuum during the injection stage. The hydrogen peroxide and water molecules diffuse throughout the chamber and come in intimate contact with the materials to be sterilized during the diffusion stage, and the plasma that is generated during the plasma phase is known as a hydrogen peroxide plasma.

There are several different methods of generating low temperature gas plasmas. The most common methods, which are reviewed by Grill (*12*) and Bell, et al. (*13*) include direct current (DC), radio frequency (RF) and microwave (MW) power applied to a gas. RF systems are the most common and provide more flexibility from a design consideration. For example, RF plasmas can be sustained with internal as well as external electrodes, while DC discharges require the electrodes to be inserted inside the reactor and be in direct contact with the plasma. RF plasmas are also characterized by higher ionization efficiencies than DC plasmas and can be sustained at lower pressures than DC or MW plasmas.

With RF induced plasma, one can utilize either a capacitively coupled system or an inductively coupled system for coupling the RF energy into the plasma. Different chamber and electrode designs can also be utilized such that samples to be processed will reside in either a high intensity or a low intensity electric field. All of these factors can potentially affect the material compatibility of devices sterilized in a plasma sterilization process.

When working with glow discharge processes, it is important that the proper safety precautions be observed. The STERRAD Sterilization Process utilizes an RF generated plasma operating at 13.56 MHz, a frequency approved by the Federal Communication Commission (FCC) for industrial application. The RF power supply used to generate the low temperature gas plasma in the STERRAD Sterilizer can only be turned on when the sterilization chamber door is closed and the chamber is under vacuum. In addition, the unit is shielded to couple with FCC class A electromagnetic interference emissions standard. The STERRAD Sterilization System also meets the IEC CISPR standard and the VDE 871 Electrical Standard, the world's most stringent standard for RF emissions.

Chemistry of Plasma Sterilization

The chemistry of a plasma sterilization process will depend upon the precursor gases or vapors from which the plasmas are generated. In the STERRAD Sterilization Process, which utilizes H_2O_2 in the pretreatment step, a limited number of decomposition products can be generated during the plasma stage due to the simplicity of the H_2O_2 molecule.

The steps involved in the STERRAD Sterilization Process are as follows: In the pretreatment phase, hydrogen peroxide is introduced into the low initial pressure of the chamber. The hydrogen peroxide evaporates and is allowed to diffuse throughout the sterilization chamber and thereby come into close proximity with the items to be sterilized. The hydrogen peroxide serves as a precursor for the generation of free radicals and other biologically active chemical species. In addition, the hydrogen peroxide vapor can also exert lethal effects of its own. In a highly simplified form, the reactions in plasma for which H_2O_2 serves as a precursor may be summarized as follows:

$$H\text{-}O\text{-}O\text{-}H \xrightarrow{\quad e^-, \ 1\text{ - }10\text{ eV}\quad} HO\cdot \ + \ \cdot OH$$

$$HO\cdot \ + \ H\text{-}O\text{-}O\text{-}H \longrightarrow HOH \ + \ \cdot O\text{-}O\text{-}H$$

$$H\text{-}O\text{-}O\text{-}H \xrightarrow{\quad e^-\quad} H\text{-}O\text{-}O\text{-}H *$$

$$H\text{-}O\text{-}O\text{-}H * \longrightarrow H\text{-}O\text{-}O\text{-}H \ + \ \text{visible and UV radiation}$$

The chain of events resulting in the low temperature hydrogen peroxide plasma is initiated by the accelerated electrons which result from the electric field imposed on the chamber by the RF system. This series of events is similar to those suggested to occur during the secondary reactions related to gamma irradiation, i.e. (14).

$$H\text{-}O\text{-}H \xrightarrow{\quad r, \ 1.17\text{ - }1.33\text{ MeV}\quad} HO\cdot \ + \ \cdot H$$

$$H\cdot \ + \ O_2 \longrightarrow \cdot O\text{-}O\text{-}H$$

$$HO\cdot \ + \ \cdot OH \longrightarrow H\text{-}O\text{-}O\text{-}H$$

$$2\,(\,\cdot O\text{-}O\text{-}H\,) \longrightarrow H\text{-}O\text{-}O\text{-}H \ + \ O_2$$

A major difference in these series of events involves the energies associated with the initiation of these reactions. In the low temperature gas plasma, electron energies are in the 1-10 eV range, while in gamma irradiation, the energies are in excess of 1 MeV. This difference in energies, which relate to the fact that low temperature plasmas are known to affect only a thin layer, a few atoms in depth, on the surface of non-metallic materials as compared to the bulk effects observed with gamma irradiation, will be discussed in section VII on effects of low temperature plasma sterilization on medical materials and devices. Other reactions also occur in the hydrogen peroxide plasma to produce reactive oxygen ($O\cdot$) and hydrogen atoms, and oxygen and water as stable by-products.

$$\cdot OH + \cdot OH \longrightarrow H_2O + O\cdot$$

$$\cdot OH + \cdot O \longrightarrow H\cdot + O_2$$

$$\cdot OH + \cdot O\text{-}O\text{-}H \longrightarrow H_2O + O_2$$

During the plasma phase, the series of reactions depicted more or less continually take place under the influence of the RF energy. At the termination of the plasma phase of the cycle, any reactive species remaining recombine to form stable chemical species, predominately $H_2O + O_2$. Both metallic and nonmetallic medical substrates exposed to the low temperature hydrogen peroxide plasma STERRAD process have shown no increase in toxicological properties after exposure to repeated sterilization cycles.

The biological activity of many of the active species generated in hydrogen peroxide plasma have been the objects of study by a number of researchers (*15-18*). The free radicals generated in the hydrogen peroxide plasma are known to be reactive with almost all of the molecules essential for the normal metabolism and reproduction of living cells, i.e., DNA, RNA, enzymes, phospholipids, etc. While an exact model of the mode or modes of action in the inactivation of microorganisms by the STERRAD Sterilizer cannot be precisely established, there exists in the hydrogen peroxide plasma process a biocidal environment which comprises many reactive species, i.e., free radicals, UV radiation, hydrogen peroxide, etc., that are capable of inactivating microorganisms by chemical interactions at multiple biologically important reaction sites.

Efficacy

The efficacy of the STERRAD Sterilization System was established, in part, by demonstrating the ability of the system to kill a broad spectrum of microorganisms selected for their known resistance to chemical and physical sterilants and to provide by universally accepted validation methods a Sterility Assurance Level (SAL) greater than 10^{-6} with highly resistant bacterial spores. Additional tests were also conducted with actual medical products to confirm the validation studies conducted with the STERRAD Sterilizer.

Spectrum of Activity. By definition, a sterilization process must have the ability to inactivate a broad spectrum of microorganisms, including resistant bacterial spores. As previously noted, the hydrogen peroxide plasma process contains a number of reactive species known to be reactive with molecules essential for the normal metabolism and reproduction of microorganisms. Because of the number of reactive species involved and the ability of these species to react at multiple reaction sites in the microorganisms, one would expect that the hydrogen peroxide plasma process would exhibit broad spectrum antimicrobial activity.

Spectrum of activity studies were conducted against vegetative bacteria (including Mycobacteria), bacterial spores, yeasts, fungi, and viruses. In general, these organisms were chosen for their resistance to hydrogen peroxide, other chemical sterilants or ionizing radiation, as has been documented in scientific literature. Those organisms currently used to monitor steam, ethylene oxide and ionizing radiation sterilization processes were included in these studies. The reason for the selection of each test organism as well as the results of these tests are presented in Table I. All of the organisms shown in Table I were found to be efficiently killed by an abbreviated STERRAD System cycle consisting of 20 minutes of diffusion with 2 mg/liter of hydrogen peroxide and 5 minutes of plasma at a power of 300 watts. By comparison, the standard STERRAD Sterilization Cycle consists of 50 minutes diffusion with a minimum of 6 mg/liter of hydrogen peroxide and 15 minutes of plasma at 400 watts. As the STERRAD System cycle conditions were further reduced, bacterial spores were found to be the most resistant organism to the process. These results are consistent with those observed with other sterilization processes in which bacterial spores have been found to be more resistant to inactivation than other microorganisms.

The two virus tested, Poliovirus Type 1 and Herpesvirus Type 1, are representative of the two major classes of viruses, hydrophilic and lipophilic viruses, respectively. Of these two classes, the hydrophilic group normally exhibits the greater resistance to chemical sterilants, and poliovirus is known to be a highly resistance hydrophilic virus. The log 10 virus titers of 3.98, 3.20 and 2.84 represent the minimum concentration of viruses in these tests. Due to the nature of the virucidal test, a minimum virus concentration is determined but the actual concentration is not. In all virucidal tests there was no infectivity obtained after exposure to the highly abbreviated STERRAD System Cycle. This shows that even an abbreviated STERRAD System Cycle is capable of inactivating both hydrophilic and lipophilic viruses.

Validation of 10^{-6} Sterility Assurance Level (SAL). Well established and universally accepted methods exist for the validation of sterilization processes. For example, in the Association for the Advancement of Medical Instrumentation (AAMI), Standards and Recommended Practices, Volume 1: Sterilization (1992) (*19*), the procedures for the validation of steam and ethylene oxide sterilization processes are specified. Implicit in these test methods is the demonstration of at least a 10^{-6} sterility assurance level (SAL) for the sterilization process.

Table I
SPECTRUM OF ACTIVITY
VEGETATIVE BACTERIA, SPORES AND FUNGI

MICROORGANISM	TYPE	INTEREST IN TESTING	CONTROL[1]	RESULTS[2]
Bacillus stearothermophilus	Bacterial Spore	H_2O_2 Resistance; Steam Indicator Organism	2.04×10^6	0/9
Bacillus subtilis var. niger (globigii)	Bacterial Spore	H_2O_2 Resistance; EtO Indicator Organism	2.69×10^6	0/9
Bacillus pumilus	Bacterial Spore	Ionizing Radiation Resistance and Radiation Indicator Organism	1.82×10^6	0/9
Staphylococcus aureus	Gram Positive	H_2O_2 Resistance; Clinical Significance	2.82×10^6	0/9
Deinococcus radiodurans	Gram Positive	Ionizing Radiation Resistance	3.10×10^6	0/9
Pseudomonas aeruginosa	Gram Negative	Clinical Significance	1.32×10^6	0/9
Escherichia coli	Gram Negative	Clinical Significance	9.23×10^5	0/9
Serratia marcescens	Gram Negative	H_2O_2 Resistance; Clinical Significance	1.85×10^6	0/9
Moroxelia osloensis	Gram Negative	Ionizing Radiation Resistance	3.14×10^6	0/9
Mycobacterium bovis	Acid Fast	Chemical Resistance; Clinical Resistance	4.20×10^6	0/9
Candida albicans	Yeast	H_2O_2 Resistance	3.95×10^6	0/9
Candida parapsilosis	Yeast	H_2O_2 Resistance; Clinical Significance	1.07×10^6	0/9
Trichophyton mentagrophytes	Filamentous Fungus	Clinical Significance	1.25×10^6	0/9
Aspergillus niger	Filamentous Fungus	H_2O_2 Resistance; Clinical Significance	1.46×10^6	0/9

1. Average titer recovered from nine samples
2. # Positive / # Tested

Continued on next page

Table I. Continued

VIRUSES

MICROORGANISM	TYPE	INTEREST IN TESTING	VIRUS TITER Log 10		INFECTIVITY
Poliovirus Type 1 (Brunhilde)	Hydrophilic	Chemical Resistance; Clinical Significance	Test 1 Test 2	$\geq$3.98 $\geq$3.98	Not detected Not detected
Herpesvirus Type 1	Lipophilic	Clinical Significance	Test 1 Test 2	$\geq$3.20 $\geq$2.84	Not detected Not detected

Sterilization is a probability function and a minimum SAL of 10^{-6} means that the probability of a bioburden microorganism surviving after exposure to the sterilization process is no greater than 10^{-6}. Or it can also be stated that the probability of having a non-sterile device after processing is less than one in one million when the sterilizer is used as directed. This definition for sterility of terminally sterilized products is well accepted in the scientific community.

Critical to the demonstration of the 10^{-6} SAL is the use of a consistent and reproducible biological monitor for evaluating the efficacy of the sterilization process. According to the AAMI guidelines the organism used in the biological challenge should be resistant to the sterilization process being monitored. The recommended biological challenge organisms for steam and ethylene oxide sterilization processes are **Bacillus stearothermophilus** and **Bacillus subtilis** var. **niger**, respectively. It is recommended that **B. stearothermophilus** be used at a population of 10^3 to 10^6 and that **B. subtilis** be used at a population of at least 10^3 and that it is typically used at a population of 10^6.

The STERRAD Sterilizer was validated by the classical overkill method of validation (AAMI Standards and Recommended Practices, Vol. 1 : Sterilization 1992). The overkill approach is based on the premise that the sterilization process will inactivate a given population of a spore challenge that is resistant to the sterilization process and provide an additional safety factor. The method is called "overkill" because the cycle conditions established to kill the resistant spore challenge, with an additional safety factor, are far more severe than those required to inactive the product bioburden. For example, for ethylene oxide sterilization, the typical spore challenge is 10^6 spore of **B. subtilis**, and an overkill cycle would provide a 6 log reduction of the microbial challenge at one-half of the sterilization cycle exposure time. This provides for a safety factor since the complete cycle would provide a theoretical 12 log reduction of the **B. subtilis** spores that are more resistant to the ethylene oxide sterilization process than normal bioburden organism.

Two methods are specified in the AAMI standards for detecting the number of organisms present on a biological indicator used in the validation of a steam or ethylene oxide sterilization process. The first method in which the number of viable organisms on the biological indicator is counted or enumerated is generally known as the survivor curve, count reduction or plate count method. The surviving number of microorganisms is evaluated using standard microbiological plating techniques, at fractional sterilization cycle exposure times. The second method, which is known as the quantal analyses, fraction-negative analyses, or sterility test method, involves placing the biological indicator in a broth media and evaluating for a growth/no growth response. Replicate units are exposed to fractional sterilization cycles, cultured, incubated, and scored for growth (positive) or no growth (negative). The data can be used to estimate the number of surviving organisms per unit using the most probable number analysis of Halvorson and Ziegler (*20*). The STERRAD Sterilization System was validated by the universally accepted overkill procedure utilizing both the survivor curve and fraction negative test methods.

Selection of Validation Organism.   Before validating the STERRAD Sterilization Cycle, studies were conducted to determine which organism exhibited the greatest resistance to the hydrogen peroxide plasma process. Since spectrum of activity tests had shown that bacterial spores were the most resistant of the organisms tested, tests were conducted on four bacterial spores used to validate comparable sterilization systems to determine their relative resistance to the STERRAD Process. In these tests the test organism was inoculated onto paper strips which presented a greater challenge than with medical material substrates because paper tends to absorb hydrogen peroxide thus competing with the spores for the available hydrogen peroxide in the system. The paper strips were placed in diffusion restricted tubular test pieces that were then placed in standardized hospital trays of instruments.

The test method used involved varying the amount of diffusion time in the cycle holding all other components of the cycle fixed at their full value. This method was used because the biological challenge was placed in the most diffusion restricted location in the validation load and diffusion, therefore, became the rate determining step in establishing the efficacy of the system. Testing was conducted by the fraction negative method in which replicates of 10 spore strips were subjected to sterility testing and the fraction negative samples plotted against diffusion time. The results of these tests, which are presented in Figure 1, demonstrate that **Bacillus stearothermophilus** was the most resistant of the spores tested.

Based on the results of these tests, **B. stearothermophilus** spores on paper strips in the diffusion restricted test configuration described above was selected as the biological challenge for the validation of the STERRAD 100 Sterilization System. The STERRAD 100 Sterilization Process was validated with both the fraction negative and survivor curve test methods.

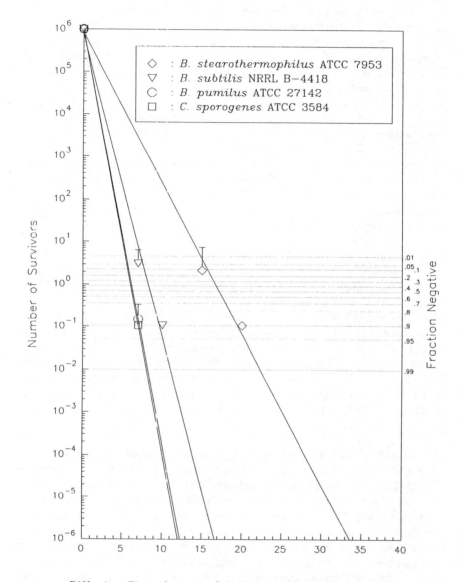

Figure 1. Kill Curves by Sterility Method (Fraction Negative) for *Bacillus stearothermophilus*, *Bacillus subtilis*, *Bacillus pumilus*, and *Clostridium sporogenes* Spores on Paper in Tubular Assemblies in The Validation Load.

Validation by Survivor Curve Test Method. The Survivor Curve Test Method involved the determination, by the direct plate count technique, of the number of viable **B. stearothermophilus** spores recovered from paper spore strips after exposure to STERRAD Sterilization Cycles with increasing diffusion times. Figure 2 contains the composite results of three sets of plate count data. Each set of data contained 10 replicates at each time point so that each point on the graph represents a total of 30 replicates. The range of variation around the points is indicated by the vertical line through the points. The linear regression line extrapolates to the 10^{-6} survivor level, on the horizontal axis at less than 30 minutes of diffusion time. The D-value calculated from the slope of the linear regression line was 1.96 minutes.

It should be noted that the linear regression line does not intersect the vertical axis at the 1×10^{6} point, the minimum initial population of spores on the paper strip. If a smooth curve was fitted to the observed averages, beginning with the starting population, a curve with an initial shoulder and increasing slope to a maximum rate of kill between 12 and 16 minutes of diffusion would be seen. This shape of kill curve is caused by the spore strips being located in a diffusion constrained load. Hydrogen peroxide must diffuse down a narrow tube to reach the spore strip. Some time in the diffusion phase of the cycle is required before the maximum achievable kill rate at the end of the plasma phase is observed. Because the eight minute survivor points are obviously still on the shoulder of the kill curve, the regression line calculated from all the 8, 12, and 16 minute data is a more conservative estimate of time to reach 10^{-6} survivors than would be the case using only a regression line from the 12 and 16 minute time points. In spite of the conservative analysis, the intersection of the regression line with the 10^{-6} survivor line provides an SAL of 10^{-6} in less than 30 minutes of diffusion with the full 15 minutes of plasma.

Validation by Fraction Negative Test Method. Tests were conducted by the Fraction Negative Method as described above utilizing 10 replicates of paper spore strips containing a minimum of 1×10^{6} **B. stearothermophilus** spores in each test set. All tests were done in triplicate. As seen in Figure 3, an extrapolation of the linear regression line, through the data points to the 10^{-6} survivor level, demonstrates that an SAL of 10^{-6} is achieved with less than 35 minutes of diffusion followed by 15 minutes of plasma. The D-Value calculated from the shape of the linear regression line was 2.79 minutes. The Fraction Negative Method provides a slightly more conservative method of determining a 10^{-6} SAL since the data are extrapolated from the initial spore population of 1×10^{6} and the actual curvature of the kill curve, due to testing in a diffusion restricted location, is not considered.

The kill curve for the STERRAD Process over the entire range of test conditions is best defined by combining the results of the survivor and fraction negative test data as presented in Figure 4. The linear regression line generated using both sets of data extrapolates to a 10^{-6} survivor end point that is in very good agreement

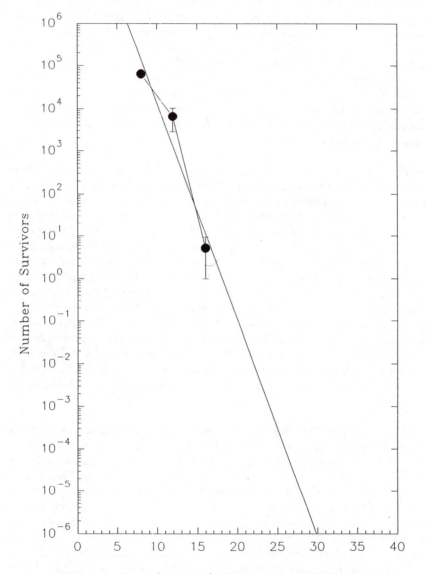

Figure 2. Composite Kill Curve by Plate Count Method for *Bacillus stearothermophilus* Spores on Paper in Tubular Assemblies in The Validation Load.

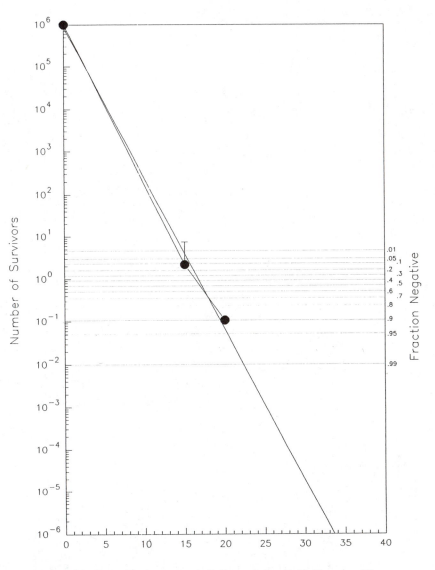

Figure 3. Composite Kill Curves by Sterility Method (Fraction Negative) for *Bacillus stearothermophilus* Spores on Paper in Tubular Assemblies in The Validation Load.

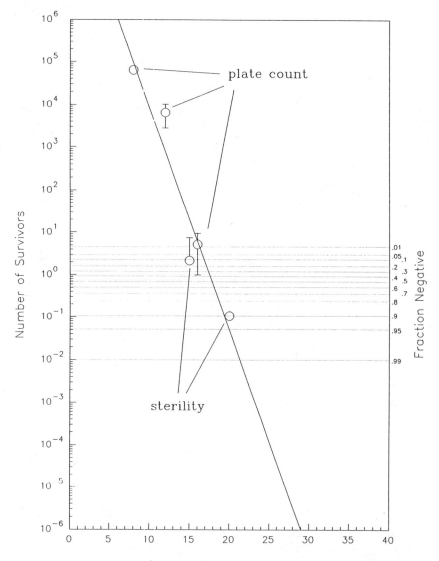

Figure 4. Composite Kill Curves by Plate Count and Sterility Methods (Fraction Negative) for *Bacillus stearothermophilus* Spores on Paper in Tubular Assemblies in The Validation Load.

with the end point obtained by the two independent test methods. This confirms that the two universally accepted test methods recommended in the AAMI standards for the validation of sterilization processes provide comparable 10^{-6} sterility survivor end points with the STERRAD Sterilization system.

Effects of Low Temperature Gas Plasma Sterilization on Medical Materials and Devices

Some of the active species formed in the hydrogen peroxide plasma are similar to those formed when r-radiation is used to sterilize medical devices. For example, both hydroxyl free radicals and hydroperoxyl free radicals are formed from the interaction of r-radiation with water molecules in the presence of oxygen. In addition, the recombination of radicals present in r-radiation also produces hydrogen peroxide. Although similar reactive species are present in both processes, the effect of the two processes on the physical properties of some non-metallic devices can be dramatically different. The high energy r-radiation is capable of passing through non-metallic materials used in medical devices and generating secondary reactions that can detrimentally affect the bulk properties of the materials. Low temperature plasmas are known to affect only a thin layer a few atoms in depth on the surface of non-metallic materials and do not affect the bulk properties of these materials. In addition, the STERRAD Sterilization System utilizes a secondary plasma that minimizes surface modification since the item to be sterilized is not exposed to the direct or primary plasma discharge.

The effect of the low temperature gas plasma process utilized in the STERRAD Sterilization Process on both metallic and non-metallic materials, as well as complex medical devices containing a combination of these materials, has been investigated. Whenever possible, the effect of the STERRAD Sterilization Process on the physical, chemical, and functional properties of these materials and devices have been compared to the effects observed after sterilization by another sterilization process, i.e., gamma radiation, steam, etc.

Metallic Device. Metallic medical devices have traditionally been sterilized in the hospital environment by steam or by dry heat. Although most metallic devices withstand the temperatures associated with these processes with minimal deterioration in performance, delicate cutting edges, such as exist in microsurgical instruments, can be rapidly degraded by repeated exposure to high temperature and high humidity sterilization processes (*21*).

Because of the sensitivity of microsurgical instruments to high temperature sterilization processes, the effect of the STERRAD Sterilization Process on microsurgical instruments were evaluated and compared to that of steam sterilization. Microsurgical scissors were exposed to a total of 50 STERRAD Sterilization cycles. The scissors were manipulated between sterilization cycles to simulate actual use of the product and tested for cutting efficiency by the manufactures specified functionality protocol after every ten sterilization cycles.

A second set of scissors were similarly exposed to repeated steam sterilization cycles under the same test protocol. The microsurgical scissors exposed to 50 STERRAD Sterilization cycles show no obvious change in appearance or any change in cutting efficiency. By contrast, after exposure to 10 steam sterilization cycles the surface of the microsurgical scissors were visually duller in appearance and the scissors did not pass the manufacturers test for cutting efficiency.

Similar results have been obtained in studies conducted in Germany on microsurgical ophthalmic instruments (Draeger, J., Universitats - Krankenhaus Eppendorf, Hamburg, Germany, personal communication, 1995), (22). Apparently, the combination of low temperature and low humidity that exists in the low temperature plasma STERRAD Sterilization Process does not adversely affect the functional properties of the metals utilized in delicate microsurgical instruments.

Non-metal Devices. As noted earlier, there is a significant similarity between the active species generated in the low temperature hydrogen peroxide plasma sterilization process utilized in the STERRAD Process and those generated during gamma irradiation sterilization. The differences in energies associated with the generation of these species are however remarkably different. For example, the accelerated electrons present in the low temperature hydrogen peroxide gas plasma process have energies in the 1 to 10 eV as compared to the gamma radiation energies that exceed 1 MeV. The low electron energies associated with low temperature gas plasma processes limit the penetration and subsequent chemical effects to the surface of non-metallic substrates where as the high energy gamma radiation is known to affect the bulk properties of these materials. To examine the difference in the effect of these two processes on a non-metallic medical substrate, a nonwoven polypropylene based fabric (Spunguard™ Heavy Duty Sterilization Wrap) was exposed to three consecutive sterilization processes by the STERRAD system and gamma irradiation. The bulk physical properties, i.e., tensile strength and elongation, of polypropylene are known to be adversely affected by gamma irradiation sterilization. The nonwoven fabric tested consists of a layer of melt blown polypropylene fibers laminated between two layers of spunbonded polypropylene fibers. The fiber diameter of the spunbonded layer is approximately 10 micron which the melt blown layer has a fiber diameter of 1 micron. The fabric therefore contains a polymer that is degraded by gamma irradiation and also has a large surface area that can chemically react with the active species generated in the low temperature gas plasma sterilization process.

Table II contains the grab tensile and elongation values for the non-woven fabric after exposure to 0 to 3 sterilization cycles with the STERRAD Sterilization Process and gamma irradiation. The results with the gamma irradiated samples are as expected. As the exposure to gamma radiation increased, i.e., from 1, 2 and 3 sterilization cycles at a dose of 2.8 Mrad per cycle, the tensile strength and elongation of the fabric decreased. The bulk properties of the polypropylene fiber are being altered by the chemical reactions occurring in the polymer. The results with the STERRAD Sterilized samples are interesting due to the lack of effect of the process on the bulk physical properties of the polypropylene substrate. All values fell within the standard deviation for the individual test data. Low temperature gas plasma is known to be a surface phenomena and only a few angstroms on the surface of even non-metallic substrates are affected by the active species generated during the plasma phase. The alteration of a few angstroms of material at the surface of the substrate has an insignificant effect on the bulk physical properties of the material being sterilized even when the diameter of the substrate is in the 1 to 10 u range. For this reason, low temperature gas plasma sterilization can provide some unique advantages for sterilization of non-metallic medical devices.

Table II

PHYSICAL PROPERTIES[1] OF NON-WOVEN POLYPROPYLENE FABRIC EXPOSED TO STERRAD AND GAMMA IRRADIATION STERILIZATION

Cycles (#)	STERRAD[2]		r - RADIATION[3]	
	Tensile Strength (lb.)	Elongation (%)	Tensile Strength (lb.)	Elongation (%)
0	22.5 ± 4.1	50.5 ± 6.7	22.5 ± 4.1	50.0 ± 6.7
1	21.9 ± 4.4	50.2 ± 9.6	13.7 ± 3.1	29.8 ± 4.8
2	20.2 ± 3.9	45.8 ± 11.2	8.9 ± 2.1	19.5 ± 3.3
3	20.3 ± 3.9	45.8 ± 11.2	4.8 ± 1.3	10.3 ± 2.6

1. All physical property measurements were an average for 12 samples.
2. STERRAD Process consisted of 50 minutes of pretreatment with 6.0 mg/liter H_2O_2 followed by 15 minutes of plasma with 400 watts of power.
3. Gamma irradiation dosage was 2.8 Mrad.

Surface Effects of Low Temperature Plasma. The effect of the active species present in low temperature gas plasma on the surface of non-metallic substrates is mainly manifested by a change in the wetting properties or contact angle observed when solutions come in contact with the substrate surface. This effect is due to the modification of the monolayer of molecules that reside at the substrate surface and their interaction with the contacting solution. Since the hydrogen peroxide plasma, utilized in the STERRAD Process, is oxidative in nature, the tendency is for hydrophobic surfaces to become more hydrophilic, i.e., contact angles with aqueous solution will decrease, after exposure to STERRAD Sterilization. This change in wetting properties is dependent on many factors, including the strength of the chemical bonds in the molecules at the surface and the hydrophobic properties of the material. For example Teflon, which is highly hydrophobic and has strong chemical bonds (C-F bond is about 110 Kcal/mole), is less affected than polypropylene, which is less hydrophobic and has lower energy bonds (C-H bond is about 90 Kcal/mole). It should also be noted that the change in contact angle observed after plasma processing is normally transient in nature and the contact angle changes with time in the direction of the original contact angle. For example, the contact angle of Teflon after exposure to the STERRAD Sterilization process changed from 98° to 89°, but within 2 days had reverted to 95°. This change in contact angle with time is reported to be due to reorientation of the molecules at the surface such that the original surface chemistry is restored (*23, 24*). This phenomena further illustrates that only a thin layer of molecules at the surface are modified by low temperature gas plasma sterilization.

Bulk Property Effects. Since plasma effects are mainly limited to the surface of non-metallic substrates, the effect on the bulk properties of these substrates are more related to the chemistry of the pretreatment or diffusion stage. In the case of the STERRAD Sterilization Process, the pretreatment stage involves the use of hydrogen peroxide in the vapor or gas phase. The potential interaction of hydrogen peroxide with the chemical structures involved in the non-metallic substrate is therefore of primary interest.

Hydrophobic non-metallic substrates, such as polyethylene, polypropylene, Teflon, etc., that have low polarity and are not capable of forming hydrogen bonds with hydrogen peroxide have excellent material compatibility with the STERRAD Sterilization Process. At the other extreme are highly hydrophilic substrates, such as cellulosic based materials, i.e., paper, linen, etc., that are highly polar and that contain multiple hydroxyl groups capable of forming hydrogen bonds with hydrogen peroxide. These materials also have large surface areas that accelerate the interaction of hydrogen peroxide with the cellulosic structure. For that reason, these materials are not recommended to be used as packaging materials in the STERRAD Sterilizer due to the high affinity of the cellulosic substrate for hydrogen peroxide and the fact that they would act as a barrier to penetration of the hydrogen peroxide into the interior of packaged items during the pretreatment stage of the process. It is also not recommended that large quantities of cellulosic based products, such as cotton, wood, etc., be sterilized in the STERRAD Sterilizer since they would act as a hydrogen peroxide sink and the proper pressures would not be achieved during the pretreatment stage. Since the STERRAD Sterilizer is software controlled, the presence of an excess of these materials in the sterilizer would result in the sterilization cycle being canceled due to low pressure in diffusion. Although the use of higher concentrations of hydrogen peroxide or longer pretreatment times would overcome these problems, these approaches would require longer plasma times to remove residual hydrogen peroxide from the substrates and would detract from the advantages of the STERRAD Sterilization Process which includes low concentrations of hydrogen peroxide and short sterilization cycles. In addition, cellulosic based materials are capable of being rapidly sterilized in almost all cases by steam sterilization.

In between the highly hydrophobic and highly hydrophilic non-metallic substrates lies a wide range of chemical structures of intermediate polarity and of varying ability to interact with hydrogen peroxide. These include silicones, polyvinyl chloride, polyvinylidene fluoride, polycarbonates, latex rubber, polyether imide, polystyrene, polysulfone, polyamides, polyurethanes, polyacetal, etc. In general, those compounds that are capable of forming hydrogen bonds with hydrogen peroxide, i.e., polyamides, polyacetal, etc., absorb more hydrogen peroxide than those materials that do not form hydrogen bonds. In addition, those compounds that absorb more hydrogen peroxide have the potential to be more affected by exposure to repeated sterilization processing than those materials that do not absorb hydrogen peroxide. However, since there are hundreds of different chemical formulations possible with a given class of compounds, i.e., polyamides, polyurethanes, silicones, etc. It is difficult to generalize the interaction of hydrogen peroxide with a specific class of compounds.

In general, the STERRAD Sterilization Process exhibits good material compatibility with non-metallic substrates and interactions with a specific polymeric structure or the effect of repeated exposure to the sterilization process on functional properties of an individual device would need to be determined on an individual basis.

Functionality of Medical Products. Ultimately, the utility of a sterilization process is determined by its ability to repeatedly sterilize complex medical devices containing mechanical, optical, or electrical components without affecting the functional properties of these devices. Laboratory tests were conducted on specific medical devices to quantify the effect of exposure to repeated STERRAD Sterilization cycles on the functional properties of the devices. The devices selected for testing represent a wide range of materials including metals, plastics, rubber and optical surfaces which must retain properties such as flexibility, optical clarity, electrical discharge, etc., after being repeated sterilized. In all tests, the devices were manipulated between sterilization cycles to simulate actual use of the product. Devices were cycled for a total of 50 cycles and were evaluated after 10, 20, 30, and 40 cycles as well as after the 50 cycles. The total of 50 cycles was chosen for devices which normally undergo repetitive sterilization and use. If no effect is seen in 50 cycles, it was considered unlikely that any adverse effect would occur as a result of additional exposures. Functionality tests were conducted according to the device manufacturer's protocol, or by a quantifiable test procedure developed specifically for that devices. The results of tests on three medical devices a resectoscope, defibrillator paddles and a flexible fiberoptic sigmoidoscope, are presented below.

Table III contains the results of testing the electrical properties of a defibrillator set after repetitive sterilization cycles compared to a control set that was manipulated to simulate actual use of the device but was not exposed to the STERRAD Sterilization Process. As noted in the footnote, the manufacturers specification indicates that the charge and discharge values should not vary by more than 15%. The data presented in Table III illustrates that there is not a significant difference in the charge and discharge values obtained with the control and test defibrillator sets, and that both sets of data are well within the manufacturers specified limits.

Table III

EFFECT OF MULTIPLE STERRAD STERILIZATION CYCLES ON
THE ELECTRICAL FUNCTIONALITY OF A DEFIBRILLATOR SET

Number of Sterilization Cycles	Electrical Test Results [1]					
	STERRAD Treated			Manipulation Control		
	Charge (Joules)	Discharge (Joules)	Difference (Joules)	Charge (Joules)	Discharge (Joules)	Difference (Joules)
0	20.0	19.7	-0.3	20.0	19.5	-0.5
	50.0	49.9	-0.1	50.0	49.9	-0.1
10	20.0	19.7	-0.3	20.0	19.8	-0.2
	50.0	49.0	-1.0	50.0	50.3	+0.3
20	20.0	20.8	+0.8	20.0	19.6	-0.4
	50.0	49.8	-0.2	50.0	49.2	-0.8
30	20.0	19.7	-0.3	20.0	19.8	-0.2
	50.0	49.9	-0.1	50.0	50.2	+0.2
40	20.0	19.7	-0.3	20.0	19.8	-0.2
	50.0	50.4	+0.4	50.0	50.6	+0.6
50	20.0	20.2	+0.2	20.0	19.8	-0.2
	50.0	50.3	+0.3	50.0	50.2	+0.2

1   The manufacturer specifies that the difference between the charge and
    discharge values not exceed 15%, i.e., for a 20 Joules charge ± 3 Joules, and a
    50 Joules charge ± 7.5 Joules.

Table IV shows the results of testing the electrical properties of a resectoscope
versus a control. The electrical performance of the test and control devices was
determined to be the best objective criterion for evaluation of this device.
Analysis of the data showed that both devices were within the 15 percent deemed
to be acceptable variability between the power supply alone and the power supply
with working elements and cutting loop.

Table V contains the results of testing mechanical and optical properties of a
flexible fiberoptic sigmiodoscope after repetitive sterilizing cycles in the
STERRAD Sterilizer. The testing was done using a control endoscope for
comparison. No change in either mechanical or optical properties could be
observed in the test sigmoidoscope as compared to the control device.

Report from clinical test sites, as well as information obtained from hospitals who
have used the STERRAD Sterilizer to repetitive sterilize medical devices for over
one year, have substantiated that the STERRAD Sterilization process is
compatible with a wide range of medical substrates and devices normally used in
the hospital environment.

Table IV

EFFECT OF MULTIPLE STERRAD STERILIZATION CYCLES
ON THE ELECTRICAL PROPERTIES OF A RESECTOSCOPE

Number of sterili-zation cycles	Power Setting	STERRAD Treated			Manipulation Control		
		Power Supply Alone (watts)	Power Supply with Working Elements and Cutting Loop (watts)	Difference between Power Supply and Cutting Loop[1] (%)	Power Supply Alone (watts)	Power Supply with Working Elements and Cutting Loop (watts)	Difference between Power Supply and Cutting Loop (%)
0	35	87	87	0	96	92	-4.2
	55	255	255	0	255	247	-3.1
	65	380	370	-2.6	350	360	+2.9
10	35	96	96	0	110	110	0
	55	262	262	0	275	280	+1.8
	65	380	390	+2.6	390	395	+1.3
20	35	101	92	-8.9	96	96	0
	55	270	255	-5.6	275	280	+1.8
	65	390	360	-7.7	380	390	+2.6
30	35	101	92	-8.9	101	101	0
	55	270	255	-5.6	270	280	+3.7
	65	390	380	-2.6	380	395	+3.9
40	35	87	96	+10.3	92	101	+9.8
	55	247	270	+9.3	255	270	+5.9
	65	350	380	+8.6	390	380	-2.6
50	35	96	96	0	96	106	+10.4
	55	262	262	0	270	270	0
	65	370	370	0	380	380	0

1   The specification for this test, which was developed in-house, states that the difference between the power supply alone and the power supply and cutting loop not exceed 15%.

Table V

EFFECT OF MULTIPLE STERRAD STERILIZATION CYCLES
ON THE OPTICAL AND MECHANICAL PROPERTIES
OF A FIBEROPTIC SIGMOIDOSCOPE

Number of Sterilization Cycle	STERRAD Treated		Manipulation Control	
	Optical Properties	Angulation Control	Optical Properties	Angulation Control
0	Unchanged	Unchanged	Unchanged	Unchanged
10	Unchanged	Unchanged	Unchanged	Unchanged
20	Unchanged	Unchanged	Unchanged	Unchanged
30	Unchanged	Unchanged	Unchanged	Unchanged
40	Unchanged	Unchanged	Unchanged	Unchanged
50	Unchanged	Unchanged	Unchanged	Unchanged

Literature Cited

1. Bell, A.T. In *Techniques and Applications of Plasma Chemistry*; Hollahan, J. R.; Bell, A. T., Eds.; Fundamentals of Plasma Chemistry; Wiley-Interscience: New York, **1974**.
2. Menashi, W. P. *Treatment of Surface*. U. S. Patent 3,383,163. **1968**.
3. Ashman, L. E.; Menashi, W. P. *Treatment of Surfaces with Low Pressure Plasmas*. U. S. Patent 3,701,628. **1972**.
4. Fraiser, S.; Gillette, R. B.; Olson, R. L. *Sterilizing and Packaging Process Utilizing Gas Plasma*. U. S. Patent 3,851,436. **1974**.
5. Fraser, S. J.; Gillette, R. B.; Olson, R. L. *Sterilizing Process and Apparatus Utilizing Gas Plasmas*. U. S. Patent 3,948,601. **1974**.
6. Boucher, R. R. *Seeded Gas Plasma Sterilization Method*. U. S. Patent 4,207,286. **1980**.
7. Bithell, R. M. *Packaging and Sterilizing Process for Same*. U. S. Patent 4,321,232. **1982**.
8. Bithell, R. M. *Plasma Pressure Pulse Sterilization*. U. S. Patent 4,348,357. **1982**.
9. Jacobs, P. T.; Lin, S. *Hydrogen Peroxide Plasma Sterilization System*. U. S. Patent 4,643,876. **1987**.
10. Addy, T. O. In *Low Temperature Plasma: A New Sterilization Technology for Hospital Application in Sterilization of Medical Products*; Morrissey, R. F; Prokopenko, Y. I., Eds.; Proceedings of the International Kilmer Memorial Conference on the Sterilization of Medical products; Polyscience Publications: Morin Hights, Canada, **1991**, Vol. V; pp 80-95.
11. Caputo, R. A.; Campbell, B. A.; Monltan, K. A. *Plasma Sterilizing Process with Pulsed Antimicrobial Agent*. U. S. Patent 5,084,239. **1990**.
12. Grill, A. *Cold Plasma in Material Fabrication from Fundamentals to Applications*. IEEE Press: New York, NY, **1994**, pp 25-45.
13. Suhr, H. In *Applications of Non-equilibrium Plasmas to Organic Chemistry. Techniques and Application of Plasma Chemistry*; Bell, A.T.; Hallahan, J. R. Eds. New York, NY, **1974**, pp 58-63.
14. Silverman, G. J. In *Sterilization by Ionizing radiation*; Block, S. Ed.; Disinfection, Sterilization, and Preservation; Lea & Febiger: Philadelphia, PA, 3rd edition. **1983**, pp 91-92.
15. Ewing, D. *J. Radiat. Biol. Relat. Study. Phys. Chem. Med.* **1983**, 43, pp 565-570.
16. Ewing, D. 1983. *Radiat. Res.* **1983**, 94, pp 171-189.
17. Ewing, D. *Radiat. Res.* **1983**, 96, pp 275-283.
18. Friedman, Y. S., Grez, N. *Acta Aliment*. **1974**, 3, pp 251-265.
19. *Association for the Advancement of Medical Instrumentation. Standard and Recommended Practice*. Vol. 1: Sterilization, Arlington, VA, **1992**.
20. Halvorson, H. O. and Ziegler, N. R. *J. Bacteriology*. **1993**, 25, pp 101-121.
21. Draeger, J.; Pruter, J. W. Klin. Monatsbl. Augenheilkd, *Clinical Monthly Ophthalmalogy*. **1990**. 197, pp 133-137.

22. Draeger, J.; Pruter, J. W.; Fortsch M. In *Infections Diseases of the Eye*; Bialasiewicz, A. A. & Schaal, K. P. Eds; AEolus Press Science Publishers, Buren the Netherlands.

23. Occhiello, E.; Morra, M.; Morini, G.; Garbassi, F.; Humphrey, P. *J. Applied Polymer Sci.* **1991**, 42, pp 551.

24. Yasuda, H.; Sharma H. K.; Yasuda, T. *J. Polym. Sci. Phy. Ed.* **1981**, 19, pp 1285.

RECEIVED September 18, 1995

Chapter 18

Radiation Sterilization of Medical Devices and Pharmaceuticals

Karen J. L. Burg[1] and Shalaby W. Shalaby[2]

[1]Department of Bioengineering, 301 Rhodes Engineering Research Center, Clemson University, Clemson, SC 29634–0905
[2]Center for Applied Technology, Poly-Med, Inc., 511 Westinghouse Road, Pendleton, SC 29670

The advantages of ionizing irradiation over traditional sterilization methods are overviewed and recent developments in the field are described. The effect of irradiation on synthetic polymers, natural materials, and pharmaceuticals is discussed, along with the advantages and disadvantages of irradiation on each of these systems. Biomedical materials have the stringent requirement of maintaining their integrity during sterilization as well as displaying short and long-term biocompatibility in the biological environment. In effect, a biomedical device or pharmaceutical product should maintain its chemical and physical properties through different stages of processing including terminal sterilization. Sterilization is a sensitive, critical step in this process.

Before implantation in the body, an implant must be sterilized to free the device of viable microorganisms. Typically, traditional metal implants are sterilized with dry or moist heat; however, polymeric implants may deform, hydrolyze, or even melt under such conditions. Sterilization with ethylene oxide is a common alternative; however, residual gases may cause post-implantation tissue necrosis (*1-2*). Additionally, a mixture of ethylene oxide and fluorinated hydrocarbons (Freon) is commonly used in small-scale hospital and industrial use to eliminate the explosion hazard of pure ethylene oxide. Fluorinated hydrocarbons face severe restrictions currently and, as a result, the sterilization industry is seeking viable ethylene oxide diluents such as carbon dioxide and nitrogen. A second sterilization alternative, particularly in the pharmaceutical industry, is aseptic processing; however, this is generally costly and hardly effective for device sterilization. Gas plasma sterilization is a relatively inexpensive option which is ideal for moisture and/or heat sensitive devices which require daily reuse (*3*); however, the surface properties of certain polymeric implants can be altered, potentially changing the device biocompatibility. Use of ionizing irradiation by gamma or electron beam (EB) for sterilization is preferred for stable

0097–6156/96/0620–0240$12.00/0

devices but so far it is not the method of choice for pharmaceuticals. The first demonstrated use of EB as a means of sterilization was on catgut surgical sutures in the late 50s (*4*). The obvious potential for this process and gamma sterilization has spawned a vast array of research. Gamma sources are cesium-137 or cobalt-60 while electron ionization is produced by an electron accelerator.

Ionizing irradiation may cause crosslinking or chain scission in certain polymers. Although these may cause undesirable changes in the mechanical properties, there can be distinct advantages to irradiation. Bone graft materials, for example, do not have the mechanical demands of load bearing implants but as a result of their radiation sterilization can display reduced immunogenicity. Ionizing radiations are also advantageous in that they are economically feasible for large-scale terminal sterilization of products in sealed packages. Accordingly, the International Atomic Energy Agency (IAEA) has supported and encouraged the development of ionizing radiation sterilization (*5-6*).

Guidelines for Sterilization

The binding energy of a covalent bond is below 12 eV while the irradiation energy is much higher. Therefore, irradiation of biomedical products may cause bond cleavage in the target microorganisms and even in the products themselves. While destroying the DNA in the microorganisms is desired as an outcome of bond cleavage, the extent of molecular changes in the product itself is critical, for they can lead to changes in the functionality/properties of the implanted product. Also, the amount of time that the biomedical device is expected to survive is a key factor in developing an appropriate sterilization protocol. The allowable effect that a physicochemical change due to irradiation has upon device performance may be radically different depending on the device's expected useful life.

The intensity of radiation decreases exponentially as it penetrates a product. Generally 5 to 10 MeV is used for electron irradiation, where 0.5 cm penetration depth is achieved per MeV for an object of 1 g/cm^3 (i.e. water) density. On the other hand, the intensity of gamma radiation is reduced by approximately 50% at 10 cm penetration depth for an object of 1 g/cm^3 density. The dose rate is much lower for gamma irradiation as compared to EB. The geometry of the product, product density, and product positioning during irradiation are key factors which are considered toward ensuring that the product receives a uniform dose.

The microorganism survival level deemed acceptable for medical sterilization was set in 1984 by the Association for the Advancement of Medical Instrumentation (AAMI) at 10^{-6} (or 1 colony forming unit per million) survivors (*7*). Certain viruses and bacterial spores were found to be the most radiation-resistant species while gram-negative rods were found to be the least resistant. The level of resistance will be affected, to a certain extent, by the radiation environment and its moisture and gas components. Regardless, the consensus is that an average dose of 25 kGy is acceptable with a wide enough margin of safety for most medical devices. Commonly used doses of different biomedical products are given in Table I. Stricter, customized regulations may need to be devised in order to include organisms with higher radiation resistance and to classify devices with respect to intended function.

The Association for the Advancement of Medical Instrumentation (AAMI) set guidelines in 1991 (*8*), requiring that audits be routinely performed to continually verify a given sterilization dose. This involves a quarterly testing of 100 product samples, in order to account for any seasonal variations, the results of which are applied to a dosage adjustment.

Table I. Examples of Gamma Irradiation Doses for Medical Products

Product	Dose (kGy)
Gloves	8-15
Swabs	8-15
Specimen Containers	8-15
Electrodes	8-15
Syringe	15-20
Gown	15-20
Catheter	15-20
Scrub Brush	25
Bandage	25
Vascular graft	25
Suture	25
Orthopaedic prostheses	25

Sterilization of Natural Materials

Bone cartilage and skin graft material is used in reconstructive surgery and the sterilization of these materials by gamma irradiation is under scrutiny. The irradiated behavior of collagen is of particular mechanical importance as applied to tendon replacement, because the irradiation process tends to decrease the elastic modulus and increase the range of fiber crimping or waviness. Sterilization by gamma irradiation at temperatures lower than $0^{\circ}C$ lowers the shrinkage temperature of collagen. Such an effect was recently verified upon irradiation at $-40^{\circ}C$ (*9*) and was attributed to chain scission, as compared with crosslinking which is thought to occur at temperatures above $0^{\circ}C$. Couplings of radicals formed by irradiation is thought to be negligible at subzero temperatures, thereby minimizing crosslinking and insolubility effects. On the other hand, irradiation of wet collagen caused a decrease in solubility which was attributed to the formation of intermolecular crosslinks. It is further suggested that the presence of water allows better motion of the molecular chains to allow radical coupling leading to crosslinking. Irradiation of dry collagen, in comparison, caused an increase in the solubility and a lower shrinkage temperature. This is attributed to polypeptide chain cleavage. There appears to be a predictable mechanical response based on the modulus and crimp geometry in the presence or absence of gamma irradiation (*10*).

Sterilization of bone allograft material is important because often the bacterial condition of the donor material is unknown. In most cases, radiation sterilization has a positive effect on the material such as decreased immunogenecity; however, it may also have negative effects in terms of mechanical changes and reduced osteoinductive capacity of the graft. Again, the irradiation temperature can affect the characteristics of the final product; this was emphasized in a recent study (*11*). Results of the study suggest that graft specimens preserved by deep freezing and then irradiated at a subzero temperature obtained approximately the same osteoinductive qualities as a nonirradiated material. Lyophilization preservation followed by irradiation at room temperature resulted in resorption of the graft with minimal osteogenesis.

Sterilization of Synthetic Material

Radiation sterilization of synthetic polymeric materials has been used successfully for many polymeric medical devices. Polyethylene terephthalate (PET) demonstrates stability at a standard single dose (*12*), even in air, perhaps due to the energy absorbing aromatic groups. This is satisfactory for "disposable" implants, but multiple sterilizations which are necessary for reusable biomedical items are not possible without compromising the microstructural integrity of the polymer. Multiple sterilization of PET can lead to a decrease in crystallinity and tensile strength. Such a change in microstructure may also affect the biocompatibility of the device. Furthermore, the polymer geometry, for example single fiber or bulk material, and its processing history can contribute to its response to ionizing radiation.

Polymeric materials which contact blood require even stricter irradiation regulation. Ionizing radiations cause both bulk and surface changes (*13*) which may affect the thermodynamics of the system and its adsorption of blood components (*14*). This can be due to a change in electrical characteristics of the surface and chemical composition, which may lead to changes in blood rheology. When the bulk of the device is affected, the sorption of plasma components may be increased which may lead to swelling of the device and dimensional changes.

Polymers tend to undergo crosslinking and/or chain scission when irradiated. Generally, polymers with high heats of polymerization tend to crosslink while those with low heats of polymerization tend to undergo chain cleavage (Table II). Sterilization may cause the release of gases, free radical formation, discoloration, or even the leaching of additives. The latter is of concern as many additives have a specific processing function and are proprietary; however, their chemical identity is critical in assessing the appropriate irradiation dose to avoid complications (*15*).

Table II. Typical Examples of the Effect of Irradiation on Polymers (*15*)

Polymer	Common Effect	Heat of Polymerization
Polyethylene	crosslink	22 kcal/mol
Polypropylene	crosslink	16.5 kcal/mol
Polystyrene	crosslink	17 kcal/mol
Polyisobutylene	chain scission	13 kcal/mol
Polymethylmethacrylate	chain scission	13 kcal/mol

Radiochemical Sterilization

Low level irradiation can be combined with controlled chemical sterilization in order to optimize the sterilization procedure, avoiding the problems of residual chemicals remaining in the implant, human exposure, related toxicity, environmental impact due to released gases, toxicity, lack of penetration into the implant, and polymer degradation. This unique method of sterilization (*16*) therefore uses subthreshold quantities of chemical and ionizing radiation to obviate the problems noted for traditional sterilization. The process calls for a powder chemical precursor, polyoxymethylene, which upon low dose gamma irradiation will dissociate a sterilant. The gamma irradiation coupled with the sterilant serve to sterilize the material. The gas generated is controlled by the irradiation dose, generally 5 or 10 kGy, and is dependent on the weight of the polymer sterilant precursor. Future areas of research may involve correlating the radiochemical sterilization variables with the concentration of gas released and the resulting sterility level.

Sterilization of Pharmaceutical Devices

Radiation sterilization of pharmaceutical devices is a more recent development, since earlier the industry was preoccupied by more complicated, costly chemical and toxicological issues which required a rigorous approval course. For instance, sterilization of bioabsorbable drug delivery systems has been investigated as microsphere technology progressed. In assessing the viability of radiation sterilization, the microsphere polymer and the drug are individually tested for reaction to gamma irradiation and then combined to test the interaction.

The absorbable materials often used in drug delivery systems, commonly the polyesters, are sensitive to gamma irradiation and tend to be undergo chain scission and/or crosslinking. This can affect the desired absorption profile of the material and consequently the desired release pattern of the drug. Sterilization under vacuum of poly(D,L-lactide-co-glycolide), for example, tends to minimize any molecular weight loss, apparently through oxygen and moisture reduction (*17*). However, the molecular weight of the polymer is increasingly reduced by an increase in gamma irradiation dose; higher initial molecular weight substrates show higher losses with increased dose. It is hypothesized, because the polydispersity index remains relatively constant with dose change, that the degradation appears to occur by random chain cleavage rather than by an unzipping reaction which would affect primarily the terminal groups of the chain.

Summary

Ionizing irradiation has the benefit of eliminating the negative aspects of chemical sterilization and heat sterilization; namely, diffusion barriers and temperature, respectively. Ionizing irradiation, however, may damage organic implants by molecular excitation, the extent of which depends on the physical state of the specimen as well as the temperature of irradiation. The presence of water and greater diffusivity potential and the higher molecular energy associated with a higher temperature appear to allow greater chemical reaction and therefore sensitize the product to sterilization side effects. The synthetic polymeric implants may also be detrimentally affected depending on the chemical structure as well as the dose rate of the sterilization. Ionizing irradiation is therefore a very critical processing step, requiring particular attention on a case specific basis.

Literature Cited

1. Christensen, E.A.; Kristensen, H. In *Principles and Practice of Disinfection, Preservation and Sterilization*; Russell, A.D.; Hugo, W.B.; Ayliffe, G.A.J., Eds.; Blackwell Scientific Publications: Oxford, Great Britain, 1992; pp 558-559.
2. Steiger, E.; Synek, J. In *Handbuch der Desinfektion und Sterilization*; Horn, V.H.; Privora, M.; Weuffen, W., Eds.; Band II. Grundlagen der Sterilization; Verlag Volk und Gesundheit: Berlin, Germany, 1973; pp 181-215.
3. Jacobs, P.T. In *Radiation Effects on Polymers: Chemical and Technological Aspects*; Shalaby, S.W.; Clough, R.L., Eds.; ACS Symposium Series; ACS: Washington, DC, 1995.
4. Artandi, C.; Van Winkle, W. *Nucleonics.* **1959**, *17*, pp 86-90.
5. International Atomic Energy Agency. *Radiosterilization of Medical Products. Proceedings of a Symposium*; International Atomic Energy Agency: Vienna, Austria, 1967.
6. Association for the Advancement of Medical Instrumentation. *Guideline for Gamma Radiation Sterilization*; ANSI/AAMI ST32-1991; AAMI: Arlington, VA, 1992.
7. Association for the Advancement of Medical Instrumentation. *Guidelines for Gamma Radiation Sterilization of Medical Devices*; AAMI: Arlington, VA, 1984.
8. Hansen, J.; Shaffer, H.; Bryans, T.; Reger, J.; Duda, D. *Med Device & Diagn Ind.* **1994**, *May*, pp 218-222.
9. De Deyne, P.; Haut, R.C. *Connect Tissue Res.* **1991**, *27*, pp 51-62.
10. Belkoff, S.M.; Haut, R.C. *J Orthop Res.* **1992**, *10*, pp 461-464.
11. Dziedzic-Goclawska, A.; Ostrowski, K.; Stachowicz, W.; Michalik, J.; Grzesik, W. *Clin Orthop and Relat Res.* **1991**, *272*, pp 30-37.
12. Nair, P.D.; Sreenivasan, K.; Jayabalan, M. *Biomaterials.* **1988**, *9*, pp 335-338.
13. Chapiro, A. *Radiation Chemistry of Polymeric Systems*; High Polymers; Wiley-Interscience: New York, NY, 1962; Vol. XV, pp 341-363.
14. Moacanin, J.; Lawson, D.D.; Chin, H.P.; Harrison, E.C.; Blankenhorn, D.H. Biomater Med Dev Artif Organs. **1973**, *1*, pp 183-190.
15. Bruck, S.D.; Mueller, E.P. *J Biomed Mater Res: Applied Biomaterials.* **1988**, *22(A2)*, pp 133-144.
16. Shalaby, S.W.; Linden, Jr., C.L. Radiochemical Sterilization. U.S. Patent 5,422,068. 1995.
17. Volland, C.; Wolff, M.; Kissel, T. *J Controlled Release.* **1994**, *31*, pp 293-305.

RECEIVED August 10, 1995

Chapter 19

Radiochemical Sterilization: A New Approach to Medical Device Processing

Shalaby W. Shalaby[1] and C. L. Linden, Jr.[2]

[1]Poly-Med, Inc., 6309 Highway 187, Anderson, SC 29625
[2]Department of Bioengineering, Clemson University, Clemson, SC 29634

Current sterilization techniques of medical products are reviewed briefly to show the need for a new process which can be particularly useful for radiosensitive polymeric materials and offers additional safe-use features. A novel approach to sterilization of medical products entailing the use of combinations of sub-optimal doses of high energy radiation and an *in situ* generated adjuvant gaseous sterilant is reported. Using unique combinations of low dose gamma radiation and radiolytically formed gaseous formaldehyde from a polymeric solid precursor, sterility was achieved in a model system. In typical experiments, small amounts of an unstabilized formaldehyde polymer were incorporated with spore strips in sealed containers and irradiated with about 0.5 to 1.3 Mrad of gamma rays from a cobalt-60 source. Sterility of the strips was tested following standard protocols and results were correlated with the amounts of formaldehyde-precursor and radiation dose used.

Toward illustrating the incentive to develop a novel radiochemical sterilization process, existing techniques are overviewed. However for a focused discussion of the topic of sterilization, it is to be first distinguished from disinfection. "Sterilization is the process or act of inactivating or killing all forms of life, especially microorganisms, and a sterilizer is any agent (physical or chemical) or process that achieves sterilization" (1). Disinfection on the other hand is a less lethal process that destroys all recognized pathogenic microorganisms on inanimate objects (2).

Traditional methods of sterilization include the use of steam, dry heat, chemicals, and radiation (3). The application of steam and dry heat require the

0097–6156/96/0620–0246$12.00/0

use of extremely high temperatures, which is not suitable for many medical products. The chemical method of sterilization incorporates both gases and aqueous solutions of highly reactive chemicals to sterilize or disinfect the specific product. Lastly, ionizing radiation is relatively the most recent and effective form of sterilization.

Heat is one of the oldest agents of destruction. Both moist and dry heat are classic sterilizing methods. The advantages of steam under pressure is that it is inexpensive and sterilizes penetrable materials and exposed surfaces rapidly. In comparison dry heat is relatively slow and requires higher temperatures of application. However, dry heat can impart sterility to materials, such as oils and petrolatum, in closed containers, that due to their impermeability are not suitable for steam sterilization (4).

The use of certain chemicals in sterilization can be accomplished in the gaseous phase as in the case of ethylene oxide, propylene oxide, formaldehyde, methyl bromide, and β-propiolactone. Most common among those is ethylene oxide gas, which is the predominant method of sterilization in hospitals today. Ethylene oxide is flammable and explosive, has been shown to be potentially carcinogenic, and requires long periods of time for it to degas or diffuse from porous products (5). Formaldehyde and its aqueous solution have also been used for some time, but it is disadvantaged by the: (a) instability of stock solutions; (b) toxicity when used for vented packages; (c) difficulty of generation and even distribution as a gas; (d) polymerization of the gaseous monomer; (e) possible need for heat and humidity to secure sterility; (f) explosiveness; and (g) a lack of penetration (6). Chemical sterilization is used for materials which are not suited for sterilization by physical methods, namely by using steam, dry heat, or radiation sterilization. While the chemical agents are basically surface sterilants, radiation can be considered a volume sterilant since it is capable of sterilizing throughout the entire volume with equal results (7). Sterilization through the use of ionizing radiation can be achieved in the presence of ultraviolet radiation, electron beam, microwave, X-rays, and gamma rays. In the sterilization with high-energy radiation, such as gamma rays and electron beam, it is hypothesized that these radiations cause production of excited atoms and free radicals. In organic compounds, free radicals can induce many effects which include, polymerization, cross-linking, gas production and the formation of double bonds (8). It is these reactions that can produce irrecoverable damage to microorganisms. It is also thought that gamma radiation may damage the cells ability to function due to damage of the cells deoxyribonucleic acid (DNA). In addition to the effects of free radicals, toxic molecules, such as hydrogen and organic peroxides are known to form (9). These effects are produced without any appreciable rise in temperature, and radiation sterilization is generally known as "cold sterilization" (10). A major shortfall to utilizing ionizing radiations, especially gamma rays or electron beam, for sterilization is the degradation of polymers exposed to an optimal sterilizing dose of about 2.5 Mrad (11). The use of ionizing radiation in the sterilization of pharmaceuticals and tissue for transplantation has met some success (12, 13, 14, 15). In modern hospitals where many of the items used are disposable in nature, the use of polymers in the form of plastics and films is a significant portion of the disposable materials. The effects of ionizing radiation on polymers is significant in that all polymers suffer modifications of their properties to some degree. The amount of degradation of the polymer is dependant on the type of polymer involved and some polymers are more resistant than others. The chemical changes that are induced by the radiation which underlie these physical modifications are limited to a few basic processes: (a) gas evolution, (b) crosslinking, (c) and degradation (16). With the aforementioned effects of gamma sterilization, it becomes obvious that the smaller the total dose, the lower the chance of material degradation, which is the object of the study, subject of this report.

The present study addresses the action of a novel radiation sterilization process which minimizes the limitations of both the chemical and radiation processes. Instead of depending on the bacteriostatic capabilities of a full dose of chemical or ionizing radiation to sterilize medical products, it is possible to use subthreshold quantities of a chemical agent and ionizing radiation that will act cooperatively to accomplish sterilization with minimal side affects.

Materials and Methods

Unstabilized Celcon M-90 powder (supplied by Hoechst-Celanese, Summit, New Jersey) was used as the polymeric precursor (prosterilant) of gaseous formaldehyde. Cobalt-60 was used as the gamma ray emitting source (Bausch and Lomb, Greenville, South Carolina), using a 0.326 Mrad/hr dose rate. For preparation of spore strips, a 3.4×10^{10} CFU/ml suspension of *B. pumilus* ATCC 27142 (Raven Biological Laboratories, Omaha, Nebraska) was used. Trypticase soy agar (TSA, BBL Microbiology Systems, Cokeysville, Maryland) was used in the microbiological assays. Vacutainer tubes (Terum Medical Corp., Elkton, Maryland) were used to test the efficacy of the radiochemical sterilization process on spore strips placed in these tubes.

Experimental Results

Confirmation of Bacillus Pumilus Concentration. An aliquot of 0.1 ml of the 3.4×10^{10} CFU/ml concentration of *B. pumilus* ATCC 27142 was added to 9.9 ml of sterile peptone broth for a 10^{-2} dilution. Three more 1/100 dilutions and a final 1/10 dilution were made. The dilutions of 10^{-8} and 10^{-9} were plated using the spiral plater method (Spiral Systems, Bethesda, Maryland; Speck 1984). Another 0.1 ml aliquot of the *B.pumilus* culture was added to 9.9 ml of peptone and heated at 70°C for 15 minutes to heat shock the spores. The sample was then diluted and plated in the same manner as the first on trypticase soy agar.

Normal Conditions	Heat Shocked
4.4×10^{10} CFU/ml	2.4×10^{10} CFU/ml

It can be concluded that the stated concentration of 3.4×10^{10} CFU/ml is accurate because the values obtained by this experiment were not significantly different.

The experiment also shows that the spores do not have to be heat-shocked to obtain a count of viable cells.

Preparation of Spore Strips and Vacutainer Tube Test System. Spore strips were prepared using aseptic technique under a sterile hood. An aliquot of 0.1 ml of the *B. pumilus* 3.4 x 10^{10} CFU/ml suspension was added to each strip using an Eppendorf pipette and sterile tips. The strips were placed in sterile Vacutainer® tubes using sterile forceps, and the rubber stoppers were replaced. A total of twelve tubes were prepared which contained *B. pumilus* spore strips and 0 to 500 mg of Celcon M-90. In Celcon-containing tubes, Celcon M-90 was placed at the bottom. Glass wool was placed above the Celcon M-90, but below the spore strip.

Spore Growth of Irradiated Systems: A Pilot Experiment. Gamma radiation from a cobalt-60 source was used. Tubes A-1 through A-4 were exposed to 1.35 Mrad gamma radiation. Twenty hours after irradiating the samples, the spore strips were aseptically removed from each tube and each strip was introduced to a 10 ml tube of trypticase soy broth (TSB). All of the tubes were incubated at 37°C until positive growth was observed in the control sample (A-1). After 24 hours there was no visible growth in any of the samples. However, at 72 hours, samples A-1 and A-2 were positive. The positive samples were diluted 1/100 and were plated using the spiral plater method to confirm that the turbidity was due to *B. pumilus*. The plates were then incubated at 37°C for 24 hours. All of the plates had heavy growth and were pure cultures. A summary of these data is given in Table I.

Table I
Spore Growth after Exposure to ∼1.35 Mrad Radiation: A Pilot Study

Sample Number	Contents	Growth at 72 Hours
A-1	No Celcon M-90	positive
A-2	10 mg Celcon M-90	positive
A-3	100 mg Celcon M-90	negative
A-4	500 mg Celcon M-90	negative

Effect of Celcon Mass on Sterility at 1.42 Mrad Radiation Dose. Ten Vacutainer tubes were prepared in a manner similar to the pilot experiment; all tubes contained *B. pumilus* with varying concentrations of Celcon M-90. The tubes were exposed to a total dose of 1.42 Mrad. The samples were plated using the spiral plater method and TSA plates. Samples B-1 and B-4 were positive and showed a high colony count. Samples B-3, B-8, B-9, and B-10 all contained one colony; and samples B-5, B-6, and B-7 were negative. The plates containing samples B-2, B-3, B-5, B-6, B-7, B-8, B-9, and B-10 were pour plated by adding 1 ml of the trypticase soy broth to brain heart infusion agar. The plates were incubated at 37°C overnight and all were found to be negative. Since these results contradict some of the spiral plater results, the remaining portions of these media, along with the spore strips, were pour plated in trypticase soy agar. The plates were incubated at 37°C overnight. All of the samples were negative. The results are summarized in Table II.

Table II
Spore Growth after Exposure to ~1.42 Mrad Radiation

Sample Number	Weight of Celcon M-90	Spiral Plate	Pour Plate Count
B-1	10 mg	7.8×10^9 CFU	---
B-2	20 mg	negative	negative
B-3	30 mg	1 CFU	negative
B-4	40 mg	6.0×10^9 CFU	---
B-5	50 mg	negative	negative
B-6	60 mg	negative	negative
B-7	70 mg	negative	negative
B-8	80 mg	1 CFU	negative
B-9	90 mg	1 CFU	negative
B-10	100 mg	1 CFU	negative

Effect of Celcon in the Presence or Absence of Gamma Rays. Sixteen tubes were prepared having spore strips and the amounts of Celcon M-90 were as follows:

Tube Numbers		Weight of Celcon M-90
C-1, C-8		None
C-2, C-9		50 mg
C-3, C-10		100 mg
C-4, C-11		200 mg
C-5, C-12		300 mg
C-6, C-13		400 mg
C-7, C-14,	C-15	500 mg
	C-16	10 mg

Samples C-15 and C-16 were not exposed to gamma radiation to observe any effects caused by Celcon M-90 alone. The total dose for each set was as follows:

Tube Numbers	Radiation (Approximate)
C-1 - C-7	0.50 Mrad
C-8 - C-14	0.75 Mrad
C-15, C-16	0 Mrad

The spore growth results of irradiated samples (C-1 to C-14) using 0.50 and 0.75 Mrad are summarized in Tables III and IV. Meanwhile, spore strips remained intact in samples C-15 and C-16, which contained Celcon M-90, but were not irradiated.

Table III
Spore Growth after Exposure to ~0.50 Mrad Radiation

~0.50 Mrad		Growth
Sample Number	Celcon M-90	At 72 Hours
C-1	None	1.8×10^6
C-2	50 mg	negative
C-3	100 mg	negative
C-4	200 mg	negative
C-5	300 mg	negative
C-6	400 mg	negative
C-7	500 mg	negative

Table IV
Spore Growth after Exposure to ~0.75 Mrad Radiation

~0.75 Mrad		Growth
Sample Number	Celcon M-90	At 72 Hours
C-8	None	3.6×10^6
C-9	50 mg	negative
C-10	100 mg	negative
C-11	200 mg	negative
C-12	300 mg	negative
C-13	400 mg	negative
C-14	500 mg	negative

Discussion and Conclusion

Results of the pilot experiment entailing a 72-hour incubation of treated samples illustrate not only the efficacy of the novel radiochemical sterilization process, but also its dependance on the radiation dose. Thus, using the available volume of the Vacutainer tube, efficacy can be achieved with 100 mg and 500 mg Celcon at 1.35 Mrad. Upon studying the effect of Celcon mass and hence, the amount of generated formaldehyde on the process efficacy at 1.42 Mrad using the Vacutainer, it was shown (Table II) that a Celcon mass of about 20 mg is sufficient to achieve sterility. To eliminate the possibility of any effect due to the Celcon alone in the absence of radiation and to determine the lower limit for the gamma dose at low Celcon mass, a set of samples were tested using 10 to 500 mg of Celcon, and exposed to gamma radiation doses of 0 to 0.75 Mrad. The data in Tables III and IV show clearly that Celcon weights of 50 to 500 mg with 0.5 to 0.75 Mrad doses achieve full sterility. On the other hand, samples C-15 and C-16, which were not irradiated, exhibited full spore viability; and hence, Celcon alone does not have any effect. This also may reflect the stability of Celcon under ambient storage conditions.

The data presented in this report reflect the viability of a novel, radiochemical sterilization process and project its potential application in many segments of the health care industry. The process calls for a gamma dose of about 0.5 to 1.0 Mrad and a small amount of powdered Celcon M-90 to generate, radiolytically, sufficient formaldehyde to augment the sterilization with such low radiation dose. More details and supporting data for the synergism associated with this process and its efficacy can be found in the patent literature (17).

Acknowledgement

The authors are indebted to Ms. Nicole M. Zirk and Dr. Susan F. Barefoot of the Food Science Department, Clemson University, for their valuable contributions to the microbiological testing.

Literature Cited

1.   Beloin, A. In *Methods of Testing For Sterility: Efficacy of Sterilizers, Sporicides, and Sterilizing Processes, In Disinfection, Sterilization and Preservation*; Block, S.S., Ed.: Lea & Febinger, Philadelphia, PA, **1983**, 3rd Ed., pp 885-917.

2.   Dempsey, D.J.; Thirucote, R.R. *J. Biomed. Appls*. **1989**, 3, pp 454-470.

3.   Dawson, J.O. *Different Methods of Sterilization of Medical Products, In Radiosterilization of Medical Products*:   International Atomic Energy Agency, Vienna, Austria, **1974**, pp 265-268.

4.   Joslyn, L.   *Sterilization By Heat, In Disinfection, Sterilization and Preservation*; Block, S.S., Ed.: Lea & Febinger, Philadelphia, PA, **1983**, 3rd. Ed., pp 3-46.

5. Caputo, R.A.; Odlang, T.E. In *Sterilization With Ethylene Oxide and Other Gases, In Disinfection, Sterilization, and Preservation*; Block, S.S., Ed.: Lea & Febinger, Philadelphia, PA, **1983**, 3rd Ed., pp 47-64.

6. Newsom, S.W.B.; Mathews, J. In *Use and Methods for Testing Formaldehyde Sterilization, In Disinfectants: Their Use and Evaluation of Effectiveness*; Collins, C.H., Ed.: Academic Press, London, **1981**, pp 61-68.

7. Dempsey, D.J.; Thirucote, R.R. J. Biomed. Appls. **1989**, 3, pp 454-470.

8. Plester, D. The Effects of Radiation Sterilization of Plastics: Duke University Press, Durham, NC, **1972**, Industrial Sterilization Symposium.

9. Swarbrick, J. In *Encyclopedia of Pharmaceutical Technology*, **1992**, 6, pp 305-332.

10. Silverman, G.J. In *Sterilization by Ionizing Irradiation, In Disinfection, Sterilization, and Preservation*; Block, S.S., Ed.: Lea & Febinger, Philadelphia, PA, 3rd Ed., **1983**, pp 89-105.

11. Alcock, H.R.; Lampe, F.R. *Contemporary Polymer Chemistry*: Prentice Hall, Inc., Englewood, NJ, **1981**; pp 95-118.

12. Phillips, G.O. *Medicines and Pharmaceutical Base Materials, In Manual on Radiation Sterilization of Medical and Biological Materials*: International Atomic Energy Agency, Vienna, **1973**, Report Ser. No. 149, pp 207.

13. Gaughran, E.R.L.; Goudie, A.J. *Sterilization by Ionizing Radiation*: Multiscience Publ. Ltd., Montreal, Canada, **1974**.

14. Silverman, J. Advance in Radiation Processing, In Trans. 2nd International Meeting Radiation Processing: Radiation Physical Chemistry, Vol 14, **1979**.

15. Masefield, J. In *Advances Made In Cobalt-60 Gamma Sterilization, In Sterilization of Medical Products*; Gaughran, E.R.L.; Morrissey, R.F., Eds.: Multiscience Publ. Ltd., Montreal, Canada, **1981**.

16. Chapiro, A. In *Physical and Chemical Effects of Ionizing Radiations on Polymeric Systems, In Sterilization by Ionizing Radiation*; Gaughran, E.R.L.; Goudie, A.J., Eds.: Multiscience Publ. Ltd., Montreal, Canada, **1974**.

17. Shalaby, S.W.; Linden, Jr., C.L. U.S. Pat. 5,422,068, (Radiochemical Sterilization) **1995**.

RECEIVED September 26, 1995

Chapter 20

Advances in Food Irradiation Research

Karen J. L. Burg[1] and Shalaby W. Shalaby[2]

[1]Department of Bioengineering, 301 Rhodes Engineering Research Center, Clemson University, Clemson, SC 29634–0905
[2]Center for Applied Technology, Poly-Med Inc., Pendleton, SC 29670

A brief background of the use of irradiation of food is given, limited to food for human consumption. Attention is directed to the effects of radiation on packages and organic food products as well as the resulting interactions. The main variables of interest in food irradiation include temperature, water content, packaging atmosphere, dose rate, package material, and food type. Certain food types release specific volatiles which may be used to monitor effectiveness of irradiation as well as dosage. Avenues of future research interest are highlighted.

The U.S. Centers for disease control and Food and Drug Administration (FDA) estimate that between 6.5 and 33 million cases of food-borne illnesses occur each year, 9,000 resulting in death, costing the U.S. economy millions of dollars (1). The 1963 FDA ruling that bacon, irradiated as a means of preservation, was satisfactory for human consumption (2) has propelled a relatively new area of food technology research designed to address this problem. This approach greatly extended the shelf life of the food and therefore reduced the quantity lost to spoilage. No radiation processed foods were commercially available prior to 1983, and this advance instigated food irradiation research on both a national and international scale (Table I).

The benefit to the consumer, in terms of reduced health care costs, and the agriculture industry, in terms of reduced product loss costs, is on the order of millions of dollars. Losses due to a food scare because of low consumption and sales can represent a substantial financial penalty to the retailers and wholesalers. Internationally, large quantities of food, particularly in developing countries, are lost to insect infestation and/or spoilage. Irradiation research is particularly important with the impending Clean Air Act legislation banning such suspect ozone depleting substances as fumigant methyl bromide that are currently used to combat spoilage and infestation (1).

Table I. Evolution of Food Irradiation Technology (*2,3,4*)

Year	Event
1950s	"Atoms for Peace" program established by President Eisenhower
1963	FDA declares irradiated bacon fit for human consumption
1963	FDA approves disinfestation of wheat and wheat flour by irradiation
1964	FDA approves irradiation as a sprout inhibitor
1965	Office of Surgeon General of U.S. Army declares foods irradiated with doses up to 56 kGy fit for human consumption
1976	Joint Food and Agricultural Organization (FAO), International Atomic Energy Agency (IAEA), and World Health Organization (WHO) committee concludes irradiated potatoes, wheat, chicken, papaya, and strawberries unconditionally safe for human consumption
1980	Joint FAO, IAEA, and WHO committee concludes irradiated foods safe up to 10 kGy
1992	U.S. Department of Agriculture (USDA) allows irradiation of raw, packaged poultry in approved facilities

The three levels of food irradiation preservation are: (a) low radiation dosages of less than about 2 kGy (0.2 Mrad) which delay sprouting of vegetables as well as aging of fruits, (b) pasteurization or medium radiation dosages of 2 to 5 kGy (0.2 to 0.5 Mrad) which do not kill all bacteria, and (e) sterilization, with higher dosages of 20 to 45 kGy (2 to 4.5 Mrad). The most common type of radiations used to achieve these doses are high energy gamma rays which are emitted from cobalt-60 or cesium-137 sources and which have relatively high penetration depth. Another important type of commonly used high energy radiation is the beta particle which is less penetrating than the gamma ray. The beta particles are high speed electrons which can be generated by a linear accelerator or a Van de Graaff generator and therefore have the advantage of directionality. X rays are generated indirectly using an electron beam. The accelerated electron may be directed through a thin metal film of high heat resistance to produce the X ray which has slightly less penetrating capacity than the gamma rays. Both X ray and electron beam are useful in irradiating large quantities of small foods, where penetration depth is not an issue (*5-6*). The electron beam source is advantageous from the standpoint that the regulatory issues necessary for a radioisotope are not encountered. The capital investment for gamma and electron irradiation appear to be approximately equivalent.

Generally, ionizing radiations are directed through the food where the energy of the particles disrupts the macromolecules of bacteria and microorganisms that might be present. Each growth phase of bacterial cells has a unique resistance to irradiation, where the stationary phase demonstrates the highest resistance and the logarithmic phase demonstrates the lowest (*7*).

Chemical Effects of Irradiation on Packaging

Typically, the food is irradiated post packaging; hence, radiation effects on the food and package individually are as critical as the packaging-food-radiation interactions. Irradiation of glass containers is not aesthetically pleasing and therefore not desirable since glass acquires a brown tint upon irradiation. Canning in metal containers is an acceptable practice for food irradiation, although a tin or aluminum container may partially shield the contents from irradiation at the lower doses used for foods. Plastic packaging offers the convenient alternative of a lightweight, space minimizing storage alternative, but may suffer damage that compromises its integrity as a barrier to microbiological contaminants. Pasteurization, for example, does not eliminate all bacteria so the plastic must not be conducive to bacterial growth. Certain plastics are damaged by high energy radiation and may not effectively form a microbial resistant heat seal after damage of this kind. The irradiation process may also lead to leaching of small particles from the plastic package into the food or emission of gases, thus changing the food quality (*8-12*).

Irradiation Stability of Common Packaging Materials. Polystyrene (PS) and polyethylene terephthalate (PET) are quite stable when irradiated, largely due to the aromatic groups which are able to absorb and dissipate the penetrating energy. Polypropylene (PP) tends to undergo crosslinking and chain scission. Polytetrafluoroethylene (PTFE) and cellulose based products also demonstrate very poor stability. Degradation or cleavage of polymeric chains can cause loss in mechanical strength which is critical to handling and storage. Polymers may influence the odor and even taste of the sealed food due to the emission of gaseous products. Certain polymers produce gases as a result of radiation-induced chemical reactions. Any such released volatiles might migrate into the food resulting in a change as minor as food odor (*13*) or as major as health hazard. The polyesters, polystyrenes, and polyamides are minimally affected by this phenomenon; however, PP packages are susceptible to irradiation damage.

Low density polyethylene may release aliphatic hydrocarbons, aldehydes, ketones, and carboxylic acids, a phenomena that may be reduced by incorporating antioxidants in the polymer. Hydrogen can also be produced by the packaging material. This does not have a negative impact on the food as the gas will typically diffuse through the plastic (*11*). Water content of the packaged food also plays an important role in volatile production, where the higher water content causes a higher hydrogen release. Moisture in the packaging materials can intensify the radiation effects.

In general, the plastics are more stable when irradiated under vacuum than in atmospheric conditions. Oxygen tends to increase main chain cleavage while inhibiting crosslinking in such polymers as PE, PP, and PS. The hydrocarbon level, however, is relatively unaffected by a change in oxygen concentration. This is attributed to the theory that the hydrocarbon formation during irradiation occurs due to short branch cleavage.

Carboxylic acid generation is much greater in the case of gamma irradiation than that of electron beam when applied to polyethylene, polypropylene, and cellulose. A low gamma dose rate (Gy/hr) results in much lower concentrations of primary radicals and therefore a lower amount of crosslinking and a higher concentration of carboxylic acid. In contrast, higher concentrations of primary radicals tend to favor recombination, thus reducing the likelihood of oxidation (*13*). Lower applied radiation energies (eV) with their weak penetration forces result in lower release of carboxylic acids and corresponding smaller changes in odor.

Migration of Additives. Plastics may contain low molecular weight compounds, namely oligomers and monomers, and additives such as heat and light stabilizers, antioxidants, UV absorbers, lubricants, and plasticizers. These are used for improved stability during processing to form articles with minimally compromised mechanical strength. Smaller chain molecules have much greater mobility and therefore potential to migrate into the food. The irradiation process may not only affect the polymer packaging itself, but it may also cause degradation of additives and migration of by-products into the food. Poly(vinyl chloride) (PVC), for example, is stabilized with organotin compounds such as dibutyltin bis-(isooctylthioglycollate) or dibutyltin bis-(isooctylmaleate). These stabilizers degrade largely to tin (IV) chloride upon exposure to gamma irradiation (*14-15*), and the degradation increases rapidly above 25 kGy (2.5 Mrad). Results indicate that the stabilizer degradation products show a far greater tendency to migrate than the intact stabilizer.

Packaging Atmosphere. Packaging under modified atmospheres have been in use since the 1800s; they found international use in the 1930s with the New Zealand and Australian lamb and beef exports which were packed in a carbon dioxide enriched atmosphere (*16*). Modified atmosphere packaging (MAP) has been developed more recently to extend shelf life and maintain food quality by combining mixtures of carbon dioxide, oxygen, and nitrogen (*17-18*). The effect of the packaging atmosphere on microorganism survival is dependent on radiation dose, packaging temperature, as well as gas composition. The modified atmosphere packaging appears particularly effective in lengthening shelf life of such low acid, perishable foods as beef, poultry, seafood and dairy products.

Generally the optimal irradiation package atmosphere for extending shelf life by inhibiting spoilage microorganisms appears to be a high carbon dioxide, low oxygen mixture; furthermore, there is a carbon dioxide threshold level above which an increase in carbon dioxide appears to have no additional inhibitory effect. Carbon dioxide was shown to increase the lag phase of many microorganisms while oxygen serves to inhibit growth of anaerobic pathogens. A high carbon dioxide environment, however, may be conducive to the appearance of toxins such as the microorganism toxin *botulinus*, produced by *Clostridium botulinum* found in meat. The presence of oxygen is also a major factor in the enzymatic degradation of lipids and subsequent changes in odor in chicken packaging (*18*). A high level of nitrogen in a package appears to have no additional inhibitory effect and, in fact, allows bacterial growth very similar to that in air.

Irradiation may decrease the rate of change in headspace gas composition by destroying the microorganisms and inactivating the enzymes which cause this change. The irradiation of oxygen may form ozone which will further inhibit microorganisms.

An increase in irradiation dose appears to be increasingly effective in a low or no oxygen environment; however, the irradiation resistance of individual strains of the same microorganisms varies.

Chemical Effects of Irradiation on Food

Irradiation will not significantly alter the elemental composition of the food; however, it may alter the food components as well as the food contaminants. The absorption of the ionizing radiation causes free radical formation, the amount dependent on the material itself and the radiation dose (*19*). The radiation dose required for sterilization may sometimes cause a change in sensory characteristics, an "off-flavor", in certain foods (*20-23*). The sensory characteristics may be protected and sterilization achieved by combining a lower irradiation dose with other preservation procedures (*4,23,24*).

The half-life of free radicals decreases with an increase in water content while calcified tissue such as pork or chicken bone may increase the half-life. Fatty acids found in foods rich in fat, pork for example, will undergo cleavage at the ester groups, yielding hydrocarbons when irradiated, the production of which increases with the dose as well as temperature of irradiation (*19*). Hydrocarbon production is predictable in chicken, beef, and pork and in fact behaves as a dosimeter. Frog legs (*25*) and chicken (*26*) containing lipids are found to release hydrocarbons and aldehydes upon irradiation. Hydrogen gas can be produced by irradiation of foods; carbohydrates tend to release more gas than fats and proteins. Hydrogen gas may be released from foods such as peppers and can be detected up to 4 months after treatment (*27*). The irradiation of chicken meat causes the triglyceride of palmitic acid to release 2-dodecylcyclobutanone which is detectable for up to 20 days postirradiation. It was suggested that this also can be beneficially used as a marker to determine satisfactory sterilization of chicken (*28*). Lipids undergo irradiation in the absence of a protective environment; for instance, eggs, milk powder, or flour may produce hydroperoxide, detectable up to six months posttreatment (*29*). The amount of oxidation may, however, depend more on the exposure of the food to oxygen or light as well as the storage temperature.

A lower water content may make the microorganisms more resistant to irradiation (*4*). Spores such as *B. cereus* and *Clostridium botulinum* have been shown to be more resistant to low dose irradiation at freezing temperatures than at refrigeration temperatures (*7*). This generalization is attributed to the theory that the mobility of the hydroxyl radical is decreased at sub-freezing temperatures. This temperature dependence appears to be case specific; since, other studies have yielded contradictory results (30). Irradiation of water causes the generation of a hydroxyl radical which may interact with the amino acids found in protein-containing meats. One of the possible products formed as a result, not naturally found in protein, is 2-hydroxyphenylalanine (*31*). This reaction depends on the dose, dose rate, and temperature of irradiation. The effect of irradiation on protein in the presence of water is lessened, possibly due to the water absorbing a portion of the energy (*32*). The irradiation of a dry protein cleaves primarily hydrogen in secondary and tertiary structures, which can distort the molecule and expose susceptible groups. The globular proteins tend to favor crosslinking as they are more tightly packed. The less globular proteins, such as collagen, tend to cleave and break into smaller units upon irradiation. Higher doses of irradiation destroy the primary structure of the amino acid; however, at the relatively low dose levels used in food irradiation, this is not relevant. The protein also produces fatty acids, mercaptans, and other sulfur compounds which become part of the food.

Polysaccharides of carbohydrates are found in various foods, e.g. sugars and grains, and may generate such products as oligosaccharide fragments and hydroxyl radicals (*33*). Shrimp, for example, releases N-acetylglucosamine oligosaccharides upon irradiation. Low molecular weight carbohydrates produce gases such as hydrogen, carbon dioxide, methane, and carbon monoxide upon irradiation. Nongaseous products such as acetone, lactones, formaldehyde, and acetaldehyde may also be produced. Carbohydrate containing foods that have high water content may undergo oxidative degradation to form an array of acid derivatives. The amount of acid produced may be increased with an increase in the presence of oxygen (*34*). This is due to the reaction of the low molecular weight sugars with radiolytic byproducts such as hydroxy radicals.

Dairy Products.   Irradiation of dairy products produces such compounds as acetaldehyde and dimethyl sulfide. Butter fat may release volatiles in the form of aliphatic hydrocarbons, acids, alcohols, aldehydes, ketones, and esters. Oxygen is soluble in the fat; therefore, irradiation in the presence of oxygen has practically no added value (*35*).

Meat and Poultry. Irradiated uncooked beef releases over 70% of its volatiles as hydrocarbons, originating from the lipids or lipoproteins. Some aromatic and sulfur groups are released from proteins.   There are also many more less-volatile components released than volatile ones, including longer chain hydrocarbons and aldehydes, diol esters, and diglycerides. There is virtually no change in the amino acid content of beef as a result of irradiation (*36*). The solubility of collagen in beef is increased due to the cleavage of peptide chains, yielding lower molecular weight units and thus tenderizing the meat, an attribute of food irradiation.

DNA molecular damage in meat, fish, and vegetables, specifically strands between bases and proteins cleaving or crosslinking, may occur as a result of irradiation (*37-38*). Damage may also occur to DNA bases or sugars.

Grains and Vegetables. Wheat and rice have high amounts of protein which remain essentially unchanged after irradiation. The carbohydrates in the grain release peroxides, which are related to an odor change. The higher water content causes radiation induced higher maltose values and depolymerization of starch whereas low moisture content is related to radical formation (*39*).

Future of Food Irradiation

Regulatory issues will depend on many factors including irradiation temperature, pH, packaging   atmosphere,   water   content,   microorganism,   specific   strain   of microorganism, all of which warrant closer examination for additional food applications to be approved. The effects of cooking or further treatment of irradiated food have not been described in the literature. Finally, studies have shown that public knowledge of the irradiation process greatly enhances the appeal of buying such food (*40*); needless to say, the dissemination of information by the regulatory agencies will be necessary to develop a successful market for irradiated food. It will not be surprising to find widespread use of food irradiation for hospital patients before the end of this decade.

Literature Cited

1. Loaharanu, P. *Food Technol.* **1994**, *January*, pp 104-108.
2. Urrows, G.M. *Food Preservation by Irradiation*; Understanding the Atom Series; U.S. Atomic Energy Commission: Oak Ridge, TN, 1968; pg 1.
3. Christensen, E.A.; Kristensen, H.; Miller, A. In *Principles and Practice of Disinfection, Preservation and Sterilization*; Russell, A.D.; Hugo, W.B.; Ayliffe, G.A.J., Eds.; Blackwell Scientific Publications: Oxford, Great Britain, 1992, Second Edition; pp 528-543.
4. Radomyski, T.; Murano, E.A.; Olson, D.G.; Murano, P.S. *J Food Prot.* **1994**, *57*, pp 73-86.
5. Shalaby, S.W.; Williams, B.L. In *Encyclopedia of Pharmaceutical Technology*; Swarbrick, J.; Boylan, J.C., Eds.; Marcel Dekker, Inc.: New York, NY, 1988, Vol. 6; pg 44.
6. Urbain, W.M. *Food Irradiation*; Food Science and Technology Series; Academic Press, Inc.: Orlando, FL, 1986; pp 4-5.
7. Thayer, D.W.; Boyd, G. *J Food Prot.* **1994**, *57*, pp 758-764.
8. Agarwal, S.R.; Sreenivasan, A. *J Food Technol.* **1972**, *8*, pp 27-37.
9. Allen, D.W.; Leathard, D.A.; Smith, C. *Chem Ind.* **1988**, *12*, pp 399-400.
10. Lambert, A.D.; Smith, J.P.; Dodds, K.L. *J Food Prot.* **1991**, *54*, pp 94-101.
11. Pratt, G.B.; Kneeland, L.E.; Heiligman, F.; Killoran, J.J. *J Food Sci.* **1967**, *32*, pp 200-205.
12. Milz, J. In *Food Product-Package Compatibility*; Gray, J.I.; Harte, B.R.; Miltz, J., Eds.; Proceedings; Technomic Publishing Co., Inc.: Lancaster, PA, 1987; pp 30-43.
13. Azuma, K.; Tsunoda, H.; Hirata, T.; Ishitani, T.; Tanaka, Y. *Agric Biol Chem.* **1984**, *48*, pp 2009-2015.
14. Allen, D.W.; Brooks, J.S.; Unwin, J. *Chem Ind.* **1985**, *15*, pp 524-525.
15. Allen, D.W.; Crowson, A.; Leathard, D.A.; Smith, C. In *Food Irradiation and the Chemist*; Johnston, D.E.; Stevenson, M.H., Eds.; Annual Chemical Congress Special Publication No. 86; Royal Society of Chemistry: Cambridge, Great Britain, 1990; pp 124-139.
16. Williams, A.C. Jr. In *Food Product-Package Compatibility*; Gray, J.I.; Harte, B.R.; Miltz, J., Eds.; Proceedings; Technomic Publishing Co., Inc.: Lancaster, PA, 1987; pp 170-177.
17. Labuza, T.P.; Breene, W.M. *J Food Process Preserv.* **1989**, *13*, pp 1-69.

18. Soffer, T.; Margalith, P.; Mannheim, C.H. *Int J Food Sci Technol.* **1994**, *29*, pp 161-166.
19. Rosenthal, I. In *Electromagnetic Radiations in Food Science*; Yaron, B.; Thomas, G.W.; Van Vleck, L.D., Eds.; Advanced Series in Agricultural Sciences; Springer-Verlag: New York, NY, 1992, Vol. 19; pp 55-59.
20. Urrows, G.M.; *Food Preservation by Irradiation*; Understanding the Atom Series; U.S. Atomic Energy Commission: Oak Ridge, TN, 1968; pp 15-16.
21. Silverman, G.J. In *Disinfection, Sterilization, and Preservation*; Block, S.S., Ed.; Lea & Febiger: Philadelphia, PA, 1983; pp 98-99.
22. Urbain, W.M. *Food Irradiation*; Food Science and Technology Series; Academic Press, Inc.: Orlando, FL, 1986; pp 126-127.
23. Thakur, B.R.; Singh, R.K. *Trends Food Sci Technol.* **1995**, *6*, pp 7-11.
24. Urbain, W.M. *Food Irradiation*; Food Science and Technology Series; Academic Press, Inc.: Orlando, FL, 1986; pp 257-263.
25. Morehouse, K.M.; Ku, Y.; Albrecht, H.L.; Yang, G.C. *Radiat Phys Chem.* **1991**, *38*, pp 61-68.
26. Meier, W.; Burgin, R.; Frohlich, D. *Radiat Phys Chem.* **1990**, *35*, pp 332-336.
27. Dohmaru, T.; Furuka, M.; Katayama, T.; Toratani, H.; Takeda, A. *Radiat Res.* **1989**, *120*, pp 552-555.
28. Boyd, D.R.; Crone, A.V.J.; Hamilton, J.T.G.; Hand, M.V.; Stevenson, M.H.; Stevenson, P.J. *J Agric Food Chem.* **1991**, *39*, pp 789-792.
29. Katusin-Razem, B.; Mihaljevic, B.; Razem, D. *Nature.* **1990**, *345*, pg 584.
30. Monk, J.D.; Clavero, R.S.; Beuchat, L.R.; Doyle, M.P.; Brackett, R.E. *J Food Prot.* **1994**, *57*, pp 969-974.
31. Karam, L.R.; Simic, M.G. *Anal Chem.* **1988**, *60*, pp 1117A-1119A.
32. Urbain, W.M. *Food Irradiation*; Food Science and Technology Series; Academic Press, Inc.: Orlando, FL, 1986; pg 48.
33. Den Drijver, L.; Holzapfel, C.W.; van der Linde, H.J. *J Agric Food Chem.* **1986**, *34*, pp 758-762.
34. Urbain, W.M. *Food Irradiation*; Food Science and Technology Series; Academic Press, Inc.: Orlando, FL, 1986; pp 38-39.
35. Urbain, W.M. *Food Irradiation*; Food Science and Technology Series; Academic Press, Inc.: Orlando, FL, 1986; pg 74.
36. Urbain, W.M. *Food Irradiation*; Food Science and Technology Series; Academic Press, Inc.: Orlando, FL, 1986; pg 67.
37. Grootveld, M.; Jain, R.; Claxson, A.W.D.; Naughton, D.; Blake, D.R. *Trends Food Sci Technol.* **1990**, *1*, pp 7-14.
38. Moseley, B.E.B. In *Food Irradiation and the Chemist*; Johnston, D.E.; Stevenson, M.H., Eds.; Annual Chemical Congress Special Publication No. 86; Royal Society of Chemistry: Cambridge, Great Britain, 1990; pp 97-108.
39. Urbain, W.M. *Food Irradiation*; Food Science and Technology Series; Academic Press, Inc.: Orlando, FL, 1986; pp 74-78.
40. Pohlman, A.J.; Wood, O.B.; Mason, A.C. *Food Technol.* **1994**, *48*, pp 46-49.

RECEIVED August 10, 1995

STABILITY AND STABILIZATION
OF POLYMERS TO IONIZING RADIATION

Chapter 21

Molecular-Level Response of Selected Polymeric Materials to the Low Earth Orbit Environment

Philip Young, Emilie J. Siochi[1], and Wayne S. Slemp

Langley Research Center, National Aeronautics and Space Administration, Hampton, VA 23681–0001

The NASA Long Duration Exposure Facility (LDEF) enabled the exposure of a wide variety of materials to the low Earth orbit (LEO) environment. This paper provides a summary of research conducted at the Langley Research Center into the response of selected LDEF polymers to this environment. Materials examined include graphite fiber reinforced epoxy, polysulfone, and addition polyimide matrix composites, films of FEP Teflon, Kapton, and several experimental high performance polyimides, and films of more traditional polymers such as poly(vinyl toluene) and polystyrene. Exposure duration was either 10 months or 5.8 years.

Flight and control specimens were characterized by a number of analytical techniques including ultraviolet-visible and infrared spectroscopy, thermal analysis, scanning electron and scanning tunneling microscopy, x-ray photoelectron spectroscopy, and, in some instances, selected solution property measurements. Characterized effects were found to be primarily surface phenomena. These effects included atomic oxygen-induced erosion of unprotected surfaces and ultraviolet-induced discoloration and changes in selected molecular level parameters. No gross changes in molecular structure or glass transition temperature were noted.

The National Aeronautics and Space Administration Long Duration Exposure Facility (LDEF) provided a novel opportunity for the aerospace community to examine the effects of long term low Earth orbit (LEO) exposure on a variety of materials. The 11-ton satellite depicted in Figure 1 was returned to Earth by the Space Shuttle Columbia in January 1990 after 69 months in orbit. It contained 57 experiments to assess the effects of the space environment on materials, living matter, and various space systems (1). The saga of this remarkable vehicle is continuing to unfold through a series of symposia, workshops, and journal articles (2-7). Perhaps as much as 90% of our first-hand knowledge of LEO space environmental effects rests with the LDEF and its contents (8).

The Langley Research Center actively pursued the chemical characterization of polymeric materials which flew on LDEF (9-22). The present paper summarizes

---

[1]Current address: Lockheed Engineering and Sciences Company, Hampton, VA 23666

almost 5 years of LDEF-related polymer research at this facility. It represents the collective efforts of a number of individuals and organizations in both assembling and analyzing a broad variety of control and exposed specimens.

Table I lists LDEF polymeric materials assembled for analysis. These materials were provided by several Principal Investigators and, depending on LDEF row and tray location, experienced somewhat different environments. Specimens were exposed for either 10 months or 5.8 years as noted. Materials ranged from early 1980 state-of-the-art graphite fiber reinforced polymer matrix composites, to space films and coatings, high performance polymer films, and more traditional polymers. Representative data obtained on these materials is given in this paper. A more complete data presentation can often be found in accompanying referenced reports. The intent of this activity is to increase our fundamental understanding of space environmental effects on polymeric materials and to develop benchmarks to enhance our methodology for the ground base simulation of those effects so that polymer performance in space can be more reliably predicted.

Experimental

Most materials identified in Table I were originally obtained from commercial sources. The fabrication, quality control, specimen preparation, and baseline testing of Langley-supplied P1700/C6000, 934/T300, and 5208/T300 composite materials is discussed in references 23 and 24. Polyimide-polysiloxane copolymer films were synthesized under NASA Grant NAG-1-343 with Virginia Polytechnic Institute and State University, Blacksburg, Virginia. Several high performance polyimides films were synthesized in-house (25-27).

As noted in the text, some specimens were exposed for only 10 months while other materials received the full 5.8-year exposure. Specimens exposed for 10 months were inside an Experimental Exposure Control Canister (EECC) (1). The EECC was closed when LDEF was launched. It opened 1 month after deployment and closed 10 months later. Various environmental exposure parameters are included with Figures 1 and 2.

Instrumental Methods of Analysis. Thermal analyses were conducted using a DuPont 9900 Computer/Thermal Analyzer to process data from a DuPont 943 Thermomechanical Analyzer operating in the expansion mode. The glass transition temperature (Tg) was obtained by noting the point of inflection from the thermogram baseline. Ultraviolet-Visible (UV-VIS) transmission spectra were scanned on a Perkin-Elmer Lambda 4A Spectrophotometer. Infrared spectra were recorded on a Nicolet 60SX Fourier Transform Infrared System (FTIR) using a diffuse reflectance technique (28). X-ray Photoelectron Spectroscopy (XPS) measurements were conducted under NASA Grant NAG-1-1186 at the Virginia Tech Surface Analysis Laboratory, VPI&SU, Blacksburg, VA. Measurements were made on a Perkin-Elmer PHI 5300 Spectrometer equipped with a Mg Kα x-ray source (1253.6 eV), operating at 15 kV/120mA. Scanning Tunneling Microscopy (STM) was performed in air on a NanoScope II instrument (Digital Instruments, Inc., Santa Barbara, CA) using a tungsten tip and G-Head accessory. Specimens were prepared by coating with 5-7 nm of gold-palladium using a Hummer IV sputtering system (Anatech, Ltd., Alexandria, VA). Transmission Electon Microscopy (TEM) analyses were conducted under NASA Contract NAS1-19656 at the Virginia Institute of Marine Science, Gloucester Point, VA. A Cambridge StereoScan 150 (Cambridge Instruments, Deerfield, IL) equipped with an EDAX S150 detecting unit (EDAX International, Inc., Prarie View, IL) performed Scanning Electron Microscopy (SEM) analyses. Various photographic techniques were used to document specimen appearance.

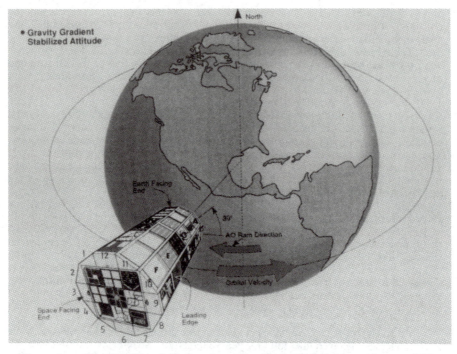

Figure 1.   The LDEF and flight orientation (Reproduced from reference 20.).

TABLE I. LDEF POLYMERIC MATERIALS

Composites:
  [a]P1700/C6000 Polysulfone
  [a]934/T300 Epoxy
  [a]5208/T300 Epoxy
  [b]PMR-15/C6000 Polyimide
  [b]LARC-160/C6000 Polyimide

[e]High Performance Polymers:
  Polyimide-Polysiloxane Copolymer
  BTDA-ODA Polyimide
  BTDA-ODA-Al^{3+} Doped polyimide
  6F-DDSO$_2$ Soluble polyimide
  6F-BDAF Soluble polyimide
  PMDA-DAF Polyimide

  [g]Polyetheretherketone (PEEK)

Films:
  [a]FEP Teflon
  [c]Silvered FEP Teflon
  [a]Kynar Fluorocarbon
  [a]P1700 Polysulfone
  [a,d]Kapton Polyimide

[f]Traditional Polymers:
  Polystyrene
  Polyvinyl toluene
  Polytetrafluoroethylene
  Polymethylmethacrylate
  Nylon
  Polyethylene terephthalate

  [g]Various Silicones
  [g]Polyurethane

Source: [a] W. Slemp, PI, Expts. A0134/S0010 (B9).
         [b] R. Vyhnal, PI, Expt. A0175 (A1 and A7).
         [c] LDEF MSIG (various LDEF locations).
         [d] J. Whiteside, PI, Expt. A0133 (H7).
         [e] W. Slemp and A. St. Clair, PI, Expt. S0010 (B9).
         [f] J. Gregory, PI, Expt. A0114 (C9/C3).
         [g] A. Whitaker, PI, Expt. A0171 (A8).

Reproduced from reference 7.

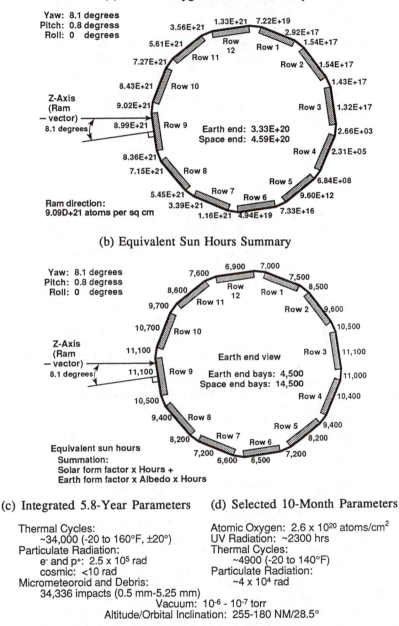

(a) Atomic Oxygen Fluence Summary

Yaw: 8.1 degrees
Pitch: 0.8 degress
Roll: 0 degrees

1.33E+21 7.22E+19
3.56E+21
2.92E+17
5.61E+21   Row 12   Row 1   1.54E+17
7.27E+21   Row 11   Row 2   1.54E+17
8.43E+21   Row 10   1.43E+17

Z-Axis
(Ram
− vector)   9.02E+21   Row 3   1.32E+17
8.1 degrees   8.99E+21   Row 9
                        Earth end: 3.33E+20   2.66E+03
                        Space end: 4.59E+20
8.36E+21   Row 4   2.31E+05
7.15E+21   Row 8
                              Row 5   6.84E+08
5.45E+21   Row 7   Row 6   9.60E+12

Ram direction:
9.09D+21 atoms per sq cm   3.39E+21
1.16E+21 4.94E+19 7.33E+16

(b) Equivalent Sun Hours Summary

Yaw: 8.1 degrees
Pitch: 0.8 degress
Roll: 0 degrees

6,900   7,000
7,600   7,500
8,600   Row 12   Row 1   8,500
9,700   Row 11   Row 2   9,600
10,700   Row 10   10,500

Z-Axis
(Ram
− vector)   11,100   Earth end view   Row 3   11,100
8.1 degrees   11,100   Row 9
                     Earth end bays: 4,500   11,000
                     Space end bays: 14,500
10,500   Row 4   10,400
9,400   Row 8
                         Row 5   9,400
8,200   Row 7   Row 6   8,200

Equivalent sun hours
Summation: 7,200
Solar form factor x Hours +   6,600   6,500   7,200
Earth form factor x Albedo x Hours

(c) Integrated 5.8-Year Parameters

Thermal Cycles:
~34,000 (-20 to 160°F, ±20°)
Particulate Radiation:
e⁻ and p⁺: 2.5 x 10⁵ rad
cosmic: <10 rad
Micrometeoroid and Debris:
34,336 impacts (0.5 mm-5.25 mm)

(d) Selected 10-Month Parameters

Atomic Oxygen: 2.6 x 10^{20} atoms/cm^2
UV Radiation: ~2300 hrs
Thermal Cycles:
~4900 (-20 to 140°F)
Particulate Radiation:
~4 x 10^4 rad

Vacuum: 10^{-6} - 10^{-7} torr
Altitude/Orbital Inclination: 255-180 NM/28.5°

Figure 2. LDEF environmental exposure parameters (reproduced from ref. 7). (a) atomic oxygen fluences at end of 5.8-year mission for all row, longeron, and end-bay locations including the fluence received during the retrieval attitude excursion (30). (b) equivalent sun hours at end of 5.8-year mission for each row, longeron, and end-bay location (31). (c) additional integrated 5.8-year exposure parameters. (d) selected 10-month exposure parameters.

Solution Property Measurements. Molecular weight measurements for polystyrene specimens were determined on a Waters 150-C Gel Permeation Chromatograph (GPC) interfaced with a Viscotek (Viscotek Corp., Porter, TX) Model 150R Differential Viscometer (DV). The general technique used to make these measurements has been previously described (11, 20, 29). Experiments were conducted in toluene at 40°C using a Waters $10^3/10^4/10^5/10^6$Å MicroStyragel HT column bank. The flow rate was 1.0 ml/min. A universal calibration curve was constructed using Polymer Laboratories (Polymer Laboratories, Inc., Amherst, MA) narrow dispersity polystyrene standards. Polyimide films were analyzed at 35°C in 0.0075M LiBr in DMAc. GPC-DV measurements on all polysulfone specimens were determined on a standard Waters Associates chromatograph at room temperature using a $10^3/10^4/10^5/10^6$Å MicroStyragel HT column bank. The solvent was chloroform. Molecular weight parameters reported are averages obtained from two or three GPC-DV analyses.

Results and Discussion

A precise orbital orientation was achieved by the LDEF spacecraft. As a result, the environmental exposure a sample received often depended on where it was located on the vehicle. Figure 1 depicts the spacecraft and its flight orientation. The 30 feet long by 14 feet in diameter gravity gradient stabilized structure had 12 sides or rows with 6 experimental trays per row. Additional trays were mounted on the Earth and space-pointing ends (1). The location of polymeric materials in this report are identified by Tray and Row. For example, B9 identifies a specimen from Tray B on Row 9.

The Row 9 leading edge nominally faced the RAM or velocity vector direction and the Row 3 trailing edge faced the WAKE direction. A detailed analysis of several factors determined that the actual RAM direction was about 8° of yaw from the perpendicular to Row 9 in the direction of Row 10, with 1° of pitch (30). This orientation was sustained throughout flight, from deployment in April 1984 until retrieval in January 1990.

Figure 2 gives additional environmental parameters. The procedure for calculating total atomic oxygen (AO) fluence and equivalent ultraviolet (UV) sun hours may be found elsewhere (31, 32). The assymetrical AO fluence around the vehicle is partly due to the 8° of yaw and a short excursion during which the vehicle received exposure while in the Shuttle payload bay after retrieval. A further discussion of orientation and environments may be found in several references (1-5, 7).

Films. The primary environments of concern for polymeric films in low Earth orbit are AO and UV. The most dominant visual effect for exposed films was AO-induced surface erosion which generally resulted in a diffuse or frosted appearance. The AO erosion of films has been described as producing a textured, carpeted, or "christmas tree" morphology (33). Figure 3 gives SEM photomicrographs of the edge of a 5 mil Kapton film specimen which flew on the space end of LDEF at H7. This specimen was oriented such that AO flow was perpendicular to the film edge and in the plane of the film surface. Figure 3a. shows the "christmas tree" morphology associated with the leading edge of this film. The trailing edge, shown in 3b, did not receive direct exposure and remained fairly smooth.

The roughened surface dramatically reduced the ultraviolet-visible transmission properties of exposed film. Figure 4 gives before and after UV-VIS spectra between 200 and 600 nm for 5 high performance polyimide films (20). Kapton was included as a reference. The molecular structure of each polyimide is identified in Table II. All films received 10 months of Row 9 exposure and, thus, experienced the environment summarized in Figure 2d. While some UV and AO degradation of the

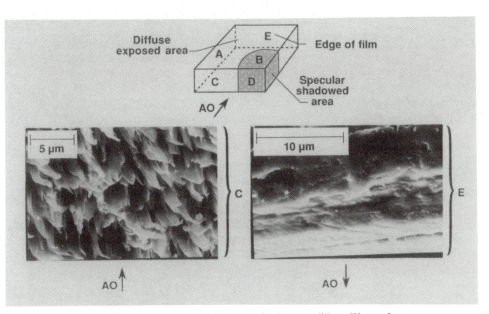

(a) leading edge   (b) trailing edge

Figure 3.   SEM photomicrographs of exposed Kapton film (Reproduced from reference 13.).

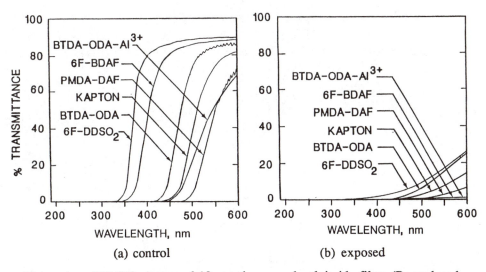

(a) control   (b) exposed

Figure 4.   UV-VIS spectra of 10-month exposed polyimide films (Reproduced from reference 20.).

TABLE II. MOLECULAR STRUCTURE OF 10-MONTH
EXPOSED POLYIMIDE FILMS

SAMPLE	STRUCTURE	COLOR
6F-BDAF		pale yellow
6F-DDSO$_2$		colorless
PMDA-DAF		reddish-yellow
BTDA-ODA		yellow
BTDA-ODA-Al^{3+}	BTDA-ODA + Al(acac)$_3$	brownish-yellow
KAPTON		yellow

Reproduced from reference 20.

polymer molecular backbone no doubt occurred, the decrease in transmittance with exposure is considered to be largely due to physical roughening of the film surface by AO. The uneven surface reflected and/or refracted the impinging radiation and, thus, less light was transmitted.

Scanning Tunneling Microscopy (STM) has proven to be an effective tool for profiling the surface of space-exposed films. Figure 5 shows STM line plots for unexposed and exposed BTDA-ODA, a high performance polymer film included in a previous study (20). The smooth surface shown in 5a. is typical of that observed for other unexposed LDEF specimens. Minor undulations in the x-direction likely resulted from the drawing procedure used during casting when the film was doctored onto a glass plate. The egg-crate-like appearance of exposed BTDA-ODA is also typical of the "christmas tree" morphology observed for several other LDEF films. The mechanism by which polymeric materials develop this conical shape upon exposure to AO is not adequately understood. VUV-induced surface crazing/crosslinking may play a role in this phenomenon.

Transmission Electron Microscopy (TEM) also proved to be an effective tool for characterizing AO erosion of polymer films. Figure 6 shows TEM photomicrographs of the five experimental polyimide films and Kapton. Prior to analysis, a segment of each film was cast into an epoxy potting resin, the resin cured, and then carefully microtomed. Irregular features emanating from the darkened films are artifacts of this potting procedure. A constant magnification was used for all specimens in Figure 6. The righthand portion of each micrograph in the figure indicates the original thickness of each film. This portion was protected from direct exposure by an aluminum retaining template which held the films in place during exposure. The diminished film thickness due to AO erosion is noted by the lefthand portion. The PMDA-DAF film was practically eroded through. The aluminum ion-containing film shows little effect at this magnification.

Figure 7 shows TEMs at two magnifications for BTDA-ODA-Al^{3+}. The 3675X micrograph given in 7a. shows the smooth back and eroded front film surfaces. The arrow to the left in the figure denotes the original uneroded surface. The textured morphology is clearly apparent in 7b. The survival of small strands of polymer in such a hostile environment is astonishing.

Composites. Unprotected composites were also affected by LEO exposure. Figure 8 shows the SEM of T300 carbon fiber reinforced 934 epoxy matrix specimens for two exposures. The 10-month and 5.8-year samples were placed adjacent to each other in the SEM to enable simultaneous analyses. The left side, low magnification photomicrograph shows both exposed surface and surface protected from direct exposure by an aluminum retaining template. The fabric-appearing pattern visible in the micrograph was transferred to the composite surface by a glass cloth peel-ply during processing. The right side photomicrograph shows a higher magnification SEM of exposed regions of two specimens. Individual carbon fibers apparent with the 10-month composite are no longer distinguishable after 5.8 years.

Table III gives XPS data for control, 10-month, and 5.8-year T300-934 epoxy composites. Surface carbon content increased in the first 10 months of exposure. This probably reflects increased carbon fiber content due to an initial preferential erosion of matrix resin. Oxygen and sulfur did not appear to change significantly. Fluorine on the control likely resulted from release agent used during processing. Fluorine was not detected on exposed composites because this outer surface was eroded away by AO. The increased silicon content with exposure is no doubt due to a well-documented LDEF contamination to be discussed later. Additional chemical characterization including FTIR, TMA, and DSC failed to detect significant differences between the two specimens.

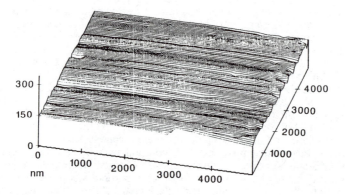

(a) BTDA-ODA unexposed

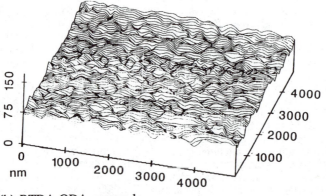

(b) BTDA-ODA exposed

Figure 5.   STM analysis of BTDA polyimide films (Reproduced from reference
            20.).

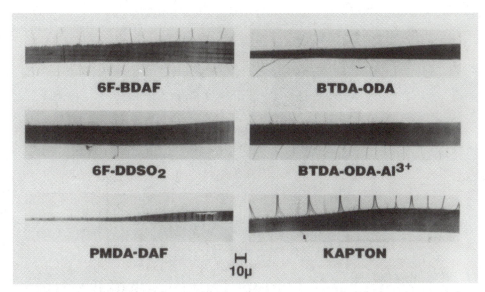

6F-BDAF

BTDA-ODA

6F-DDSO$_2$

BTDA-ODA-Al^{3+}

PMDA-DAF

KAPTON

10μ

Figure 6.   TEM analysis of exposed polyimide films (770X) (Reproduced from reference 7.).

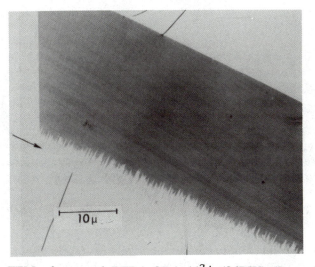

10μ

Figure 7a.   TEM of exposed BTDA-ODA-Al^{3+} (3675X) (Reproduced from reference 7.).

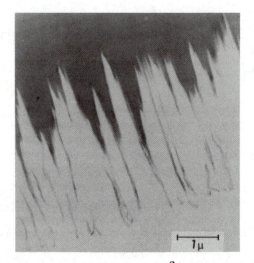

Figure 7b.   TEM of exposed BTDA-ODA-Al^{3+} (25,200X) (Reproduced from
reference 7.).

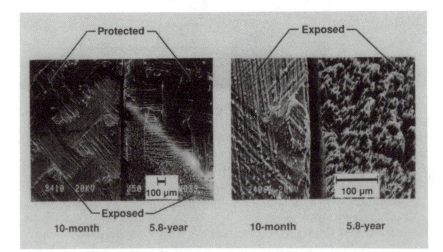

Figure 8.   SEM photomicrographs of 934/T300 epoxy/graphite
composites after 10-month and 5.8-year LDEF exposures
(Reproduced from reference 17.).

PMR-15 and LARC™-160 are two similar addition polyimide resins of considerable promise in the early 1980's when LDEF experiments were being designed. Celion 6000 graphite fiber reinforced composites of these polymers were flown on a Row 1 and Row 7 experiment (34). Specimens of these two materials were made available to Langley for chemical characterization. Among other analyses, exposed and unexposed specimens were examined by DR-FTIR spectroscopy. Spectra given in Figure 9 suggest little difference. A new absorption band at $1650cm^{-1}$ was anticipated for the exposed specimen. This band would have been indicative of oxidation of a methylene group in the polymer molecular backbone. This band along with an accompanying band at $930cm^{-1}$ had been detected in the spectrum of thermally aged LARC™-160 composites (28). The two bands are missing in the spectra shown in Figure 9. The band at $1684cm^{-1}$ is due to the anhydride portion of the polymer backbone and is supposed to be present. Similar FTIR analyses of other LDEF-exposed composites have failed to detect significant molecular level differences as a result of exposure.

One of the unsolved mysteries concerning materials on LDEF were "stripes" and/or "gray ash" associated with selected epoxy matrix composites. This phenomenon was investigated in some detail. Figure 10 gives a photograph of a striped 5.8-year exposed 934/T300 epoxy composite specimen and also SEM photomicrographs of a sample of the gray ash. Projections rising from the composite surface were apparently caused by contamination protecting underlying material from AO attack. The righthand photograph, obtained by overlaying three individual micrographs, shows graphite fiber presumedly sheared off by AO. The gray ash in question is visible at the base of the finger-like projection.

Figure 11 provides additional information on the 934/T300 flight specimen. The upper righthand portion shows a high resolution SEM of the ash. The residue appears to contain crystals on the order of $0.1\mu m$ in diameter. EDS and XPS analyses on these crystals, given in Figure 11, revealed sulfur to be a significant component. This was an unexpected result. However, sulfur is present in the diaminodiphenylsulfone (DDS) cured epoxy matrix resin. Similar-appearing residues have been noted for DDS-cured 5208 epoxy composites. The exact chemical nature of this sulfur-containing species has not been established; sodium may be a counter ion. XPS data in Table III for another epoxy flight specimen shows no unusual sulfur content. Apparently this analysis was not conducted on an ash-rich portion of the exposed composite.

Figure 12 contains information on a striped 5208/T300 epoxy composite. Optical and SEM micrographs are given at the top of the figure. The slightly recessed dark stripe is on the order of a tow wide. XPS analysis of a $1mm^2$ spot size failed to note significant differences between white and dark areas. Why one tow of the epoxy composite behaved differently from adjacent tows has not been adequately explained.

The composites addressed in this paper were uncoated materials. They were intentionally left uncoated in order to maximize the effects of long term LEO exposure. Identical materials protected with thin coatings, such as 1000Å of nickel followed by 600Å of silicon dioxide, exhibited outstanding resistance to surface erosion (9, 16, 17). Several additional inorganic coatings were found to be effective in preventing surface degradation (18).

Glass Transition Temperature. Table IV summarizes glass transition temperature (Tg) measurements conducted on a series of composites and films which flew on LDEF along with selected data obtained on films flown on STS-8 and STS-46 Space Shuttle experiments. Exposure duration, row, and tray location are identified in the table. A careful inspection of Tg values for control and flight

TABLE III. XPS ANALYSIS OF 934/T300 COMPOSITES

Photopeak	Control	10-Month Exposed	5.8-Year Exposed
C 1s B.E.[a] (eV)	285.0 ... 292.3[c]	283.6 ... 289.7	283.9 ... 288.5
A.C.[b] (%)	68.9	73.3	72.1
O 1s B.E. (eV)	531.5/532.7/533.9	531.3 ... 534.0	531.1/532.5/534.8
A.C. (%)	18.1	18.8	19.7
S 2p B.E. (eV)	168.4	168.6	170.0
A.C. (%)	1.1	0.8	0.9
N 1s B.E. (eV)	399.9	399.6	400.6
A.C. (%)	3.4	5.5	0.8
Si 2p B.E. (eV)	103.2	103.7	104.0
A.C. (%)	1.0	0.9	6.4
Na 1s B.E. (eV)	1072.2	——	
A.C. (%)	2.0	NSP[d]	
F 1s B.E. (eV)	689.3	——	——
A.C. (%)	5.5	NSP	NSP

[a] Binding Energy.   [b] Atomic Concentration.   [c] Multiple Peaks.
[d] No Significant Peak.

Reproduced from reference 17.

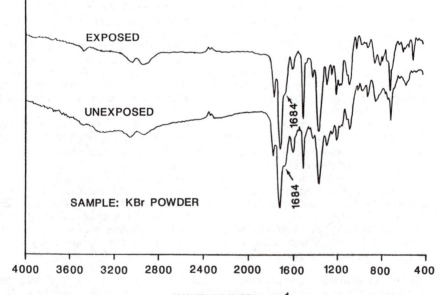

Figure 9.   DR-FTIR Spectra of PMR-15/C6000 polyimide composites
(Reproduced from reference 14.).

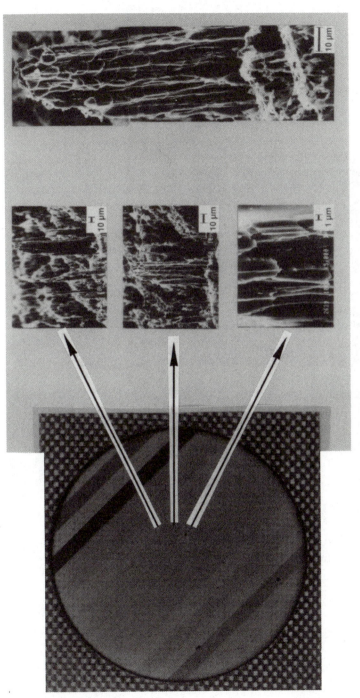

Figure 10.   Photograph and SEM photomicrographs of 5.8-year exposed 934/T300 epoxy composite (Reproduced from reference 17.).

EXPOSED COMPOSITE

Scanning Electron Microscopy

Energy Dispersive Spectroscopy

X-Ray Photoelectron Spectroscopy

Photopeak	Binding Energy, ev	Control Atomic Conc., %	Exposed Atomic Conc., %	Gray Ash Atomic Conc., %
C 1s	285.0	68.8	72.0	32.5
O 1s	532.6	18.1	19.6	39.6
N 1s	399.9	3.4	0.8	
Na 1s	1072.2	2.0		13.1 (Na^+)
S 2p	168.4	1.0	0.8	7.4 (SO_4^{2-})
F 1s	689.3	5.5		0.5 (F^-)
Si 2p	103.2	1.0	6.3	6.5
Sn $3d^5$				0.4 (SnO)

Figure 11. Characterization of 934/T300 epoxy composite (Reproduced from reference 7.).

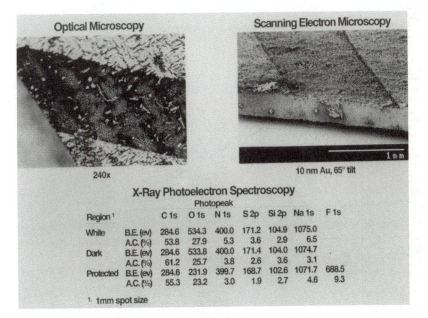

Region [1]		C 1s	O 1s	N 1s	S 2p	Si 2p	Na 1s	F 1s
White	B.E. (ev)	284.6	534.3	400.0	171.2	104.9	1075.0	
	A.C. (%)	53.8	27.9	5.3	3.6	2.9	6.5	
Dark	B.E. (ev)	284.6	533.8	400.0	171.4	104.0	1074.7	
	A.C. (%)	61.2	25.7	3.8	2.6	3.6	3.1	
Protected	B.E. (ev)	284.6	231.9	399.7	168.7	102.6	1071.7	688.5
	A.C. (%)	55.3	23.2	3.0	1.9	2.7	4.6	9.3

[1.] 1mm spot size

Figure 12. Characterization of 5208/T300 epoxy composite (Reproduced from reference 7.).

TABLE IV. GLASS TRANSITION TEMPERATURE*

Composite	Exposure Condition	Tg, °C Control	Tg, °C Exposed	Film/Resin	Exposure Condition	Tg, °C Control	Tg, °C Exposed
P1700/C6000	1	164	164	P1700	1	184	184
P1700/C6000	2	171	171	P1700	5	185	185
PMR-15/C6000	3	343	342	PET	6	85	86
LaRC-160/C6000	4	356	357	PEN-2,6	7	131	131
5208/T300	2	214	215	6F-BDAF	1	259	255
934/T300	1	202	209	6F-DDSO$_2$	1	270	273
				PMDA-DAF	1	331	—
				BTDA-ODA	1	271	269
				BTDA-ODA-Al^{3+}	1	295	290
				PEEK	8	164	166
				Polystyrene	5	99	97
				Polystyrene	6	92	93
				Polyvinyl toluene	6	85	85
				Polymethyl-methacrylate	6	119	119

Exposure Conditions:

1. 10-Month LDEF, Row 9, Tray B
2. 5.8-Year LDEF, Row 9, Tray B
3. 5.8-Year LDEF, Row 7, Tray A
4. 5.8-Year LDEF, Row 1, Tray A
5. STS-46 EOIM-III, 40 hr RAM
6. 5.8-Year LDEF, Row 3, Tray C
7. STS-8 EOIM-II, 40 hr RAM
8. 5.8-Year LDEF, Row 8, Tray A

*Thermomechanical Analysis

Reproduced from reference 7.

specimens suggests no significant change as a result of exposure. Tg effects do not appear to be an issue for polymeric materials in LEO, at least for polymers exposed for 69 months or less.

Solution Property Measurements. Selected solution property measurements have been conducted on several polymeric materials which flew on LDEF. The most extensively studied material is a thermoplastic polysulfone resin designated P1700. Both film and graphite fiber reinforced composites have been examined. Table V summarizes molecular weight data for P1700 film and composite specimens and Figure 13 gives typical molecular weight distribution curves for 10-month exposed film. All data was obtained by Gel Permeation Chromatography-Differential Viscometry (GPC-DV). Several points concerning these data are evident.

A decrease in solubility was noted in testing control film, then a template-protected specimen cut from around the yellowed edge of a flight specimen, and finally a directly exposed center-cut specimen. This decreased solubility with exposure along with a significant decrease in number average molecular weight (Mn) and increase in weight- and z-average molecular weights (Mw and Mz) is evidence for chain scission plus crosslinking. This behavior confirms predictions derived from ground-based simulation of space environmental effects on this material (35-39).

Table V also includes molecular weight data obtained on the top ply of 4-ply composites flown on Row 9 and exposed for 10 months or 5.8 years. Data for the 5.8-year sample reflects no molecular weight change when compared with the control composite while the 10-month exposed specimen suggests the same molecular level trends observed with the film sample. This potentially contradictory observation is best understood by considering the orbit of the spacecraft during its flight. LDEF was deployed in an essentially circular orbit of 257 nautical miles on April 7, 1984 (40). It was retrieved 69 months later at an altitude of 179 nautical miles. Only about 2 months of orbit lifetime remained at retrieval. The atomic oxygen fluence differs greatly at these two altitudes, with significantly higher AO flux being associated with the lower altitude.

The 10-month specimens were exposed early in the mission when AO fluence was at a minimum. The 5.8-year specimens received significant AO near the end of the mission. The molecular level effects observed after 10 months, primarily related to changes in surface chemistry, had most likely been eroded away by the time the satellite was retrieved. An earlier retrieval from a higher orbit may have provided different results.

About a 25% reduction in ultimate tensile strength and tensile modulus was noted for P1700 composite specimens as the result of 5.8 years of exposure on LDEF (16). This reduction in mechanical properties was attributed to the AO-induced loss of essentially one ply of the 4-ply specimens.

Subtractive FTIR spectroscopy gave additional insight into the molecular level response of P1700 film to LEO exposure. Since the LDEF specimen was too thick for good quality transmission studies, somewhat poorer quality spectra of control and exposed specimens were obtained by diffuse reflectance (DR). Differences between the two spectra were difficult to establish until they were subtracted. Figure 14 is the result of subtracting the DR-FTIR spectrum of the control film from that of the exposed. A downward inflection in the curve is indicative of a larger amount of a particular species in the exposed spectrum.

The band centered around 3400cm^{-1} is most likely due to -OH. Bands at 1485 and 1247cm^{-1} may also be associated with that group. Reports in the literature have noted the 3400cm^{-1} -OH band for polysulfone film exposed to UV (41) and also to 3-MeV protons (38). Additionally, the loss of the 1385cm^{-1} methyl band was noted in at least one study (41). Methyl does not appear to have been lost in the present

TABLE V. MOLECULAR WEIGHT OF LDEF-EXPOSED P1700 SPECIMENS

Film	[1]Solubility	[2]M_n	M_w	M_z	M_w/M_n	[3]IV
Control	100	18,100	53,600	92,600	2.97	0.48
10 month, edge	96	12,700	73,500	183,000	5.77	0.47
10 month, center	87	12,500	90,900	326,000	7.27	0.49

Composite	% Resin					
Control, top ply	30.3	15,800	57,400	100,600	3.62	0.43
10 month, top ply	27.5	14,300	61,200	115,600	4.28	0.39
5.8 year, top ply	32.9	15,400	57,300	99,800	3.71	0.45

[1] Solubility in Chloroform, %

[2] All molecular weight in grams/mole

[3] Intrinsic viscosity, dL/g

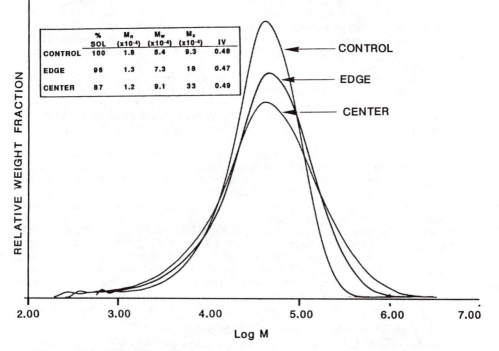

Figure 13.   GPC-DV molecular weight distribution of 10-month exposed polysulfone film (Reproduced from reference 12.).

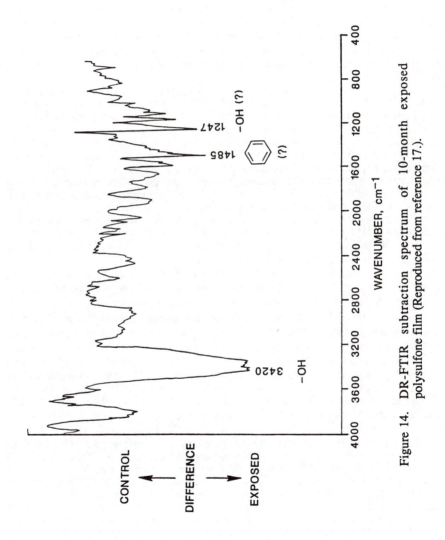

Figure 14. DR-FTIR subtraction spectrum of 10-month exposed polysulfone film (Reproduced from reference 17.).

study. A diminished -CH$_3$ content would have resulted in an upward inflection in the subtraction spectrum at 1385cm^{-1}; no band is present around 1385cm^{-1} in Figure 14. The presence of -OH has been explained by cleavage of the ether oxygen in the backbone of the polymer followed by abstraction of a proton (37), or by a photo Claisen rearrangement of the ether oxygen to produce an ortho-hydroxy substituted biphenyl linkage (36). The origin of the -OH group was not determined in this study.

Solution property measurements were also conducted on polystyrene specimens exposed for 5.8 years on LDEF Row 9 and Row 3. Table VI gives various molecular weight parameters for these two specimens. Only a slight reduction in M$_n$ was observed with exposure while M$_w$ and M$_z$ increased dramatically. This behavior is indicative of crosslinking and is the predicted response to UV for polystyrene.

Additional solution property measurements were conducted on two polyimide films exposed for 10 months on Row 9. The molecular structure of the two polyimides, 6F-BDAF and 6F-DDSO$_2$, is included with structures given in Table II. The two 6F-anhydride polymers were the only potentially soluble polyimides flown on LDEF. Table VII summarizes molecular weight data determined by GPC-DV. Analyses were conducted on a control film, the edge of a flight specimen shielded from direct exposure, and a 10-month exposed specimen cut from the center of the film. All samples contained residual insoluble material. The insoluble gel was recovered by filtration and dried to constant weight. Analyzed concentrations were then adjusted to account for the insoluble portion.

An inspection of data for 6F-BDAF suggests that various molecular weight parameters were not affected by the 10-month LEO exposure provided by LDEF. In contrast, 6F-DDSO$_2$ exhibited significant changes at the molecular level. Solubility decreased from 94.5% to 60.9% with exposure. The number average molecular weight (M$_n$) remained fairly constant while the z-average molecular weight (M$_z$) doubled. Changes in the polydispersity ratio (M$_w$/M$_n$) are also noted for this material. These observations, particularly solubility and M$_z$ behavior, are considered evidence that the 6F-DDSO$_2$ structure crosslinked during exposure. The z-average molecular weight (M$_z$) is probably a neglected parameter for evaluating crosslinking in environmentally exposed polymers (42). Figure 15 gives GPC-DV molecular weight distributions for 6F-DDSO$_2$. They show the broadening with exposure and shift to higher molecular weight documented in Table VII. The molecular weight distributions for control and exposed 6F-BDAF were virtually superimposable. The change in molecular weight for 6F-DDSO$_2$ is considered primarily a result of VUV damage to the -SO$_2$- group in the polymer backbone.

Additional Studies. *Polyimide-polysiloxane copolymers.* The chemical characterization of a series of polyimide-polysiloxane copolymers which received 10 months of Row 9 exposure also provided molecular level insight into LEO space environmental effects. In general, the films discolored somewhat with exposure but did not exhibit significant weight loss. Table VIII gives detailed XPS results for three different siloxane-containing copolymers. Data for both control and exposed film is included in the table. Several points are made concerning these data. Note the decrease in surface carbon content with exposure and concurrent increase in surface oxygen content. Note also the increase in silicon and that the silicon to oxygen ratio is about 1:2 after exposure. Finally, note the increase in the silicon 2p electron binding energy with exposure from approximately 102 electron volts to 103 eV. This data is consistent with the following interpretation. Upon AO exposure, the surface of these siloxane-containing copolymers eroded to expose silicon atoms. These atoms were initially present as an organically-bound silicon, as evidenced by the 102

TABLE VI. MOLECULAR WEIGHT OF LDEF-EXPOSED POLYSTYRENE

Experiment	Origin	Sample	[1]Solubility	[2]M_n	M_w	M_z	M_w/M_n	[3]IV
LDEF, Row 3	[4]UAH, 30 mil	Control	95	79,900	289,900	754,200	3.63	0.68
		Exposed	95	65,700	410,500	1,422,000	6.25	0.77
LDEF, Row 9	[4]UAH, 30 mil	Control	100	69,300	310,000	1,099,900	4.47	0.74
		Exposed	90.6	57,600	474,500	1,903,000	8.24	0.85

[1] Solubility in Toluene, %
[2] All molecular weight in grams/mole
[3] Intrinsic viscosity in dL/g
[4] University of Alabama in Huntsville, John C. Gregory, PI.

TABLE VII. MOLECULAR WEIGHT OF LDEF-EXPOSED POLYIMIDE FILMS

SAMPLE	[1]SOLUBILITY	[2]M_n	M_w	M_z	M_w/M_n	[3]I.V.
6F-BDAF						
Control	98.8	85,600	218,000	710,000	2.54	0.829
		87,500	218,000	659,000	2.49	0.816
[4]Edge	98.7	76,500	212,000	767,000	2.77	0.846
		84,000	225,000	800,000	2.68	0.805
[5]Center	96.8	80,500	219,000	651,000	2.72	0.824
6F-DDSO$_2$						
Control	94.5	66,900	181,000	438,000	2.70	0.715
		46,500	176,000	407,000	3.79	0.732
[4]Edge	86.8	80,700	302,000	904,000	3.74	1.048
		76,200	309,000	966,000	4.06	1.042
[5]Center	60.9	41,800	306,000	1,110,000	7.32	0.619
		52,200	274,000	802,000	5.25	0.598

1 Solubility in DMAc, %
2 All molecular weight in grams/mole
3 Intrinsic Viscosity, dL/g
4 Shielded from direct exposure
5 10-month direct exposure

Reproduced from reference 20.

TABLE VIII. XPS ANALYSIS OF 10-MONTH EXPOSED POLYIMIDE-POLYSILOXANE COPOLYMERS

PHOTOPEAK	PIPSX-6		PIPSX-9		PIPSX-11	
	CONTROL	EXPOSED	CONTROL	EXPOSED	CONTROL	EXPOSED
C 1s B.E.[a] (eV)	285.0/287.6/288.7	285.0/286.1	284.7	284.6	285.0/286.5/288.6	283.7 ... 287.8[c]
A.C.[b] (%)	57.5	15.5	54.4	16.8	69.2	19.0
O 1s B.E. (eV)	532.6	532.9/533.8	532.5	533.0	532.2/533.7	532.7/533.8
A.C. (%)	23.4	53.8	23.7	52.4	19.2	53.0
N 1s B.E. (eV)	400.1	—	—	—	400.4	—
A.C. (%)	1.6	NSP[d]	NSP	NSP	2.5	NSP
Si 2p B.E. (eV)	102.6	103.8	102.2	103.4	102.2	103.6
A.C. (%)	17.6	27.1	21.6	30.8	9.2	28.0
F 1s B.E. (eV)	—					
A.C. (%)	NSP				NSP	NSP
Na 1s B.E. (eV)		1073.1				
A.C. (%)		2.1				
Cl 2p B.E. (eV)		200.1				
A.C. (%)		1.3				

[a] Binding Energy   [b] Atomic Concentration   [c] Multiple Peaks   [d] No Significant Peak

Reproduced from reference 7.

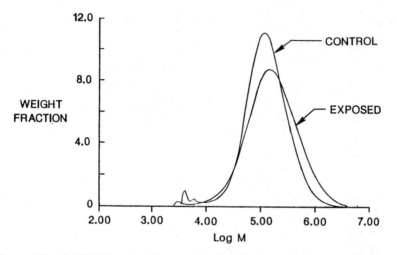

Figure 15.   GPC-DV molecular weight distribution of 10-month exposed
              6F-DDSO$_2$ film (Reproduced from reference 20.).

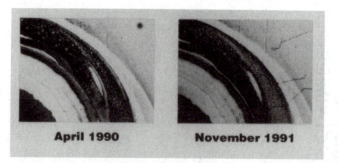

Figure 16.   Photographs of micrometeoroid impact on Ag/FEP teflon
              thermal blanket (100X) (Reproduced from reference 14.).

eV binding energy. Upon exposure to AO, organic silicon (102 eV B.E.) was oxidized to inorganic silicon (103 eV B.E.), most likely an SiO_2 type of silicate. At this point, further AO erosion was retarded. Inorganic silicates are known to be effective barriers to AO erosion. These observations suggest an exciting potential for designing AO protection into the backbone of certain polymers. This protection could likely be achieved by periodic incorporation of siloxane groups into the molecular structure of the host polymer.

Contamination. As documented in numerous LDEF publications, much of the vehicle surface was coated with varying amounts of a molecular contamination film referred to as the "nicotine stain." While several sources undoubtedly contributed to this stain, a commonly reported component was organic and/or inorganic silicon. The contamination layer likely had an effect on the behavior of polymeric materials on LDEF. UV probably contributed to the discoloration of contaminated specimens. Upon exposure to AO, the silicon-containing contamination oxidized from silicone to silicate. The resultant AO-resistant protective layer likely affected the manner in which the material eroded. Contaminated samples probably performed differently than had they not been contaminated.

Post Exposure Effects. The possibility that some LDEF polymeric materials changed or degraded after the spacecraft was returned to Earth should be addressed. Two thin films flown on STS-8 in 1983 changed significantly in appearance since they were initially examined (13, 14). One of the films, an experimental polyimide designated PMDA-DAF, was reflown on LDEF. By visual inspection, the exposed area of this film is noticeably more opaque now than when the experiment was deintegrated from LDEF in Spring 1990.

Selected Ag/FEP Teflon thermal blanket specimens have also changed in appearance with time. Figure 16 shows photographs of a micrometeoroid impact on a thermal blanket flown on Row 11. Both photographs were obtained on an optical microscope at the same magnification. The photograph on the left was taken in April 1990, shortly after the analysis of LDEF materials began at the Langley Research Center. The same specimen area was located and photographed again in November 1991. A careful inspection of the two photographs in Figure 16 will reveal that cracks on the vapor deposited silver side of the thermal blanket material had continued to form and intensify. The aged specimen had also become duller in appearance.

Thus, some polymeric materials no doubt continued to age after environmental exposure. Quantified information is needed to define the chemical and/or mechanical mechanisms associated with this phenomenon. The prudent analyst must be aware that certain characterized effects may have become exaggerated during the interval between when the specimen was returned to Earth and when it was analyzed. In retrospect, Electron Spin Resonance (ESR) should have been used to investigate the formation and lifetime of free radicals which likely contributed to various observed post exposure effects.

A Perspective. A significant amount of fundamental information has been generated on several polymers which received 10 months or 5.8 years of LEO exposure. For example, Table V gives molecular weight data on P1700 polysulfone film and composite specimens. Table VI gives similar data for polystyrene specimens flown on Row 3 and Row 9 of LDEF. Additional unreported molecular weight measurements were conducted on 5.8-year exposed Row 3 poly(vinyl toluene). Polystyrene and poly(vinyl toluene) are two well-characterized polymers. Finally, Table VII contains molecular weight data for two 10-month exposed experimental polyimide films.

All polysulfone, polystyrene, and polyimide specimens were extensively characterized by UV-VIS and FTIR spectroscopy, thermal analysis, SEM, STM, and in most cases, XPS. Thus, a thorough understanding has been obtained of the molecular level response of these polymers to extended LEO exposure. Additional polysulfone and polystyrene data has been obtained on samples flown on recent Space Shuttle experiments (43).

Can this data be used as a benchmark to calibrate the ground-based simulation of LEO space environmental effects? If effects on materials described in this report can be simulated, then the same conditions can be used to simulate the effects of LEO exposure on new and emerging polymeric materials of current interest for space application. Synergistic and accelerated effects may also then be better understood. Such understanding will significantly enhance our ability to predict the long-term performance of polymeric materials in low Earth orbit.

Concluding Remarks

Current studies indicate LDEF to be the definitive source for long-term exposure verification of LEO environmental effects on polymeric materials. A wide variety of materials has been characterized. Exposure to atomic oxygen led to changes in the physical appearance of most flight specimens and reductions in selected mechanical properties. Other environmental effects are considered to be primarily surface phenomena. Changes in various molecular weight parameters, attributed to ultraviolet degradation, were documented for several soluble polymers. Many environmental effects for specimens located on or near Row 9 were lost to increased AO erosion near the end of the LDEF mission. A ubiquitous silicon-containing contamination likely affected the behavior of many polymeric materials. Finally, the possibility that selected LDEF polymers have changed since they returned to Earth in January 1990, was addressed.

The Long Duration Exposure Facility provided a once-in-a-career opportunity for the first-hand examination of effects of long-term space exposure on a variety of polymeric materials. As a result, research on space environmental effects has taken a forward leap past pre-LDEF levels of analytical procedures, data aquisition, modeling, and understanding of degradation mechanisms in low Earth orbit.

Acknowledgment

A significant portion of this research was made possible by the generous contribution of LDEF specimens by a number of individuals. The authors express appreciation to John C. Gregory, The University of Alabama in Huntsville, James B. Whiteside, Grumman Aerospace Corporation, Richard F. Vyhnal, Rockwell International Corporation, Michael G. Grote, McDonnell Douglas Astronautics Company, Ann F. Whitaker and James M. Zwiener, NASA Marshall Space Flight Center, H. Gary Pippin, Boeing Defense and Space Group, Anne K. St. Clair and Bland A. Stein, NASA Langley Research Center.

REFERENCES

1. Clark, L. G.; Kinard, W. H.; Carter, D. J. and Jones, J. L., Eds.: The Long Duration Exposure Facility (LDEF). NASA SP-473, 1984.
2. Levine, A. S., Ed.: LDEF-69 Months in Space. First Post Retrieval Symposium, Kissimmee, FL, June 2-8, 1991. NASA CP-3134, Parts 1, 2, and 3.
3. Stein, B. A. and Young, P. R., Comp.: LDEF Materials Workshop '91. Langley Research Center, Hampton, VA, November 18-21, 1991. NASA CP-3162, Parts 1 and 2.

4. Levine, A. S., Ed.: LDEF-69 Months in Space. Second Post Retrieval Symposium,San Diego, CA, June 1-5, 1992. NASA CP-3194, Parts 1, 2, 3, and 4.
5. Gregory, J. C., Ed.: LDEF Materials Results for Spacecraft Applications Conference, Huntsville, AL, October 27-28, 1992. NASA CP-3257.
6. Murr, L. E. and Kinard, W. H.: *Amer. Sci.*, **1993**, 81, 152.
7. Levine, A. S., Ed.: LDEF-69 Months in Space. Third Post Retrieval Symposium, Williamsburg, VA, November 8-12, 1993, NASA CP-3275, Parts 1, 2, and 3.
8. Stevenson T.: *Aerospace Comp. and Matls.*, **1990**, 2(2), 12.
9. Young, P. R., Slemp, W. S., Witte, W. G. and Shen J. Y.: *SAMPE Int.Symp.*, **1991**, 36(1), 403.
10. Young, P. R. and Slemp, W. S.: NASA TM 104096, 1991.
11. Young, P. R., Slemp, W. S. and Gautreaux, C. R.: *SAMPE Int.Symp.*, **1992**, 37, 159.
12. Young, P. R., Slemp, W. S., Siochi, E. J. and Davis, J. R. J.: *SAMPE Intl. Tech. Conf.*, **1992**, 24, T174.
13. Young, P. R. and Slemp, W. S.: NASA CP-3134, **1991**, Part 2, 687.
14. Young, P. R. and Slemp, W. S.: NASA CP-3162, **1992**, Part 1, 357 (1992).
15. Young, P. R. and Slemp, W. S.: Radiation Effects on Polymeric Materials, Reichmanis, E. Frank, C. W., O'Donnell, J. H., Eds., ACS, Washington, DC, **1993**, Book Series 527, 278.
16. Slemp, W. S., Young, P. R., Witte, W. G. and Shen, J. Y.: NASA CP-3134, **1991**, Part 2, 1149.
17. Young, P. R., Slemp, W. S. and Chang, A. C.: NASA CP-3194, **1993**, Part 3, 827.
18. Slemp, W. S. and Young, P. R. : LDEF Thermal Control Coatings Post-Flight Analysis. NASA CP-3194, **1993**, Part 3, 1093.
19. Young, P. R., Slemp, W. S. and Stein, B. A.: NASA CP-3257, **1992**, 125.
20. Young, P. R., St. Clair, A. K. and Slemp, W. S.: *SAMPE Int.Symp.*, **1993**, 38(1), 664.
21. Kalil, C. R., and Young, P. R.: *39th Intl. Instrument Symp.*, **1993**, ISA 39, 445.
22. Young, P. R. and Slemp, W. S.; *Polymer Preprints*, **1990**, 31(2), 353.
23. Witte, Jr., W. G.: NASA TM 87624, 1985.
24. Witte, Jr., W. G.: NASA TM 89069, 1987.
25. St. Clair, A. K., St. Clair, T. L. and Shevet, K. I.: *Proceedings of the ACS Division of Polymeric Materials: Science and Engineering*, **1984**, 51, 62.
26. Aluminum Ion-Containing Polyimide Adhesives. U.S. Patent 4,234,461 to NASA, Aug. 13, 1981.
27. Bell, V. L.: *J. Polym. Sci.: Polym. Chem. Ed.*, **1976**, 14, 225.
28. Young, P. R., Stein, B. A. and Chang, A. C.: *Natl. SAMPE Symp.*, **1983**, 28, 824.
29. Young, P. R., Davis, J. R. J. and Chang, A. C.: *SAMPE Int.Symp.*, **1989**, 34(2), 1450.
30. Peters, P. N., Whitehouse, P. L. and Gregory, J. C.: NASA CP-3194, **1993**, Part 1, 3.
31. Bourassa, R. J. and Gillis, J. R.; NASA CR 189627, Contracts NASI-18224 and NASI-19247, May 1992
32. Bourassa, R. J. and Gillis, J. R.: NASA CR 189554, Contract NASI-18224, February 1992.
33. Koontz, S. L.: NASA CP-3035, **1989**, Part 1, 241.
34. Vyhnal, R. F. : NASA CP-3194, **1993**, Part 3, 941.
35. Brown, J. R. and O'Donnell, J. H.: *J. Appl. Polym. Sci.*, **1975**, 19, 405.

36. Brown, J. R. and O'Donnell, J. H.: *J. Appl. Polym. Sci.*, **1979**, 23, 2763.
37. Kuroda, S., Nagura A., Horie, K., and Mita, I.: *Eur. Polym. J.*, **1989**, 25(6), 621.
38. Coulter, D. R., Smith, M. V., Tsay, F. and Gupta, A.: *J. Appl. Polym. Sci.*, **1985**, 30, 1753.
39. Santos, B. and Sykes, G. F.: *SAMPE Tech. Conf.*, **1981**, 13, 256.
40. O'Neal, R. L. and Lightner, E. B.: NASA CP-3134, **1991**, Part 1, 3.
41. Gesner, B. D. and Kelleher, P. G.: *J. Appl. Polym. Sci.*, **1968**, 12, 1199.
42. Hill, D. J. T., Lewis, D. A., O'Donnell, J. H., Pomery, P. J., Winzor, C. L., Winzor,D. J. and Whitaker, A. K.: Pacific Polym. Preprints, **1989**, 1, 67.
43. Young, P. R., Slemp, W. S., and Siochi, E. J.: *SAMPE Intl. Symp.*, **1994**, 39, 2243.

RECEIVED August 10, 1995

Chapter 22

Effects of Gamma-Ray Irradiation on Thermal and Tensile Properties of Ultrahigh-Molecular-Weight Polyethylene Systems

M. Deng, R. A. Johnson, R. A. Latour, Jr., and Shalaby W. Shalaby

Department of Bioengineering, 301 Rhodes Research Center, Clemson University, Clemson, SC 29634–0905

UHMW-PE has long been used in articulating components for total joint reconstruction. However, the long-term performance (wear and creep) of the polymer has been of some concern. Attempts to improve the mechanical properties, and thus the survival rate, of the prostheses fabricated from this material have entailed crosslinking in the presence of gamma radiation. However, these attempts were associated with limited success. This study addresses the effects of gamma radiation on the thermal and tensile properties of compression-molded UHMW-PE (GUR405) at a dose ranging from 1 to 5 Mrad in air, nitrogen, acetylene and vacuum. Available DSC and tensile data indicate that both the radiation dose and type of environment affect the thermal and tensile properties of UHMW-PE in several modes. Crosslinking appears to dominate in the acetylene environment. Changes in tensile properties, melting and oxidation temperatures, as well as crystallinity are presented. Preliminary data on the effect of radiation on UHMW-PE fiber-reinforced composites are reported.

Since radiation crosslinking improves polyethylene film (*1*), its effects on medical grade ultrahigh molecular weight polyethylene (UHMW-PE) have been investigated in attempts to improve the mechanical properties and thus the survival rate of its prostheses (*2,3*). At present, many UHMW-PE joint components are sterilized by gamma irradiation. The main advantages of this process are the relatively high effectiveness and safety factor involved. However, radiation sterilization can also cause certain undesirable changes in polymer properties at the traditional sterilization dose of 2.5 Mrad (*4*). Earlier studies have addressed, mainly, the effect of radiation dose. The present study was conducted to investigate the effects of gamma radiation on the thermal and tensile properties of UHMW-PE using 1 to 5 Mrad radiation in different gas environments. A key goal of the study was to determine a means of reducing radiation sterilization damage and achieve long-term stability of UHMW-PE medical devices.

Materials and Methods

A commercial UHMW-PE, GUR405 (same as GUR415, but without calcium stearate), was provided by Hoechst Celanese Co., USA. The polymer had a molecular weight of

0097–6156/96/0620–0293$12.00/0

about 6 million and is available as a fine powder. In the present study, virgin powder GUR405 was melted and compression-molded into sheets using a rectangular metal frame placed between two stainless steel plates at a temperature of 180°C and a pressure of 7 MPa. A Carver Laboratory Press (Model C) was used for molding under an ambient laboratory environment. The polyethylene fibers were used to reinforce UHMW-PE and the effect of gamma irradiation on the tensile properties of resultant composites was studied. The polyethylene fiber used in this study was ultrahigh strength and modulus gel-spun UHMW-PE, Spectra 1000, from Allied Signal Inc. The fibers were procured as 650 deniers yarn consisting of 120 filaments. The polyethylene matrix was UHMW-PE, GUR405, from Hoechst Celanese Co. In this study, unidirectional composite laminates were made.

The polymer sheets and composite laminates were converted into standard dumbbell-shaped tensile test specimens using a metal cutting die. The specimen measures 70 X 15 X 1.3 mm (length X width X thickness) with a narrow width of 5 mm, a modification of ASTM standards. Figure 1 shows the specimen geometry. Gamma irradiation was conducted at 1.25, 2.5 and 5.0 Mrad, using a ^{60}Co source, in three different gas environments, namely, air, nitrogen and acetylene at room temperature. The experiment was also run under vacuum. The nitrogen and acetylene environments were established by repeatedly evacuating and purging with the desired gas through a two-way ground joint vessel containing the samples. The duration for each step was about 15 minutes and the process was repeated four times prior to sealing. To achieve vacuum, the vessel was evacuated below 1 mm Hg before sealing. Prior to irradiating, the acetylene gas pressure inside the vessel was established at 2.5 psi above atmospheric pressure. The nitrogen pressure was atmospheric.

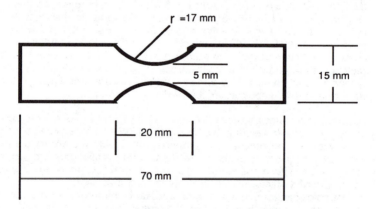

Figure 1. Tensile specimen geometry

Differential scanning calorimetry (DSC) was used to study the thermal properties of UHMW-PE. A TA Instrument 2000 thermal analyzer was used for this purpose. Samples weighing approximately 4 mg were placed in a sealed, aluminum pan and heated at 10°C/min from room temperature to 300°C, in air. The weight change of the polymer after gamma-irradiation was obtained by weighing samples before and after treatment using a balance with an accuracy of 0.01 mg. The thermal properties and weight change were measured only for samples irradiated at 2.5 Mrad. An average of 3 samples was used for DSC analysis and at least 8 were considered for weight measurements. Considering the long-term effects of gamma radiation, the properties were examined just after irradiation and four months thereafter. During this study, the samples were placed in glass bottles and shielded from light. The glass bottles were occasionally opened to atmosphere when the measurement was necessary. The tensile properties of irradiated specimens just after gamma irradiation were evaluated using an Instron universal mechanical tester (Model 1125), at room temperature. A loading rate of 20 mm/min and a gauge length of 20 mm were used. At least 4 samples were used for each case. Additionally, selected samples of new fiber-reinforced UHMW-PE composites were studied under the conditions similar to those described above.

Results and Discussion

Thermal properties. DSC thermograms were used to study two events, melting and thermal oxidation, in terms of peak melting temperature (T_m) and peak oxidation temperature (T_o), respectively, as illustrated graphically in Figure 2. The thermograms were also used to determine the heat of fusion (ΔH). Using the ΔH data, the percent crystallinity of the polyethylene was estimated by dividing ΔH by the heat of fusion of fully crystalline polyethylene, reported as 289.3 J/g (*5*).

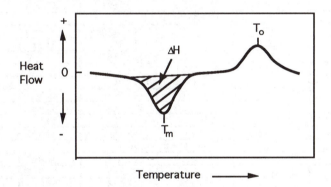

Figure 2. Graphical illustration of a DSC plot

The DSC results are summarized in Table 1, which are also illustrated in Figures 3-5. It is clear that the thermal properties of compression-molded UHMW-PE seldom change during normal storage conditions and their properties are consistent, considering that the 600-day old sample was compression-molded at least one year earlier than the rest of the samples. Following gamma irradiation, however, the long-term properties varied. First, gas environments affected the thermal properties of the polymer. The data in Table 1 show that gamma irradiation increased the melting temperature of UHMW-PE in all four conditions. The acetylene environment resulted in the highest increase, while the other three conditions led to limited increases. This may indicate that radiation-induced molecular changes resulted in recrystallization and can be associated with chain scission in the amorphous region, which is followed by recrystallization. Such events can also be affected by crosslinking. Hence, the change in crystallinity will be determined by relative amounts of crosslinking and chain scission in the prevailing environments. The results of the shift in T_o support this assumption. T_o, in all but acetylene environments, showed decrease after gamma irradiation. Following gamma irradiation, the crystallinity of UHMW-PE increased. The longer the post-irradiation, the higher the crystallinity. This suggests that irradiated polyethylene continues to change with time, possibly due to trapped free radicals. It seems that gamma irradiation for the sterilization of UHMW-PE products in an acetylene environment is desirable, considering the increase both in T_m and T_0.

Table 1. Thermal Data of UHMW-PE after Gamma Irradiation at 2.5 Mrad
(mean ± 95% confidence interval; 3 samples per group)

Time (days)	T_m (°C)	T_0 (°C)	ΔH (J/g)	Crystallinity (%)
Control (not irradiated)				
10	134.7±0.2	228.5±0.4	129.1±1.8	44.6±0.6
130	134.2±0.2	224.4±0.5	128.5±2.3	44.4±0.8
250	134.3±0.5	226.7±3.0	122.4±3.3	42.3±1.2
360	132.9±1.1	224.4±0.6	136.2±24	47.1±8.4
610	134.2±0.4	227.6±1.1	132.0±3.7	45.6±1.3
Irradiated in acetylene				
10	139.8±0.6	228.8±3.0	131.8±3.0	45.6±1.0
130	139.6±0.3	227.7±0.6	144.4±0.2	49.9±0.1
250	139.1±0.9	227.2±1.5	137.5±8.5	47.5±2.9
360	138.4±0.4	226.9±1.5	150.4±23	52.0±8.2
Irradiated in air				
10	135.6±0.3	223.1±0.8	144.3±1.8	49.9±0.6
130	136.4±0.2	223.4±2.5	151.3±6.1	52.3±2.1
250	135.5±0.5	223.7±2.5	147.5±6.7	51.0±2.3
360	134.8±0.9	222.8±5.5	156.1±40	54.0±14
Irradiated in nitrogen				
10	135.7±0.5	223.9±0.5	142.6±3.7	49.3±1.3
130	135.8±0.4	222.3±1.4	154.6±8.2	53.4±2.8
250	135.3±0.5	223.9±2.7	145.8±3.2	50.4±1.1
360	135.5±0.5	223.7±1.7	165.7±11	57.3±3.6
Irradiated in vacuum				
10	136.2±0.6	223.3±2.9	135.4±8.6	46.8±3.0
130	136.0±0.2	223.9±0.9	150.2±5.5	51.9±1.9
250	135.3±0.4	224.6±2.6	148.5±8.5	51.3±2.9
360	134.9±0.7	223.5±5.9	162.3±8.3	56.1±2.9

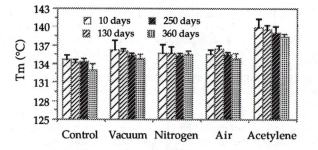

Figure 3. Effect of gamma irradiation on melting temperature of compression-molded UHMW-PE (error bar = 95% confidence interval)

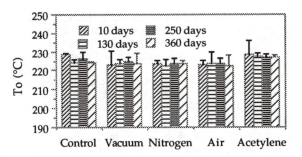

Figure 4. Effect of gamma irradiation on oxidation temperature of compression-molded UHMW-PE (error bar = 95% confidence interval)

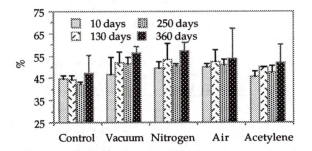

Figure 5. Effect of gamma irradiation on percent crystallinity of compression-molded UHMW-PE (error bar = 95% confidence interval)

Tensile properties. Tensile tests were run following gamma irradiation. No attempts were made to examine the post-irradiation effects. From tensile testing, the following parameters were determined: yield stress (σ_y), ultimate stress (σ_u), sample modulus (E), ultimate elongation (ε_u) and fracture energy (W_u). Table 2 lists the results (percentage of parameter as compared with the control), which are also illustrated in Figures 6-10. For the control the mean ± 95% confidence interval were: $\sigma_y = 24.7 \pm 0.6$ MPa, $\sigma_u = 47.9 \pm 1.7$ MPa, $E = 578 \pm 13$ MPa, $\varepsilon_u = 176 \pm 12\%$ and $W_u = 145 \pm 12$ lbs-in/in. It appeared that both radiation dose and type of gas environment affect the tensile properties of UHMW-PE. Statistical Analysis (ANOVA) of the result shows that (1) at the same dose, the type of gas environment significantly changes σ_y, E and ε_u ($p<0.025$); (2) in the same gas environment, the radiation dose significantly affects σ_y, σ_u, E, ε_u and W_u ($p<0.025$) except in nitrogen where σ_u, ε_u and W_u are not significantly affected ($p>0.2$). At doses of 1.25 and 2.5 Mrad, yield stress and sample modulus displayed higher increase in air and acetylene environments than in nitrogen. In fact, the sample modulus dropped in nitrogen. At these two doses, crosslinking appears to dominate in air and acetylene, while chain scission prevails in presence of nitrogen. The highest increase recorded for σ_y and E were 8 and 20%, respectively. At 5 Mrad, all three gas environments showed an increase in σ_y and E. Decreases in ultimate elongation at 5 Mrad suggest that the crosslinking began to dominate. As shown later, preliminary results on the effect of radiation on the tensile properties of self-reinforced UHMW-PE composites, using ultrahigh strength fiber (Spectra 1000), indicate that irradiation causes both chain scission and crosslinking, and the extent of which may depend on the dose and/or environments. In the absence of oxygen, C-C crosslinks are presumably formed. If oxygen is present, the crosslinking may also be associated with the formation of peroxide linkages. The later are thermally and chemically labile. Crosslinking in the presence of acetylene can result in stable C-C crosslinks. Results of the present study showed that σ_y and E decreased when the air-irradiated samples were treated with N,N-dimethyl aniline, a known catalyst for peroxide dissociation. Reduction in ultimate elongation of samples irradiated in acetylene increased as the dose increased, suggesting that the C-C crosslinking dominates in these samples, with acetylene participating as a reactant. This was consistent with the recorded increase in specimen mass after irradiation at 2.5 Mrad (Table 3). At 5 Mrad, ultimate elongation generally decreased, which is most noticeable in acetylene environment where a reduction of greater than 50% was recorded. It may be expected that a higher dose of irradiation will largely decrease the toughness of UHMW-PE and thus make the polymer more brittle, an undesirable characteristic for orthopedic implants.

Table 2. Percentage of Tensile Properties as Compared to Control
(mean ± 95% confidence interval; 5 samples per group)

	σ_y, MPa	σ_u, MPa	E, MPa	ε_u, %	W_u, lb-in/in
1.25 Mrad					
N_2	104.5±0.8	114.7±7.3	97.7±7.7	126.3±13	139.7±23
Air	105.7±1.4	106.0±5.5	121.8±17	135.0±13	136.4±15
C_2H_2	105.3±1.8	105.8±4.9	108.7±12	86.4±11	88.0±8.7
2.5 Mrad					
N_2	102.3±2.2	113.4±30	94.1±14	118.0±52	133.4±71
Air	108.2±1.0	111.2±3.7	123.4±5.0	148.8±10	155.7±23
C_2H_2	106.7±2.3	127.9±10	107.2±14	83.5±11	94.3±17
5 Mrad					
N_2	106.2±1.7	106.4±8.7	121.7±13	93.5±16	99.7±22
Air	108.8±2.2	102.0±6.1	135.5±10	97.0±9	101.8±14
C_2H_2	107.5±1.7	90.8±10	119.0±6.8	46.6±9	43.4±11

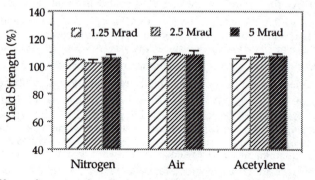

Figure 6. Effect of gamma irradiation on yield strength of compression-molded UHMW-PE (error bar = 95% confidence interval)

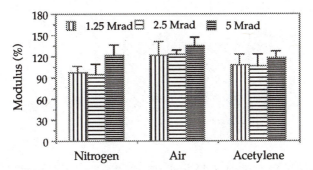
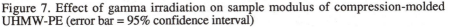

Figure 7. Effect of gamma irradiation on sample modulus of compression-molded UHMW-PE (error bar = 95% confidence interval)

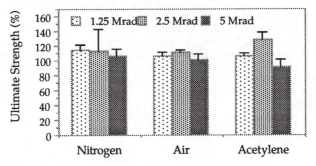

Figure 8. Effect of gamma irradiation on failure strength of compression-molded UHMW-PE (error bar = 95% confidence interval)

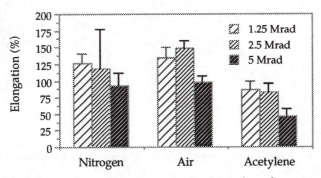

Figure 9. Effect of gamma irradiation on ultimate elongation of compression-molded UHMW-PE (error bar = 95% confidence interval)

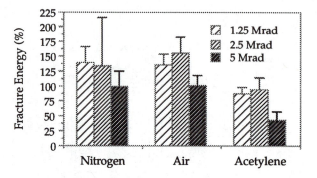
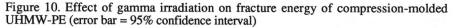

Figure 10. Effect of gamma irradiation on fracture energy of compression-molded UHMW-PE (error bar = 95% confidence interval)

Mass Change. Samples for weight measurement had the same geometry as the tensile specimens (weighing about 1.2 g). The weight changes of UHMW-PE gamma-irradiated at 2.5 Mrad under different gas environments are illustrated in Figure 11. The results suggest that gamma irradiation causes mass increases under all four conditions, with acetylene yielding highest weight gain. Although the mass increase of specimens treated in vacuum was minimal, this may be due to the fact that the environment under vacuum was not completely free of oxygen. We can also understand that post-irradiation continuously caused weight gain due to the long-lived macromolecular radicals. Irradiation in oxygen-containing environment produces peroxides, and can also lead to weight gain. The weight gain in nitrogen may be due to the factor that nitrogen can react with polyethylene under high energy ionizing irradiation.

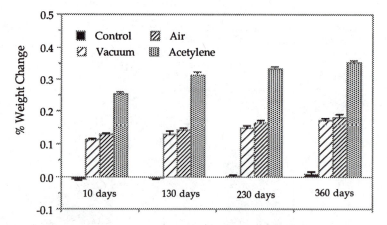

Figure 11. Percent weight change of UHMW-PE after gamma irradiation (error bar = 95% confidence interval)

Composites. To determine the effect of gamma radiation on the properties of self-reinforced composites, the tensile specimens were irradiated at 2.5 Mrad in air, nitrogen and acetylene environments. The tensile test was run following gamma irradiation. For each testing point, 5 specimens were used. The data are presented in Table 3. It can be seen that gamma irradiation increased transverse strength and modulus of the composites, but decreased longitudinal strength, particularly in air environment, while longitudinal modulus was little affected.

Table 3. Tensile properties of UHMW-PE composites irradiated at 2.5 Mrad

	Control	Air	Nitrogen	Acetylene
Yield strength (MPa)				
[0/0]s	117.1±18	83.7±6.1	110.3±6.9	107.1±7.8
[90/90]s	24.7±0.5	26.3±0.8	25.1±1.0	25.3±0.7
Modulus (MPa)				
[0/0]s	2652±479	2663±355	2362±557	2651±202
[90/90]s	694.1±72	872±63.2	753.8±74	745±87.0

Conclusions

The present study showed that gamma radiation and gas environment affect the thermal and tensile properties of compression-molded UHMW-PE sheets and self-reinforced composites. At 2.5 Mrad, the melting temperature of UHMW-PE increased more than 5°C following gamma irradiation in an acetylene environment, but less than 2°C under the other three conditions. Oxidation temperature decreased in all but acetylene. After irradiation, crystallinity continuously changed. At 1.25 and 2.5 Mrad, air and acetylene environments promote crosslinking, and thus increase in yield stress and modulus. At 5 Mrad, all three gases showed increase in yield stress and modulus, but decrease in ultimate elongation. A fraction of the crosslinks formed due to oxygen can be viewed as transient and may dissociate upon aging, particularly in the presence of reducing agents. Gamma irradiation in acetylene predominantly causes crosslinking and at low dose ranges it can be used as a means to improve the performance of UHMW-PE products. Gamma irradiation of self-reinforced UHMW-PE composites decreased longitudinal tensile properties while transverse properties were increased.

Literature Cited

1. Andreopoulos, A. G.; Kampouris, E. M. J. Appl. Polym. Sci., 1986, 31, p1061.
2. Grobbelaar, C. J.; du Plessis, T. A.; Marais, F. J. Bone Joint Surg., 1978, 60-B, p370.
3. Streicher, R. M. In Ultra-High Molecular Weight Polyethylene as Biomaterial in Orthopedic Surgery, Willert, H. G., Ed., Hogrefe & Huber Publishers, Toronto, 1991, p66.
4. Boggan, R. S., Transactions of 19th Annual Meeting of the Society for Biomaterials, XVI, Birmingham, AL, 1993, p329.
5. Wunderlich, B.; Cormier, C. M. J. Polym. Sci., 1967, 5, p987-988.

RECEIVED June 20, 1995

Chapter 23

Effects of Ionizing Radiation on the Optical Properties of Polymers

Julie P. Harmon[1], Emmanuel Biagtan[2], Gregory T. Schueneman[3], and E. P. Goldberg[2]

[1]Department of Chemistry, University of South Florida, Tampa, FL 33620–5250
[2]Department of Materials Science and Engineering, University of Florida, Gainesville, FL 32611
[3]Department of Polymer Science and Engineering, University of Massachusetts, Amherst, MA 01003

The UV/Visible transmission spectra of glassy, optical polymers are greatly affected by ionizing radiation. The effect of gamma radiation on the transparency of styrene polymers is studied. Modifications on the styrene ring can reduce or enhance the radiation induced discoloration. The color center population is monitored in a fundamental study via UV/VIS spectroscopy. This fundamental study is used to help interpret the behavior of polymeric scintillator exposed to gamma radiation.

The effect of ionizing radiation on polymer properties has been extensively studied for many years. Most of the attention has focused on mechanical properties and on the chemical reactions responsible for radiation induced changes in these properties. Reference 1 is an excellent review of this subject matter. More recently, radiation effects on optical properties of polymers have commanded attention. This is due to applications involving optical polymers that are used in high radiation environments. When transparent biomedical devices are sterilized via gamma irradiation, doses as low as 2.5 Mrads induce discoloration of the articles. Particle accelerators use polymeric scintillator to detect radioactive species formed during collisions. Polymeric scintillator is made up of polystyrene or poly(4-methyl styrene) doped with fluorescent dyes. In the new generation of high energy accelerators, scintillator will see doses of radiation in the Mrad range. These doses are sufficient to induce color center formation which diminishes the sensitivity of the scintillator. These applications triggered an interest in understanding the nature of the radiation induced color center formation and in designing polymers which resist color center formation.

Radiation effects on polymer structure vary with the energy or type of source (alpha, gamma, neutron, X-ray or electron beam) and with the dose. In addition, radiation induced reactions in the presence of oxygen differ significantly from those which occur in an oxygen-free environment. Irradiation induces a number of complicated effects on polymer optical properties ranging from decreases in refractive index(2) to the production of color centers. Earlier literature sites many examples of the effects of ionizing radiation on the production of permanent and transient color centers in polymeric materials (3-8). Recent work of Clough and Wallace defines two types of color centers, "annealable" and "non-annealable"(9). "Annealable" color centers are associated with reactive species which disappear during or after irradiation. Non-annealable color centers are associated with a permanent change in chemical structure. Wallace, Sinclair, Gillen and Clough(*10*) define the annealable color centers as free radicals that anneal in the presence of and in the absence of air. While air annealing effects are attributed to oxygen quenching of free radicals, annealing in the absence of oxygen is the result of radical recombination. Rates for both types of annealing increase with annealing temperature. The "colored" nature of free radicals has been well documented in non-polymeric organic molecules. Phenoxy radicals and their derivatives prepared by flash photolysis in gaseous and liquid systems and by photolysis in glassy solutions at low temperatures have absorption maxima between 270 and 615 nm(*11*).   Similarly, gamma irradiation induces the formation of free radicals in 3-methylpentane glasses at 77°K, and these absorb in the UV region of the electromagnetic spectrum(*12*).   Again, certain enzymes form stable, delocalized free radicals that absorb in the UV/VIS region(*13*).

Non-annealable color centers resulting from radical-radical reactions are thought to be due to the formation of conjugated double bonds(*10*). Possible permanent, colored oxidation products for gamma-irradiated polystyrene are suggested (conjugated ketones, aromatic ketones and quinones).   It has also been noted that when polystyrene is irradiated in air, diffusion limited oxidation bands appear on the outsides of the samples(*14*). Analysis of bleaching rates in samples allows the quantitative determination of oxygen consumption rates.

We conducted a series of studies aimed at identifying ways of altering the chemistry of macromolecules to make them more resistant to radiation induced discoloration (*12-21*). We investigated increasing the flexibility and permeability of macromolecules in order to increase annealing and radical recombination rates. A flexible polymer matrix enhances the probability of radical recombination and a more flexible matrix is likely to exhibit enhanced oxygen diffusion. The nature of the group or additive used to enhance flexibility may be chemically prone to or resistant to irradiation, or the enhancer may alter oxygen solubility. Furthermore, oxygen permeability may be increased via the incorporation of stiff bulky groups which deter bulk flexibility i.e.. t-butyl groups. Finally, attempts to alter permeability or recombination rates may result in the incorporation of moieties which exhibit different degrees of radical stability. The first part of this study addresses this issue by focusing on the effect of styrene substituents on radiation induced color center formation. The second part of this study focuses on dye doped polystyrene and poly(4-methyl styrene) scintillator as they   respond optically to radiation. Scintillators emit light when exposed to ionizing radiation. The light is detected by photomultiplier tubes and compared to light production by a standard. This provides a comparative measure of photons is termed light output, LO. Ionizing radiation diminishes LO due to the formation of permanent and transient color centers which absorb the light emitted by the scintillator. Additional loss in LO occurs when radiation alters the dye chromophore. The effect of dose and dose rate on scintillator light output is reviewed.

Experimental

Styrene and p-methyl styrene monomers were obtained from Scientific Polymer Products, Inc. in Ontario, NY. T-butyl styrene was obtained from Monomer-Polymer Laboratories in Windham, NH. P-ethyl styrene and p-propyl styrene were synthesized as described in reference 22. Inhibitor, t-butyl catechol, was removed from the commercial monomers via an activated alumina column. Styrene, methyl styrene and t-butyl styrene were initiated with 0.2 wt % (1,1-di(t-butylperoxy)-3,3,5-trimethylcyclohexane). Ethyl and propyl styrene were initiated with 0.2% AIBN (*23*). Samples were polymerized under nitrogen for 12 hrs. at 85 °C. The polymers are coded: polystyrene (PS), poly (4-methyl styrene) (PMS), poly (4-ethyl styrene) (PES), poly(4-n-propyl styrene) (PPS) and poly (4-t-butyl styrene) (PtBS). Six tenths cm thick discs were compression molded between plates with optical surfaces. Glass transition temperatures were characterized with a 2910 Du Pont Instruments DSC. Samples were scanned at 10 °C under nitrogen. T_gs were determined from the second scan. No further characterization was undertaken. Oxygen permeability was determined on discs machined to 0.25-0.35 mm thick. A Createch Model 201T Permeometer was used following the method of Fatt (24). Permeability is characterized by the Dk value which is in units of (cm^2/sec) (ml O_2/ml x mm Hg). Densities were determined via water displacement or with a Quantachrome Corp. MPY-1 micropyncnometer. UV/VIS spectra were recorded on a Hewlett Packard 8452A Diode Array Spectrophotometer. Samples were irradiated with a ^{60}Co source to a total dose of 10 Mrads at a dose rate of 0.039 Mrads/hr in air.

Scintillators, SCSN-38, SCSN-81 and Bicron BC-499-35, were obtained from Kurarey Company Ltd. and Bicron Corporation. Four mm thick samples were machined into 1 inch OD disks. Compositions and pre irradiation relative light outputs are in Table I. All light outputs are relative to an unirradiated sample of SCSN-38.

Table I. Commercial Polymer Scintillators

Scintillator	Base polymer	Primary Dye	Secondary Dye	Pre irradiation Light Output[a] ± 5%
SCSN-38	Cross-linked Polystyrene	1% b-PBD	0.02% BDB	100%
SCSN-81	Cross-linked Polystyrene	1% PTP	0.02% BBOT	95%
Bicron BC-499-35	Uncross-linked Poly(methylstyrene)	PTP	BBOT	105%

[a]Light outputs are relative to unirradiated sample of SCSN-38.

Each sample was irradiated with a ^{60}Co at a constant dose rate in air. Dose rates were 1.5, 0.35, 0.14, 0.04, 0.013, 0.0069, 0.0045 and 0.0023 Mrads/hr. LOs were measured immediately after irradiation, and at intervals afterwards. Measurements were made by a THORN EMI type 9124B PMT (11 stages, 1100V) located in a light-tight box. Excitation occurred via a 1.0 micro Curie Am-241 alpha-source mounted on the scintillator. Further details are described in reference 25. UV/VIS transmission spectra were recorded as above. ESR spectra were recorded on bars of the Bicron sample, 76 mm x 10 mm x 3mm which were gamma irradiated to 10 Mrads at 1.5, 0.14, and 0.04 Mrads/hr. Radical intensities were measured with a Bruker ER 200D SRC Electron Spin Resonance Detector operating at 9500 MHz and scanning from 2000 to 4000 Gauss.

Results and Discussion

Polystyrene and alkylated Polystyrene study. Figure 1 shows the UV/VIS transmission spectra for the polystyrenes before and at intervals after irradiation at 0.039 Mradshr to a total dose of 10 Mrads. In figure 2, the wavelength at 50% transmission before and after irradiation is summarized for the styrene polymers. At 0.039 Mrads/hr, PS and PMS exhibit a significant amount of post irradiation

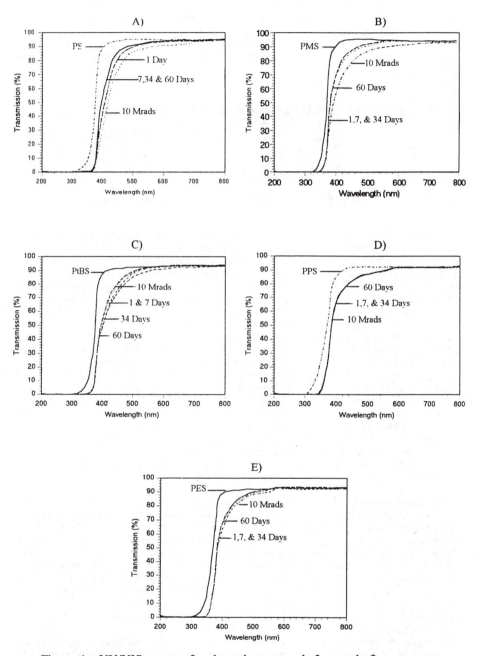

Figure 1: UV/VIS spectra for the polystyrenes before and after gamma-irradiation

recovery, while PES, PPS and PtBS do not. PES and PPS have the shortest wavelengths at 50% transmission at all times after irradiation. The 50% transmission wavelength is used for the comparative purpose of relating structure to color center formation. It must be noted, however, that all samples exhibited a decrease in transmission that extends to much longer wavelengths at transmission percentages greater that 80%.

Table II. Glass Transition Temperatures, Densities and Permeabilities

Polymer	T_g °C	Density (gm/cm^3)	Dk $(10^{11} \times (cm^2//sec)(ml\ O_2\ x\ mm\ Hg))$
PS	94	1.050	3.10
PMS	111	1.020	5.70
PES	76	1.008	10.20
PPS	48	0.997	9.10
PtBS	140	0.963	15.80

Table II shows the measured T_gs, densities and oxygen permeabilities for the polystyrenes. There is a trend showing an increase in permeability as density decreases, except for PPS which has a lower permeability than expected. Para substituted ethyl and propyl groups decrease the glass transition temperature, while methyl and t-butyl groups increase the glass transition temperature. One might reason that the flexibility of the side group could effect radical recombination, but the situation is actually too complex to sort out such effects. We anticipated that increasing the permeability would result in an increase in recovery rate. Indeed this is the case. PPS, PES and PtBS (Figure 2) show little or no recovery after irradiation, an indication that recovery occurred during irradiation. PtBS showed a slight, unexplained increase in the 50% transmission wavelength between 7 and 60 days after irradiation. If PtBS is not considered, there is a decrease in permanent damage as determined by the 50% transmission point that accompanies an increase in the Dk value. This indicates that increases in free volume which enhance permeability may also increase radical recombination rates. We are currently exploring the use of ESR to investigate this issue. PtBS exhibits the largest permanent bathochromic shift. This may be the result of preferential cleavage of the phenyl-t-butyl bond during irradiation due to the stability of the resulting t-butyl radical. This may result in the formation of a higher population of free radicals in PtBS as compared to the other polymers.

Scintillator study. Alpha particles excite the samples near the surface and light produces by this surface scintillation is transmitted through the 4 mm thick sample to the photomultiplier tube. Therefore, decreases in LO are due to degradation of the scintillating and emitting centers in the penetrated surface and to color center formation throughout the entire sample thickness. Figure 3 is a plot of the light outputs for the Bicron BC-499-35 scintillator during the seven day period immediately after irradiation. There is less recovery in the light output as the dose rate for irradiation decreased. The Bicron samples in particular displayed a wide range of recovery behavior depending on the dose rate. For all scintillators, LOs measured 2 to 3 months after irradiation showed no additional increases over readings that were taken within three to seven days after irradiation.

Figure 4 is a representative semilogarithmic plot of the light outputs for the Bicron scintillator immediately after irradiation and after full recovery versus the irradiation dose rate. As the dose rate for irradiation decreased logarithmically, the immediate LO either decreased (Bicron BC-499-35), stayed constant (SCSN-81), or increased (SCSN-38), while the final LO consistently decreased. In all cases, the

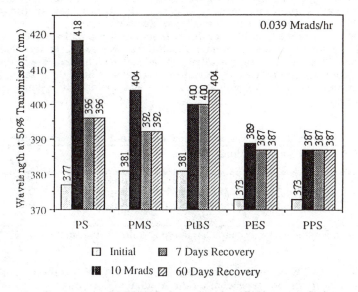

Figure 2: Summary of UV/VIS data showing 50% transmission versus polymer.

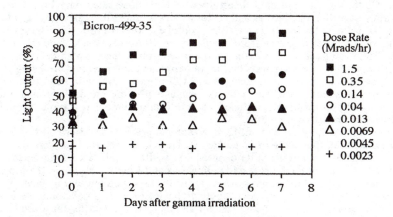

Figure 3: Bicron-499-35 light output after gamma-irradiation to 10 Mrads at different dose rates. Values relative to unirradiated SCSN-38. Reproduced with permission from ref. 25.

post-irradiation recovery, which is the difference between the immediate and final LOs, decreased as the dose rate decreased. The immediate and final LOs converged until below a certain dose rate, no recovery was observed. The dose rate where the two LOs converged was different for each scintillator. More importantly though, the final LOs below the convergence dose rate still follow the semilogarithmic relationship established above the convergence dose rate.

In figure 5, just the final LOs for the three scintillators are plotted versus the dose rate. The respective semilogarithmic equations based on regression analysis are presented in Table III. This suggests that one can estimate the decrease in final LO at low dose rates of irradiation by irradiating samples at high dose rates, fitting a semilogarithmic equation to the data, and then extrapolating the equations to the lower dose rates.

Table III. Semilogarithmic Equations Relating Final LOs to Dose Rate

Scintillator	Equation of line[a]	Correlation Coefficient
SCSN-38	Final LO = 60.028 + 12.233*Log(dose rate)	.988
SCSN-81	Final LO = 74.095 + 15.398*Log(dose rate)	.975
Bicron BC-499-35	Final LO = 87.251 + 26.247*Log(dose rate)	.994

[a]Dose rate in Mrads/hr. Light outputs relative to unirradiated sample of SCSN-38.

Bicron BC-499-35 had the highest log constant of the three scintillators, indicating that it is more dependent on the irradiation dose rate than the other two. Furthermore, note that at high dose rates, the Bicron scintillator had the highest final LO, while at low dose rates it had the lowest. Thus the order of radiation stabilities between different scintillators at high dose rates may not be the same at lower dose rates. Radiation damage data determined at high dose rates may be incorrect in estimating scintillator stabilities and useful lifetimes under real operating conditions. Figure 6 is the UV/VIS transmission spectra for the Bicron scintillators before irradiation and after recovery from irradiation to 10 Mrads at selected constant dose rates. There is a gradual red shifting of the transmission spectra with decreasing dose rate of irradiation. This suggests that more permanent color centers are being created at the lower dose rates. The decrease in LO is more dramatic than that seen in transmission spectra shifts. This is because polymer and fluor degradation diminished the initial light produced by scintillation before it travels through the remainder of the damaged matrix.

Figure 7 is a plot of the ESR derivative curves for Bicron-499-35 taken immediately after they were irradiated to 10 Mrads at the selected constant dose rates. The ordinate is in arbitrary units of intensity, while the abscissa is in Gauss. The curves best represent the most stable radical in poly(4-methyl styrene), the p-methyl-benzyl radical (26). The primary benzyl radical is not ruled out, but its presence may be obscured by the more prevalent radical species or it may not dominate at these irradiation conditions (27). The radical intensity decreased as the dose rate for irradiation decreased. Thus, irradiation at high dose rates creates a large concentration of radicals within the scintillator, while irradiation at lower dose rates creates a smaller concentration. It is important to note that the final light output, the recovery in light output and the radical concentration all decreased as the constant dose rate for irradiation decreased.

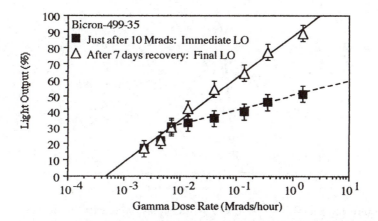

Figure 4: Bicron-499-35 immediate and final light outputs versus dose rate after gamma-irradiation to 10 Mrads. Values relative to unirradiated SCSN-38. Reproduced with permission from ref. 25.

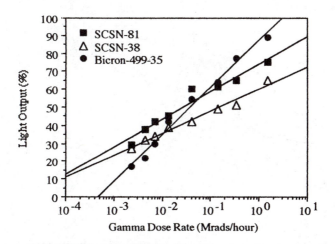

Figure 5: Final light outputs versus gamma dose rate for all three scintillators. Values relative to unirradiated SCSN-38. Reproduced with permission from ref. 25.

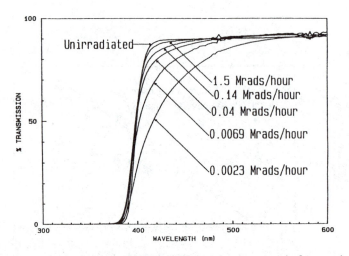

Figure 6: Bicron-499-35 UV/VIS transmission spectra before and after gamma irradiation to 10 Mrads at different dose rates. Normalized to 93% transmittance at 800 nm. Reproduced with permission from ref. 25.

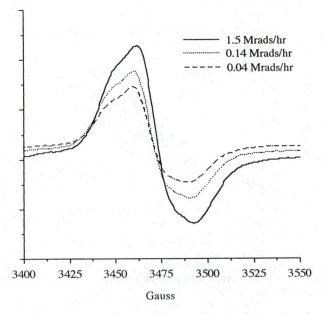

Figure 7: Bicron-499-35 ESR spectra immediately after gamma irradiation to 10 Mrads at the indicated dose rates.

Summary

The results for the study on PS and the alkylated PS samples indicate that controlling the population of color centers in irradiated samples is a complex endeavor. Increasing the oxygen permeability does increase the annealing rate. However, the substituents used to increase the oxygen permeability effect color center formation in an additional way in that they may enhance or decrease permanent color center formation. The highly permeable PtBS sample annealed rapidly, but exhibited the largest permanent bathochromic shift, presumable due to the susceptibility of the t-butyl group to radiation induce cleavage. The ethyl and propyl substituted samples exhibited the smallest permanent bathochromic shifts. This may be due to the effect of these flexible groups on radical recombination. Future research on these polymers will correlate ESR results with optical results at different radiation doses and dose rates.

The scintillator observations can be explained by assuming that radiation-induced oxidation occurs during irradiation to create effective light absorbing centers and that some of the unreacted radicals after irradiation convert into non-light absorbing or less absorbing centers.

High radiation dose rates create large radical concentrations in the scintillator within a short time period. Before the irradiation is completed, radicals react with all of the dissolved oxygen to create effective light absorbing centers. The concentration of dissolved oxygen is quickly depleted and not readily replenished by oxygen diffusion, so many radicals are left unreacted. Since the permeability of dissolved oxygen in polystyrene and poly(4-methyl styrene) is low, very few permanent light absorbing centers are created. Immediately after irradiation, there are many unreacted radicals, hence there is a high ESR intensity, and eventually high LO recovery and high final LOs.

Lower radiation dose rates create fewer radicals within a given time period. The absorbed oxygen is never depleted and may be replenished by diffusion. Oxidation degradation during irradiation is sustained so more light absorbing centers are created. Immediately after irradiation, there is a lower concentration of radicals, hence there is a lower ESR intensity and eventually less light output recovery.

Acknowledgments

The authors would like to thank Dr. K. Williams, Dr. H. Hanrahan, Dr. Talham and Dr. H. Byrd for their help in this research and for giving us access to their equipment;. The authors would also like to thank the Bicron Corporation and Fermilab for providing scintillators. Support was provided by the U.S. Department of Energy grant DE-FG05-86ER40272 and by the Texas National Research Commission grant RGFY93-281.

References

1. Clough, R. L., *In Encyclopedia of Polymer Science and Engineering, vol. 13: Radiation Resistant Polymers* ; Bikales, N., Ed; John Wiley and Sons, New York, **1983,** 667.
2. Darraud, C., Bannamane, B., Gagnadre, C., Decossas, J. L., Vareille, J. C., *Polymer;* **1994,**11, 2447.
3. Day., M. J., Stein, G., *Nature,* **1951,** 168, 644.
4. Day., M. J., Stein, G., *Nature,* **1951,** 168, 645.
5. Charlesby, A., *Nucleonics,* **1954,** 12, 18.
6. Fowler, J. F., Day, M. J., *Nucleonics,* **1955,** 13, 52.
7. Boag, J.W., Dolphin, G. W., Rotblat, J., *Radiation Res.,* **1958,** 9, 589.
8. Barker, R. E., *J. Polym., Sci.,* **1962,** 58, 553.

9. Clough, R. L., , Wallace, J. S., *In Symposium on Detector Research and Development for the Superconducting Super Collider: Radiation Effects on Organic Scintillators: Studies of Color Center Annealing.* T. Dombeck, V. Kelly, and G. Yost Eds., World Scientific, **1990**, 661.

10. Wallace, J. S., Sinclair, M. B., Gillen, K. T., Clough, R. L., *Radiat. Phys. Chem,* **1993,** 41, 85.

11. Land, E. J., Porter, G., Strachan, E., *Trans. Faraday Soc.,* **1962,** 57, 1885.

12. Neiss, M. A., Willard, J. E., *J. Phys. Chem,* **1975,** 79, 783.

13. Atkin, C. L., Thelander, L., Reichard, P., Lang, G., *J. Biol. Chem.,* **1973**, 248.

14. Gillen, K. T., Wallace, J. S., Clough, R. L., *Radiat. Phys. Chem,* **1993,**41, 101.

15. Harmon, J. P., Jhaveri, T., Gaynor, J., Walker, J. K., Chen, Z., *J. Apply. Polym. Sci.,* **1992,** 44, 1695.

16. Harmon, J., Gaynor, G., Feygelman, V., Walker, J., *Nucl. Instrum. and Meth. in Phys. Res.,* **1991,** B53, 309.

17. Harmon, J. P., Taylor, A. G., Schueneman. G. T., Goldberg, E. P., *Polym. Deg. and Stab.,* **1993**, 319.

18. Harmon, J. P. Gaynor, J. F., *J. Polym. Sci. Part B: Polym. Phys.,* **1993**, 31. 235.

19. Taylor, A. G., Harmon, J. P., *Polym. Deg. and Stab.,* **1993,**41, 9.

20. Gaynor, J., Fischer, V., Walker, J., Harmon, J. P., *Nucl. Instrum. and Meth. in Phys. Res.,* **1992,** B69, 332.

21. Harmon, J. P., Gaynor, J. F., Taylor, A. G., *Radiat. Phys. Chem,* **1993,** 41, 153.

22. Schueneman, G. T., "Radiation Stability of Polymers for High Energy Radiation Detectors" Ph. D. Thesis, University of Florida, **1994**, 14.

23. Davies, T. E., **British Plast.**, **1959,** 19, 283.

24. Fatt, I., *Int. Cont. Lens Clinic,* **1975,** 11, 179.

25. Biagtan, E., Goldberg, E. P., Stephens, R., Harmon, J. P., *Nucl. Instrum. and Meth. in Phys. Res.,* **1994,** B 93, 296.

26. Parkinson, W., Keyser, R., *The Radiation Chemistry of Macromolecules, Editor*: M. Dole, Academic Press, **1973**,, II, 57.

27. Herod, T., Johnson, K., Schlenoff, J., *Radiat. Phys. Chem.,* **1993**, 41, 1/2, 65.

RECEIVED December 1, 1995

Chapter 24

Temperature Effects in Gamma-Ray Irradiation of Organic Insulators for Superconducting Magnets for Fusion Reactors

H. Kudoh, N. Kasai, T. Sasuga, and T. Seguchi

Takasaki Radiation Chemistry Research Establishment, Japan Atomic Energy Research Institute, Takasaki, Gunma 370–12 Japan

The gamma-radiation-induced degradation at 77 K for glass fiber reinforced plastic (GFRP) was examined by flexural tests and gas analysis, and compared with room temperature irradiation results. The decrease in flexural strength at break (measured at 77 K) was much less in the case of 77 K irradiation than for RT irradiation. The evolution of CO and CO_2 was also depressed at 77 K. The temperature dependence of the degradation closely relates to the local molecular motion of the matrix resin during irradiation. Results for other polymers such as polymethylmethacrylate (PMMA) and polytetrafluoroethylene (PTFE) are also reported in terms of change in mechanical properties and molecular weight.

Polymer and organic composite materials are planned to be used as the insulator for super conducting magnets in fusion reactors. The radiation resistance of the materials must be evaluated in cryogenic environments. There is only limited research work on the temperature dependence of radiation effects on polymer and composite materials (1-4). In particular, little work on temperature effects in radiation-induced mechanical property changes is available (5-7). Wilski has pointed out that real understanding of radiation resistance in these materials will require further measurements of mechanical property changes (8). Therefore, low temperature gamma-ray irradiation effects on glass fiber reinforced plastic (GFRP) were studied in terms of changes in mechanical properties and gas evolution, and the results were compared with those obtained by irradiation at room temperature (RT). The results on PMMA and PTFE are also reported in terms of changes in mechanical properties and molecular weight.

0097–6156/96/0620–0313$12.00/0

Experimental

Materials used in this work are commercially available bis-phenol A type epoxy resin GFRP (G10CR; Glass fiber/ diglycidyl ether of bis-phenol A hardened with dicyanodiamide) of 2 mm thick specimens (6 mm width, 70 mm length), PMMA of 3 mm thick (10 mm width, 100 mm length) and PTFE of 0.1 mm thick sheet. They were exposed to Co-60 gamma rays at 77K in liquid nitrogen and RT in flowing nitrogen gas atmosphere at a dose rate of 30 kGy/h at 77 K and 10 kGy/h at RT. PTFE was irradiated also at 4 K and 30 kGy/h. The irradiation at 77 K and 4 K was carried out with low temperature irradiation test equipment at JAERI Takasaki (9,10). Samples of GFRP and PMMA for gas analysis were irradiated in evacuated glass ampules.

Mechanical properties were measured by three-point flexural tests at 77 K and RT, with span length 50 mm and crosshead speed 2 mm/min for GFRP and PMMA, and by tensile test for PTFE at RT with dumbbell shaped specimens and crosshead speed 200 mm/min.

Decomposed gas accumulated in the sample tube from GFRP and PMMA was analyzed by gas chromatography (GC) at RT. Total gas evolution was determined by the pressure rise. H_2, CO and CO_2 were measured by GC.

The molecular weight of PMMA was measured by gel permeation chromatography (GPC). For PTFE, molecular weight was determined by differential scanning calorimetry (DSC) using the method developed by Suwa et al (11). The heat of recrystallization was obtained by cooling from the melt. Molecular weight was calculated as $Mn=2.1 \times 10^{10}(Hc)^{-5.16}$, where Hc is the heat of recrystallization. DSC was also applied to measure the glass transition temperature (Tg) of GFRP from changes in the curvature in the thermogram. The DSC apparatus used was a Perkin Elmer type 7, and all DSC thermograms were taken at a heating and cooling rate of 20 K/min. The relaxation spectrum of GFRP was obtained by visco-elasticity measurement. The instrument used was a RHESCA RD-1100 torsion pendulum type visco-elastometer. The measurement was performed in the temperature range of 110 to 393 K with 1.25 K/min and the frequency range of 0.5 Hz to 2 Hz.

Results and Discussion

Irradiation Effects on GFRP (Glass/bisphenol A epoxy/dicyanodiamide).

Mechanical Properties. Figure 1 shows the change in flexural strength at break of GFRP as a function of dose. Flexural strength at break of GFRP decreases with dose, and reaches half of the original strength at 25 MGy for 77 K irradiation and at 1 MGy for RT irradiation (All mechanical measurements were performed at 77 K). The poor radiation resistance of this GFRP to RT radiation perhaps comes from the aliphatic hardener. For specimens irradiated and measured at 77 K, the strength scarcely changed upon the annealing at room temperature. These facts indicate that the radiation-induced reactions are extremely depressed at 77 K, and almost complete during irradiation.

The degradation of GFRP is mainly attributed to the matrix resin. Figure 2 shows the change in Tg of GFRP determined by DSC. The Tg of GFRP corresponds to that of epoxy resin. It decreases with dose, indicating the destruction of network structure of the epoxy resin through chain scission. The behavior of Tg to dose is the same with flexural strength; the degradation upon 77 K irradiation is much less than that upon RT irradiation, which implies a lower probability of chain scission at 77 K.

Gas Evolution. Figure 3 shows gas evolution from GFRP, indicating total gas and component gases of H_2, CO and CO_2. The total gas evolved from GFRP by irradiation increases linearly at low dose, then seems to level off above 1 MGy. The total gas evolution is less at 77 K than at RT, but it does not show as large a temperature dependence as flexural strength shows. H_2 evolution is almost the same at 77 K and at RT. However, CO and CO_2 evolutions are much less at 77 K than at RT. The evolution of CO and CO_2 well reflect the irradiation temperature dependence of flexural strength.

Irradiation Temperature Dependence. Most of the studies on the temperature dependence of polymer radiation effects refer to glass transition temperature (*1,2*). Though glass transition would play a most significant role in temperature dependence, other factors must be taken into consideration. The irradiation temperature dependence found in this study can not be explained by the molecular mobility above/below Tg, because both irradiations are carried out below Tg. We measured the relaxation spectra of unirradiated GFRP by visco-elasticity measurement. The gamma transition was found around 200 K corresponding to local molecular mobility(*12*). The local molecular motion is restricted at 77 K and allowed at RT. The formation of CO and CO_2 come from the cleavages of >C=O and -CO- bonds in bis-phenol A epoxy resin. The probability of cleavages depends on the mobility of >C=O and -CO- groups, and is reduced at 77 K. On the other hand, the H_2 evolution from the cleavage of the C-H bond is less dependent on the irradiation temperature because the molecular motion of C-H bond is allowed even at 77 K.

Irradiation Effects on PMMA.
Mechanical Properties. Figure 4 shows the change in flexural strength of PMMA measured at 77 K as a function of dose. A large temperature effect is observed; the degradation at 77 K irradiation is much less than that at RT irradiation.

Molecular Weight. Figure 5 shows the number average molecular weight

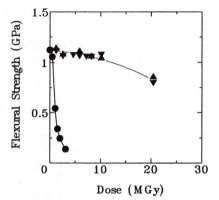

Figure 1 Flexural strength of GFRP measured at 77 K
          irradiated at 77 K without warming(▲),
          irradiated at 77 K after warming(▼),
          irradiated at RT(●).

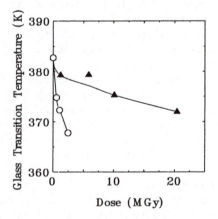

Figure 2 Glass transition temperature (Tg) of GFRP
          irradiated at 77 K(▲), and at RT(○).

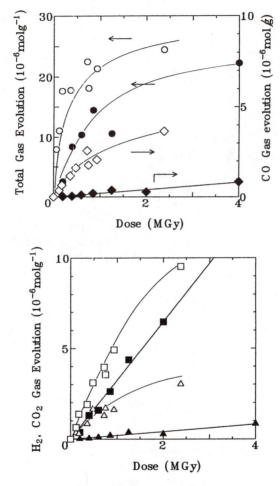

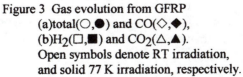

Figure 3 Gas evolution from GFRP
(a)total($\bigcirc$,$\bullet$) and CO($\diamond$,$\blacklozenge$),
(b)H_2($\square$,$\blacksquare$) and CO_2($\triangle$,$\blacktriangle$).
Open symbols denote RT irradiation,
and solid 77 K irradiation, respectively.

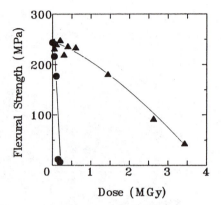

Figure 4 Flexural strength of PMMA measured at 77K
irradiated at 77 K(▲), and at RT(●).

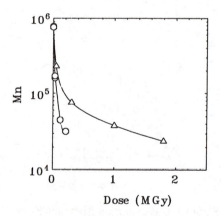

Figure 5 Number average molecular weight of PMMA
irradiated at 77 K(△), and at RT(○).

of PMMA measured by GPC. Molecular weight decreases with dose and the reciprocal of Mn increased linearly with dose, indicating that chain scission occurs. The chain scission probability is less at 77K than at RT. The G value of chain scission obtained is 1.7 at RT and 0.24 at 77 K, from Figure 5. The probability at 77 K is one seventh of that at RT, in agreement with the observation that the dose to half strength is seven times larger at 77 K than at RT in the case of mechanical properties measured at RT (*10*).

Gas Evolution. Figure 6 shows the gas evolution from PMMA. The same tendency with the case of GFRP is observed, that is, total gas evolution is less at 77 K but does not show so large a temperature dependence as does the strength. H_2 evolution is almost the same at 77 K and RT, while CO and CO_2 evolutions are much less at 77 K than at RT.

Irradiation Temperature Dependence. The irradiation temperature dependence relates to local molecular mobility, because both irradiations were performed below Tg. Since the transition temperature of the ester side group of PMMA is around 270 K, the molecular motion of the ester branch is restricted at 77 K and allowed at RT. This transition would relate to the large difference in flexural strength, molecular weight and gas evolution between 77 K and RT irradiations.

Irradiation Effects on PTFE.
 Mechanical Properties. Figure 7 shows the change in elongation at break of PTFE measured at RT as a function of dose. The degradation by 77 K irradiation is much less than that by RT irradiation, and it should be noted that the degradation by 4 K irradiation is the same as that by 77 K irradiation. The dose at equivalent change is 5 times larger at 4 K and 77 K than at RT.

Molecular Weight. Figure 8 shows the number average molecular weight of PTFE measured by DSC and Suwa's method. Molecular weight decreased with dose, which means that chain scission takes place. The chain scission probability is the same at 4 K and 77 K, and less than at RT. The dose at equivalent decrease is 5 times larger at 4 K and 77K than at RT, which agrees with the elongation behavior. The irradiation temperature dependence is related to molecular mobility as described above. The difference between 77 K and 4 K irradiation is not observed for PTFE. Considering that there is no transition in molecular motion related to the chain scission between 4 K and 77 K, the degradation at 4 K would be the same as that at 77 K, and this should be true of other polymers. This indicates that though the radiation resistance of candidate materials for fusion reactors should be evaluated at 4 K, 77 K irradiation experiments may be substituted.

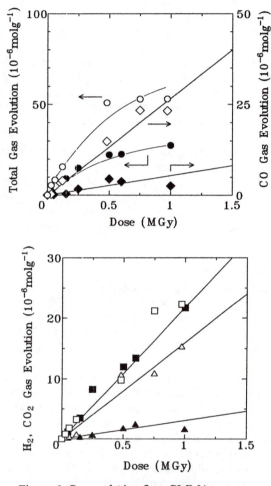

Figure 6 Gas evolution from PMMA
(a)total($\bigcirc$,$\bullet$) and CO($\diamondsuit$,$\blacklozenge$),
(b)H_2($\square$,$\blacksquare$) and CO_2($\triangle$,$\blacktriangle$).
Open symbols denote RT irradiation,
and solid 77 K irradiation, respectively.

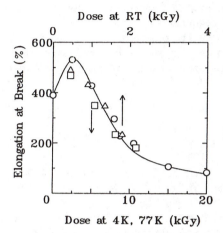

Figure 7 Elongation at break of PTFE measured at RT
irradiated at 4 K(□), 77 K(△) and RT(○).

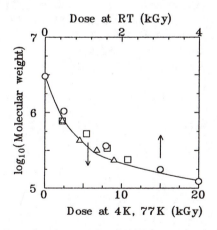

Figure 8 Molecular weight of PTFE measured with DSC
irradiated at 4 K(□), 77 K(△) and RT(○).

Conclusion

The irradiation temperature dependence on the degradation of GFRP was studied by irradiation at 77 K and RT by measuring the changes in flexural strength and in the evolved gas. The degradation in mechanical properties at 77 K was much less than at RT, and the evolution of CO and CO_2 follows the same pattern. The degradation is related to the local molecular motion of the epoxy resin in GFRP. The probability of scission decreases with decreasing molecular motion at low temperature, but would be constant below 77 K. Thus, the radiation resistance at 4 K can be evaluated by 77 K irradiation.

Literature Cited

1. Hill D. J. T., O'Donnell J. H., Perera M. C. S. and Pomery P. J., *Radiat. Phys. Chem.*, **1992**, *40*, 127
2. Wundlich K., *J. Polym. Sci. Polymer Phys. Edn.*, **1974**, *11*, 1293
3. Kempner E. S. and Verkman A. S., *Radiat., Phys. Chem.*, **1988**, *32*, 341
4. Garrett R. W., Hill D. J. T., Le T. T., Milne K. A., J. H., O'Donnell J. H., Perera M. C. S. and Pomery P. J., In *Temperature Dependence of Radiation Chemistry of Polymers, Radiation Effects on Polymers*; Clough R. L. and Shalaby S. W., Eds.; ACS Symposium Series 475; American Chemical Society: Washington, D. C., 1991; 146-155
5. Yamaoka H. and Miyata K., *J. Nucl. Mater.*, **1985**, *133/134*, 788
6. Coltman R. R. and Klabunde C. E., *J. Nucl. Mater.*, **1981**, *103/104*, 717
7. Takamura S. and Kato T., *J. Nucl. Mater.*, **1981**, *103/104*, 729
8. Wilski H., *Radiat. Phys. Chem.*, **1987**, 29, 1
9. Kasai N. and Seguchi T., **1990**, JAERI-M 90-155
10. Kudoh H., Kasai N., Sasuga T. and Seguchi T., *Radiat. Phys. Chem.*, **1994**, *43*, 329
11. Suwa T., Takehisa M. and Machi S., *J. of Appl. Polym. Sci.*, **1973**, *17*, 3253
12. Sasuga T. and Udagawa A., *Polymer*, **1991**, *32*, 402

RECEIVED September 8, 1995

Chapter 25

Radiation-Induced Effects in Ethylene–Propylene Copolymer with Antioxidant

Stefania Baccaro

Department of New Technology, Technological Services, Italian National Agency for New Technology, Energy, and the Environment, Research Center Casaccia, Via Anguillarese 301, 00060 S. Maria di Galeria, Rome, Italy

Infrared absorption spectroscopy and electron spin resonance have been used to investigate the gamma radiation effect on ethylene-propylene co-polymer loaded with an antioxidant characterized by an -NH functional group. The shape of the infrared oxidation profiles is not affected by the antioxidant content, and the dependence on thickness and dose rate is in good agreement with the Gillen and Clough model. The infrared profiles are also used to investigate the distribution of the grafted chains across the plate of ethylene-propylene co-polymer without antioxidant. The influence of dose rate and oxygen has been studied by electron spin resonance. The interactions of polymeric free radicals with the antioxidant lead to the formation of R-NO° stable radicals. The post-irradiation time evolution and the behaviour of the oxidation products have been studied.

The fundamental process of polymer degradation and stabilization has been extensively studied (1-3). Infrared absorption spectroscopy (IR) is a commonly used technique to investigate polymer degradation and stabilization induced by gamma radiation (2-8). Inhomogeneous oxidation effects in polymers exposed to ionizing radiation have been often found and extensively studied by using different experimental methods (9-11). The most widely used technique to obtain the oxidation profile is the measurement of the carboxylic content by IR absorption spectroscopy (8, 12, 13). Usually the inhomogeneity is attributed to the diffusion-limited oxidation, and in order to take into account such effects several theoretical models have been developed (12, 14, 15, 16). Up to now the largest part of these studies have been performed on pure polymers. Less attention has been devoted to polymers which are crosslinked and contain added antioxidant, in spite of their relevance for a wide variety of applications (3). A preliminary study (7) performed on an ethylene-propylene co-polymer (EPR) loaded with antioxidant suggested that the observed inhomogeneity effects can be explained (16) with the Gillen and Clough model, even if the model does not take explicitly into account the presence of antioxidant. In order to confirm this hypothesis, a more systematic study of the inhomogeneous oxidation on the EPR rubber has been

0097–6156/96/0620–0323$12.00/0

performed. Oxidation profiles have been measured as a function of thickness, dose rate and antioxidant content. Moreover, the study of the post-irradiation storage behaviour by Electron Spin Resonance (ESR) could allow, in some cases, one to clarify the role of the radical species produced by gamma radiation in the degradation processes (*17*). The ionizing radiation induced effects on an ethylene-propylene loaded with antioxidant characterized by a -NH functional group have been previously investigated using ESR spectroscopy (*17-19*). In this way it was possible to ascertain that ionizing radiation produces free radicals on the polymer that, upon interaction with the antioxidant, lead to the formation of R-NO° stable radicals. The R-NO° production starts during irradiation and lasts for about 100 hours. A stable level is reached afterwards which was found to depend on dose, antioxidant content and temperature (*18-21*). The post-irradiation time evolution of the observed effects, and the influence of dose rate and oxygen on the degradation processes, have been studied by ESR (Baccaro, S., B. Caccia, S. Onori, M. Pantaloni, *Nucl. Instr. Method. B,* in press). Measurements were performed at different temperatures in the (77-333) K range to clarify the dynamical evolution of the system, and to identify the radicals and their activation energy (*19*).

Experimental methods

The material studied (DUTRAL C0-034) is an ethylene-propylene copolymer (30% propylene) and was supplied by Montedison. All the samples were manufactured by *Pirelli Cavi S.p.A.* Company adding an antioxidant with the -NH functional group, and crosslinked with dicumilperoxide in the final form of thin plates (10 x 10 cm^2).

Samples for infrared measurements.   Two different series of samples have been studied: the first were plates 2.2 mm thick with 3 different concentrations of antioxidant (0.5, 0.8 and 1.5 mol%) which will be indicated as samples A, B and C respectively; the second series were plates 6 mm thick with 0.8 mol% antioxidant (samples D). The irradiation of the samples has been performed in air at room temperature in the ^{60}Co irradiation plant "Calliope" at ENEA Casaccia Center (Rome). Fricke dosimetry was used to measure the absorbed dose. Samples A, B and C have been irradiated at 100, 200 and 400 kGy using two different dose rates, 0.21 Gy/s for samples A_i, B_i, C_i (for i=1,3) and 1.31 Gy/s for samples A4, B4 and C4. Samples D have been irradiated to 150 kGy at three different dose rates (0.21, 0.5 and 0.9 Gy/s for samples D1, D2, D3 respectively). In order to determine the oxidation profile we cut from the center of the plates a slice 0.25 mm thick orthogonal to the surface. IR spectra of the slices were recordered using a FTIR 1720 Perkin Elmer equipped with Spectra Tech mod. Plan 11 microscope, the field of view was 300 x 300 μm. We obtained about 20 points for the oxidation profiles of samples D (6 mm thick) and only 7 for samples A, B and C (2.2 mm thick).

Samples for ESR measurements.   The samples used to investigate the free radicals were irradiated in air at 100 and 600 kGy with dose rates ranging from 0.09 up to 374 Gys^{-1} with ^{60}Co source and a 12 MeV linear accelerator. To clarify the post-irradiation behaviour, measurements were performed on some samples stored at different temperatures in the (77-333) K range. The measurements were performed at room temperature with a Bruker ESP 300 spectrometer equipped with a standard rectangular X-band TE102 microwave cavity.

Infrared results and discussion

Infrared measurements on EPR with antioxidant. The infrared spectra have been analyzed using the procedure discussed in a previous work (*7*) in order to obtain the "differential spectra", i.e. the spectral changes due to the radiation. The integrated absorption of the C=O band centered about 1720 cm^{-1} has been assumed to give a quantitative evaluation of the radiation induced oxidation. The differential spectra for the D1 sample at four different depths are reported in Fig. 1. The inhomogeneity of the oxidation is evident. The oxidation as a function of the depth is obtained by integration of the spectra in the 1850 - 1650 cm^{-1} region assuming a parabolic baseline. The results obtained for the sample D1 are presented in Fig. 2. Similar profiles are obtained for samples D2 and D3, and we have found that the oxidation depth is reduced and the profile becomes steeper with increasing dose rate; at the highest dose rate are employed (0.9 Gy/s), radiation induced oxidation is detectable only within 600 μm from the surface. As far as the A, B and C samples are concerned their thickness was small enough to enable us to measure the total absorption spectra by the usual IR technique before slicing. For all the samples, the total oxidation increases linearly with the absorbed dose in the dose range up to 400 kGy. The dose rate dependence is evident and it can be attributed to inhomogeneous oxidation effects. The oxidation decreases linearly with the antioxidant content; moreover its relative effect seems to be independent from the dose rate, thus suggesting that the antioxidant content does not change the inhomogeneity of the oxidation. We tried to obtain the parameter γ, i.e. the order of oxidation with respect to the dose rate (*16*), from our data. Since γ relates the oxidation rate in homogeneous oxidation conditions R to the dose rate I through the relation

$$R \propto I^\gamma, \qquad (1)$$

the total oxidation after the irradiation time t=D/I (where D is the absorbed dose) is given by

$$A \propto DI^{(\gamma - 1)} \qquad (2)$$

As previously noted, at the dose rates used our samples do not present homogeneous oxidation, but we can assume surface oxidation as representative of the homogenous oxidation process. The oxidation at the surface versus dose rate for samples D1, D2 and D3 is presented in Fig.3. The pure unimolecular (γ = 1) or bimolecular (γ = 0.5) processes do not properly fit the experimental results, which are well described assuming γ = 0.7 in the equation I$^{(\gamma - 1)}$. This value agrees with the results found in the literature for similar materials (*4*), and indicates that the oxidation is due either to the superposition of unimolecular and bimolecular processes or to bimolecular processes involving products partially originating from a single spur. In order to model the

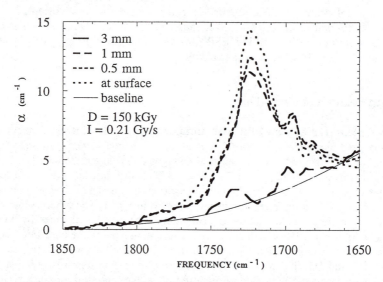

Fig.   1 - Differential spectra of sample D1 at four different depths. (Reproduced with
            permission from *Radiat. Phys. Chem.* **1993**, <u>42</u>, 211, Copyright 1993
            Pergamon).

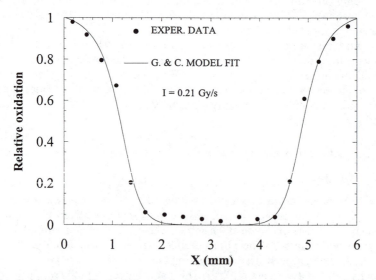

Fig.   2 - Oxidation profile for D1 sample. The best fit parameters are α=1100 and
            β=20. (Reproduced with permission from *Radiat. Phys. Chem.* **1993**, <u>42</u>, 211,
            Copyright 1993 Pergamon).

oxidation profile we have used the Gillen and Clough expression, even if it has been derived only for pure unimolecular or bimolecular processes. Indeed, as previously said, there are indications that Gillen and Clough model (GCM) is suitable to describe the total oxidation effects we observed in our material (K. T. Gillen, Sandia National Laboratories, private communication, 1991). In the GCM the oxidation profile $R(\xi)$ is given by

$$R(\xi) = \frac{\vartheta(\xi) \cdot (1+\beta)}{(1+\beta \cdot \vartheta(\xi))} \qquad (3)$$

where ξ is the normalized depth ($\xi = X/L$), L is the sample thickness and $\theta(\xi)$ is the oxygen profile concentration given by the differential equation

$$\frac{d^2\vartheta(\xi)}{d\xi^2} = \frac{\alpha \cdot \vartheta(\xi)}{(1+\beta \cdot \vartheta(\xi))} \qquad (4)$$

Assuming α and β as free parameters we have performed the best fit of the experimental oxidation profile as measured for sample D1 with relation (3). The result is shown in Fig.2. The best fit has been obtained for $\alpha = 1100$ and $\beta = 20$. From the values of the parameters α and β obtained for sample D1 we can calculate the corresponding quantities for sample B4 which has the same antioxidant content of D1. Indeed the difference between D1 and B4 are the thickness L, the dose rate I and the absorbed dose D; the absorbed dose changes only the absolute intensities because we found a linear dependence of oxidation on the dose. The parameter β is not affected by these quantities while the changes of α can be calculated using the relation

$$\alpha \propto L^2 I^\gamma \qquad (5)$$

Thus for sample B4 we have $\alpha = 512$ and $\beta = 20$. The agreement is excellent between the oxidation profile computed with this values of α and β and the experimental result for the B4 sample (Fig. 4). Moreover, the same profile scaled by a proper factor fits quite well the A4 and C4 experimental data, thus confirming that the antioxidant content do not change the oxidation profile but only its absolute value.

Infrared measurements on additive free-EPR. The spatial distribution across the additive-free EPR plate thickness of the resulting grafted chains was evaluated by adopting the same procedure, based on the IR analysis of very thin sample section [Anelli, P., S. Baccaro, M. Carenza, G. Palma, *Rad. Phys. Chem.* in press]. Graft copolymerization of hydrophilic monomers (2-hydroxyethyl acrylate (HEA), N,N-dimethylacrylamide (DMMA) and N,N-dimethylaminoethyl methacrylate (DMAEM)) onto thin plates of EPR rubber was carried out by the direct irradiation method . Experimental conditions ensuring a surface grafting reaction were ascertained. For all the grafted samples, IR analysis showed a sharp absorption band, not present in the ungrafted EPR spectrum, which is centered around 1730 cm^{-1} and is ascribed to the

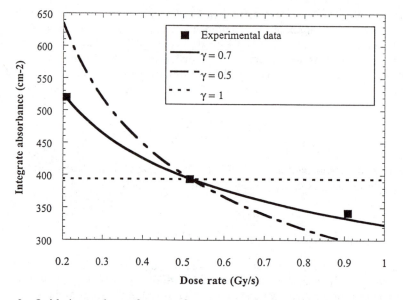

Fig.   3 - Oxidation at the surface as a function of the dose rate for D1, D2, D3
        samples. (Reproduced with permission from *Radiat. Phys. Chem.* **1993**, <u>42</u>,
        211, Copyright 1993 Pergamon).

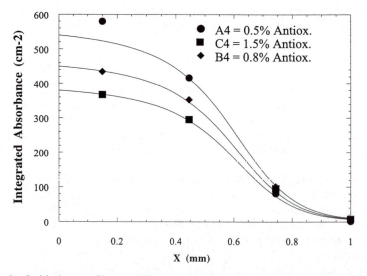

Fig.   4 - Oxidation profiles at different antioxidant content. The curves are the
        oxidation profiles computed with $\alpha = 512$ and $\beta = 20$. (Reproduced with
        permission from *Radiat. Phys. Chem.* **1993**, <u>42</u>, 211, Copyright 1993
        Pergamon).

stretching of C=O. The relative carbonyl integrated absorbance, as a function of depth (X), are reported in Fig. 5 for 6 mm EPR-g-DMAEM samples grafted to different extents. This behaviour can be explained by the slow diffusion of the monomer which initially reaches mostly the surface layers, and subsequently penetrates the sample inner regions. The inner regions are more compatible with the monomer phase as more grafted chains are produced. The effect of the monomer nature (DMAA, HEA and DMAEM) has been investigated, and the integrated absorption of the C=O band for 2 mm EPR samples grafted showed that the inhomogeneity increases in the order: HEA>DMAEM>DMAA. These results suggest that this method can give an evaluation of the grafting distribution inside the sample.

ESR results and discussion

The ESR spectrum of samples recorded after irradiation in air (Fig.6a) can be interpreted as the superposition of two signals (*19, 20*). The first is due to antioxidant R-NO° radical (g = 2.007) and the second to polymer free radicals P° (g = 2.003). During post-irradiation storage in air two different processes take place: i) the partial disappearance of the polymer radicals, and ii) the start of a bimolecular process leading to the formation of the R-NO° radical. For each dose rate at about 100 hours after irradiation, the sample reaches a stable condition and the ESR spectrum is dominated by the R-NO° component. The R-NO° concentration, measured in the stable phase, decreases with dose rate (Fig.7). It is well known (*17, 22*) that antioxidant prevents the ageing of irradiated polymers. Since the R-NO° formation is antagonist to polymer degradation, the result of Fig.7 confirms that the oxidative degradation strongly depends on dose rate (*23*). To better investigate the oxidative degradation, we have performed some measurements on samples irradiated and stored under vacuum. Fig.6b shows that the ESR signal shape is different from the one in air (Fig.6a): as a matter of fact the R-NO° component is not clearly evident in the sample irradiated under vacuum, and the shape seems to be dominated by the polymer component P°. Moreover, while the signal shape remains unchanged, the signal amplitude decreases with storage time in vacuum and reaches a plateau after about 100 hours. For a sample irradiated under vacuum the radical concentration depends on dose rate. A set of samples were irradiated under vacuum with a given dose rate at the same dose value. At different times the sample holders were broken and the samples exposed to air. Two effects were detected: i) the shape of the signal gradually changes from the shape typical of a polymer radical (see Fig. 6b) to the one typical of R-NO° radical (see Fig. 6a). This indicates that the presence of oxygen, in the model rubber, leads to the formation of R-NO° stable radicals; ii) after exposure to air, the overall radical concentration (P° and R-NO°) as measured by the signal intensity, still decreases with storage time. Fig. 8 shows the ESR spectrum of polymer irradiated and measured in air soon after irradiation (Fig. 8a) and after 100 hours storage in air (Fig. 8b). The complex spectrum can be analyzed as the superposition of two different signals. The time evolution indicates that the component at g = 2.007 is stable while the component at g = 2.003 decays in about 100 hours (Fig. 9). In the hypothesis that the peak at the lowest magnetic field is due to immobile radicals while the peak at midfield is due to mobile ones (*24*), we assume that the decay

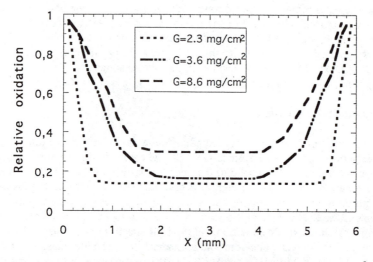

Fig. 5 - C=O band profile for EPR-g-DMAEM at the grafting extent (mg/cm^2) of 2.3 (----), 3.6 (— ---) and 8.6 (— —). (Reproduced with permission from *Radiat. Phys. Chem.* in press).

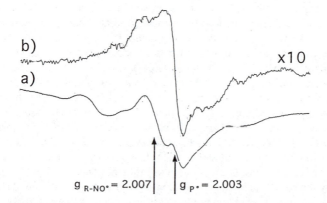

Fig. 6 - ESR spectra of samples measured soon after the irradiation at 100 kGy (dose rate 0.3 Gy/s) in air (a) and in vacuum (b). (Reproduced with permission from *Nucl. Instr. Method. B,* in press).

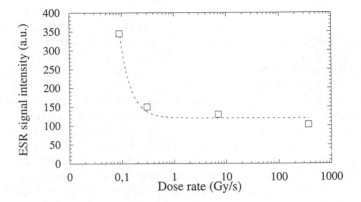

Fig. 7 - ESR signal amplitude studied as a function of the dose rate. The samples were
irradiated at 300 kGy and stored in air. (Reproduced with permission from
Nucl. Instr. Method. B, in press).

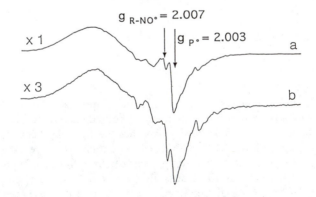

Fig. 8 - ESR spectra of polymer samples irradiated in air at 100 kGy (dose rate 0.3
Gy/s): a) just after the irradiation; b) after 100 h storage time at room
conditions. (Reproduced with permission from *Nucl. Instr. Method. B,* in
press).

of the component g = 2.003 is due mainly to the mobile radicals and that the immobile radicals, at the same measurement conditions, are stable.

To describe the radical behaviour, a semiquantitative model has been proposed for samples irradiated up to 600 kGy at 0.3 Gy/s (S.Baccaro, ENEA-Casaccia, Italy, unpublished data). The relative R-NO° radical concentration after 1000 hours storage time as a function of dose is reported in Fig. 10. The result suggests that the process involved in the R-NO° production is non linear with dose. Indeed at each dose value the ESR signal amplitude has been measured, and it increased linearly with the antioxidant content. Figure 11 reports the experimental data on time evolution of R-NO° radical and Fig. 12 of polymer P° radicals. The reported results evidence that in the presence of oxygen: i) the recombination and production processes of the involved radicals are concomitant and ii) the R-NO° vs dose production is nonlinear with the involved radical concentrations. Moreover it is well known that when antioxidant (A) is present in the material, the polymeric radicals react each other and with the antioxidant producing radicals in the antioxidant itself and termination products (17, 25). Referring only to long-lived free radicals, the simplified reactions can be used to describe the process. On the hypothesis that correlated nonlinear chemical reactions take place, after gamma irradiation, it is possible to describe the processes of the three involved species with the following rate-equations

$$\frac{dN_C}{dt} = -k_1 N_C N_B^{\alpha} - k_2 N_C$$

$$\frac{dN_B}{dt} = -k_1 N_C N_B^{\alpha} \qquad\qquad (6)$$

$$\frac{dN_A}{dt} = -k_2 N_C$$

N_A, N_B and N_C are the concentrations of A, B and C species and α is different from 1 to take into account the nonlinear processes involved. In the studied material the antioxidant contains the -NH functional group, thus the N_A radical concentration of the scheme can be related to the R-NO° radical concentration. Moreover, the ESR spectra reported in Fig.8 suggest that the two components can be assigned to N_B and N_C radical concentrations of the scheme. It is possible to give a theoretical prevision of the behaviour of R-NO° ESR signal amplitude as a function of the dose (see Fig. 10), and the time evolution of the R-NO° and P° radicals (see Figs. 11 and 12) by solving the equations with the initial condition: $N_C(0)$ and $N_B(0)$ proportional to the absorbed dose. The best agreement between experimental and theoretical data was obtained for $\alpha=2$. It is possible to hypothesize that B and C are species of type PO° and $PO_2°$ which allow the following reactions (17):

$$PO_2° + \quad A \rightarrow AH° + \text{stable products}$$
$$PO_2° + PO° \rightarrow \text{stable product}$$

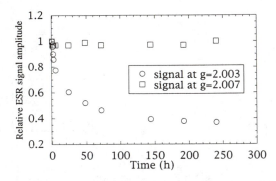

Fig. 9 - Relative ESR signal amplitude, for the component at g = 2.003 and g = 2.007, studied as a function of time. (Reproduced with permission from *Nucl. Instr. Method. B,* in press).

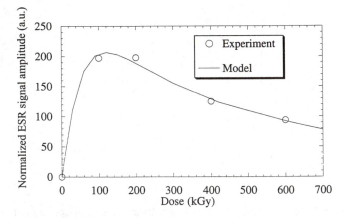

Fig. 10 - Normalized ESR signal amplitude for R-NO° vs dose (dose rate 0.3 Gy/s). The solid line is the model prevision of the free-radical response as a function of dose.

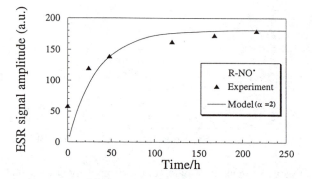

Fig. 11 - ESR signal amplitude vs. storage time for R-NO° radical irradiated at 100 kGy. The solid line represents the model prevision of the post-irradiation behaviour.

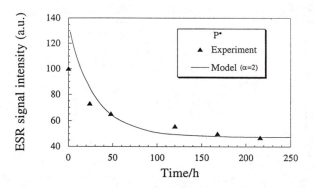

Fig.12- ESR signal amplitude vs. storage time for P° radical irradiated at 100 kGy. The solid line represents the model prevision of the post-irradiation behaviour.

In this hypothesis the number of radicals detectable on the polymer is always the sum of N_B and N_C. The proposed model well describes the R-NO$^\circ$ behaviour with dose, while some discrepancies are evident at t=0 in the R-NO$^\circ$ and P$^\circ$ time evolution. This effect can be easily understood considering that the model does not take into account the contribution of the radicals created during irradiation. In fact, the P$^\circ$ experimental value is lower than the theoretical estimate because of the recombination processes which take place during irradiation. On the contrary the R-NO$^\circ$ experimental value is higher than the theoretical estimate since the process leading to the R-NO$^\circ$ formation also works during irradiation. These discrepancies suggest the improvement of the proposed model by taking into account the processes which occur during irradiation.

Activation energy of R-NO$^\circ$ and P$^\circ$ components.   In order to evaluate the activation energy of the two processes, the time evolution of both components as a function of the storage temperature has been studied (*19*). In the temperature range of 273-333 K no thermal degradation effects were observed, moreover, for temperatures lower than 273 K the time evolution was so slow that no changes in ESR spectrum were detected in the first 200 hours. The time evolution of the process is strongly affected by temperature. In order to derive a characteristic time as a function of the temperature, a scaling factor τ of the time axis was choosen for the temperatures different from 333 K. The amplitude of the ESR signal as function of the scaled time is shown in Figs. 13 and 14 for both polymeric and R-NO$^\circ$ components. The values of the scaling parameters for each component arranged in an Arrhenius plot (ln τ vs 1/T) (*2*) exhibit a linear behaviour within the experimental errors (Fig.15); the slopes of the two linear best fits give the activation energy of the two processes. The values for the activation energies are $E_{R\text{-}NO^\circ}$ = 16 kCal/mol for the process leading to the R-NO$^\circ$ formation, and E_{P° = 6 kCal/mol for the P$^\circ$ disappearance. Indeed, a similar decay is also present in pure ethylene-propylene. The more interesting feature is the large after irradiation growth of the R-NO$^\circ$ component. Following the hypothesis discussed on our previous work on the formulated rubber (*20*), this effect can be attributed to the decay of a metastable species created by γ radiation in presence of oxygen, and not detectable with ESR technique. The well defined activation energy found for the R-NO$^\circ$ creation indicates that it is related to a single process; nevertheless the time evolution of the R-NO$^\circ$ component is far from a pure exponential as evident from Fig. 14. Those two facts are not in contradiction if we affirm that the decay of the metastable species with creation of R-NO$^\circ$ radicals is a bimolecular process. Indeed, with this assumption we have that the concentration C_X of the metastable species as a function of time is given by $C_X(t) = C_0/(1+ KT)$ where K is the reaction rate and C_0 the initial concentration. From this relation, the time dependance of NO$^\circ$ concentration normalized to its value at T = 0 is obtained as

$$C_{R\text{-}NO^\circ}(t)/\ C_{R\text{-}NO^\circ}(0) = [\ 1+ KT \times (\ C_{R\text{-}NO^\circ}(\bullet)/\ C_{R\text{-}NO^\circ}(0)\ ] \times 1/\ (1+KT) \qquad (7)$$

We have used expression (7) to fit our experimental data; the resulting best fit is shown in Fig. 14, the values obtained for the two parameters are

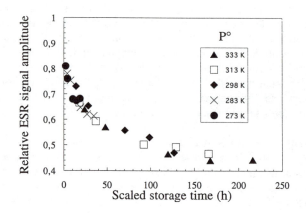

Fig. 13 - Relative ESR signal amplitude vs. scaled time t / τ for P° component. (Reproduced with permission from *Radiat. Phys. Chem.* **1993**, <u>42</u>, 241-244. Copyright 1993 Pergamon).

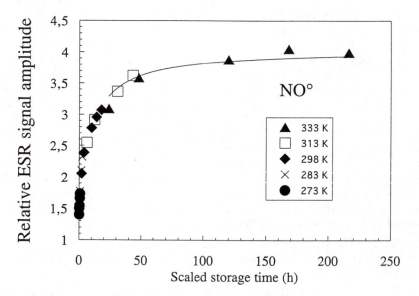

Fig. 14 - Relative ESR signal amplitude vs. scaled time t / τ for R-NO° component. The solid curve was obtained as the result of a best fit analysis using eq. (7). (Reproduced with permission from *Radiat. Phys. Chem.* **1993**, <u>42</u>, 241-244. Copyright 1993 Pergamon).

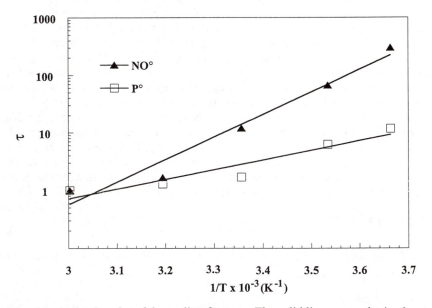

Fig. 15 - Arrhenius plot of the scaling factor τ . The solid lines were obtained as a results of a linear regression. (Reproduced with permission from *Radiat. Phys. Chem.* **1993**, <u>42</u>, 241-244. Copyright 1993 Pergamon).

$$C_{R-NO^\circ}(\bullet)/C_{R-NO^\circ}(0)=4 \qquad\qquad 1/K = 8 \text{ hours}$$

The results support the hypothesis that the process responsible for the $R-NO^\circ$ formation is bimolecular.

Conclusions

The infrared and ESR spectra show that the radiation induced effects on the investigated polymer-based rubber are strongly influenced by oxygen and dose rate. In particular, the oxidative degradation, increases at low dose rate values due to increased time for oxygen diffusion. The IR profile is a very good method to describe the oxidative degradation and to evaluate the grafting distribution inside the sample. The good agreement between infrared experimental data and theoretical prevision proposed by Gillen and Clough confirms the same scheme of reactions for a polymer added with antioxidant. The values of the α and β parameters are independent of the antioxidant content. This suggests that the antioxidant is oxidized in place of the polymer, but it does not change the reactions involved in the oxidation process and the oxygen distribution on the sample. Moreover, the analysis of the different components in the ESR spectrum of ethylene-propylene (with antioxidant irradiated in presence of oxygen) confirms that during post-irradiation storage in air two different processes take place: i) the partial disappearance of the polymer radicals, and ii) the initiation of a bimolecular process leading to the formation of the $R-NO^\circ$ radical. Further, the recombination and production processes of the involved radicals are concomitant, and the $R-NO^\circ$ production as function of the absorbed dose is nonlinear with the involved radical concentrations.

ACKNOWLEDGMENTS

The author is grateful to Prof. U. Buontempo for his fundamental help in proposing and discussing the radical behaviour model.

REFERENCES

1.   Charlesby A., *Atomic Radiation and Polymers*, Pergamon Press, Oxford, **1960**.
2.   Dole M. (Ed.) (**1972**, Vol. I and **1973**, Vol. II), *The Radiation Chemistry of Macromolecules*, Academic Press, New York.
3.   Grassie, N. and G. Scott, *Polymer Degradation and Stabilization*, Cambridge University Press, Cambridge, **1985**.
4.   Decker, C., F.R. Mayo and H. Richardson , *J. Polym. Sci. Polym. Chem. Ed.*, 11, 2879, **1973**.
5.   Decker C. , *J. Polym. Sc. Polym. Chem. Edn.* 15, 781, **1977**.

6. Bower, D.I. and W.F. Maddams, *The Vibration Spectroscopy of Polymers*, Cambridge University Press, Cambridge, **1989**.
7. Baccaro, S. and U. Buontempo , *Rad. Phys. Chem.*, <u>40</u>, 175, **1992**.
8. Baccaro, S., U. Buontempo, P. D'Atanasio , *Rad. Phys. Chem.*, <u>42</u>, 211, **1993**.
9. Gillen, K. T., R. L. Clough and N. J. Dhooge , *Polymer*, <u>27</u>, 225, **1986**.
10. Gillen, K. T., R. L. Clough and C. A. Quintana , *Polym. Degrad. and Stab.*, <u>17</u>, 31, **1987**.
11. Morita, Y., T. Yagi and W. Kawakami, *Proc. Int. Symp. Radiation Degradation of Polymers and the Radiation Resistant Materials*, Takasaki, p. 81, **1989**.
12. Cunliffe, A. V. and A. Davis , *Polym. Degrad. and Stab.*, <u>4</u>, 17, **1982**.
13. Papet, G., L. Jirackova-Audonin, J. Verdu , *J. Radiat. Phys. Chem.*, <u>29</u>, 65, **1987**.
14. Seguchi T., Hashimoto S., Arakawa K., Hayakawa N., Kawakami W. and Kuriyama I. *Radiat. Phys. Chem.,* <u>17</u>, 195, **1981**.
15. Papet, G., L. Jirackova-Audonin, J. Verdu, *J. Radiat. Phys. Chem.*, <u>33</u>, 329, **1989**.
16. Clough, R. L. and K. T. Gillen, In Radiation Effects on Polymers; R. L. Clough and S. W. Shalaby Eds.; ACS Symposium Series. <u>475</u>, 457, **1991**.
17. B.Ranby and L.F.Rabek, in: *ESR Spectroscopy in Polymer Research*, Springer-Verlag, Berlin, **1977**.
18. Onori S., S. Baccaro , B. Caccia , P. D'Atanasio , P.L. Indovina , M. Pantaloni , E. Petetti and G. Viezzoli , *Radiat. Protect. Dos.*, <u>34</u>, 299, **1990**.
19. Baccaro S., U. Buontempo, B. Caccia , S. Onori, M. Pantaloni, *Radiat. Phys. Chem.*, <u>42</u>, 241, **1993**.
20. Baccaro S., U.Buontempo, B.Caccia, S.Onori, M.Pantaloni, *Appl. Rad. and Isot.*, <u>44</u>, 331, **1993**.
21. Baccaro S., U.Buontempo, B.Caccia, S.Onori, M.Pantaloni, *Proc. Polymex 93 - Cancun (Mexico)*, 1-5 Nov., 201, **1993**.
22. Kamiga Y. and E. Niki, in *Aspects of Degradation and Stabilization of Polymers*, ed. H.H. Jellinek , Elsevier, **1979**.
23. Clough R. L. and K. T. Gillen, in *Oxidation Inhibition in Organic Materials*, eds. P.Klenchuk and J.Pospisil, CRC Press, **1988**.
24. Kashiwabara H. and T. Seguchi, in *Radiation Processing of Polymers*, eds. A. Singh and J. Silverman, Hanser Publishers, **1992**.
25. Scott G., *Atmospheric Oxidation and Antioxidants*, Elsevier Publishing Company, Amsterdam, **1965**.

RECEIVED October 23, 1995

Chapter 26

Development of an Accelerated Aging Method for Evaluation of Long-Term Irradiation Effects on Ultrahigh-Molecular-Weight Polyethylene Implants

D. C. Sun, C. Stark, and J. H. Dumbleton

Pfizer Hospital Products Group, Howmedica, Inc.,
Rutherford, NJ 07070–2584

A general scheme for developing an accelerated aging method for irradiated biomaterials is proposed. Using UHMWPE implants as an example, an accelerated thermal diffusion oxidative aging (ATDOA) method has been developed. The method requires an optimum initial heating rate and an optimum aging temperature to accelerate oxidation reactions. Based upon oxidation-induced material property changes (crystallinity by DSC, tensile properties by ASTM D638 tensile test, oxidation index by FTIR, and low molecular weight fraction by GPC), correlations between accelerated aging time and post-radiation shelf aging time were obtained. The new ATDOA method allows a rapid evaluation of long-term irradiation effects on the material properties of UHMWPE implants.

Ultra high molecular weight polyethylene (UHMWPE) has been used as an orthopaedic bearing material for more than two decades with satisfactory clinical results. The current research effort is aimed at prolonging its performance *in vivo*. As a virgin polymer, UHMWPE is chemically inert and stable. However, upon the forming and fabrication process (compression molding, ram extrusion, etc.) and high energy beam radiation (such as γ-ray, x-ray, or electron beam irradiation, a step required for sterility), free radicals are formed causing the UHMWPE to become reactive. Oxidation can occur during irradiation, post-irradiation shelf-aging, or *in vivo* in the presence of oxidants. Accompanying the oxidation reaction are slow but progressive material property changes in UHMWPE which have been unobserved in the past, due to the lack of sensitive evaluation tools. In this study, we first evaluated shelf-aged UHMWPE specimens to obtain correlations between material properties and aging time. An accelerated aging method was then developed to simulate the long-term oxidation effects in a short period of time. Finally, material property changes of clinical retrievals reported in the literature were discussed.

0097–6156/96/0620–0340$12.00/0

Sample Preparation and Evaluation Methods

UHMWPE samples (acetabular cups, tibial inserts, and blocks, all made of GUR resins) were γ-ray irradiated in air at a nominal dose of 2.5 Mrads and stored in air at room temperature for various time periods prior to evaluation. Unirradiated UHMWPE samples stored during the same time periods were also evaluated for comparison. A new experimental UHMWPE with a superior oxidation resistance was included in the study. This stabilized UHMWPE sample differed from a regular medical grade UHMWPE sample in that it was packed and irradiated in nitrogen followed by an inert annealing step at an elevated temperature to eliminate free radicals through crosslinking (*1*). For DSC thermal analysis and FTIR analysis, thin slices (ca. 50 to 100 microns) were either obtained by microtoming (when the sample had a flat surface) or by a razor blade (performed manually) from the surface zone of UHMWPE components or samples. The depth of the thin slices was kept between 0 and 2 mm from the component surface. Crystallinity and melting behavior were measured by a Perkin-Elmer 7-Series DSC thermal analysis system. All runs were carried out with nitrogen purge in the sample chamber. Sample weight was kept between 3.5 and 3.8 mg for consistency. The heating rate used was 10°C/min. The crystallinity was determined by the ratio of the heat of melting of the sample to 68.4 cal/g which is commonly used as the heat of melting for a perfect polyethylene crystal. Repetitive runs using an UHMWPE control sample (unirradiated) showed the accuracy of crystallinity to be within $\pm$ 1.5%. A Nicolet 710 FTIR microscopy system was used to determine the extent of oxidation as caused by the γ-ray radiation. The oxidation index was calculated as the ratio of the peak area between 1650 and 1800 cm^{-1} (C=O or C-O groups) to the 1468 cm^{-1} peak area (CH_2 and CH_3 groups) so that the sample thickness dependence was removed. The so-determined oxidation index had an accuracy of $\pm$ 0.01 as derived from repetitive runs on an irradiated UHMWPE control sample. Tensile tests followed ASTM D638 and D1708 procedures. Tensile specimens of 1 mm in thickness were machined from the surface zone of UHMWPE components (ca. 0 to 2 mm depth). Four to six specimens per sample condition were tested and the averaged results were reported. A crosshead speed of 25.4 mm/min was used. From repetitive tests of an unirradiated UHMWPE control, the testing variations were 25% for elongation at break (or ca. 8% on percent basis), 110 psi for tensile yield strength (ca. 4% on percent basis), and 510 psi for ultimate tensile strength (ca. 8% on percent basis), respectively. Low molecular weight fraction was determined by extraction with hot trichlorobenzene (TCB) followed by molecular weight analysis for the soluble portion by GPC. The UHMWPE samples were heated in TCB (with antioxidant N-phenyl-2-napthylamine) at 170°C for 6 hours followed by hot filtering using a Waters High Temperature Filtration Apparatus. For molecular weight analysis, a Waters 150C gel permeation chromatogram with a Jordi Gel Mixed Bed column at a column oven temperature of 145°C was used. The definition of low molecular weight was arbitrarily chosen to be below 100,000. Note that for a typical UHMWPE, the average molecular weight, Mw, is in the range of 1000,000, or ten times higher than 100,000 chosen above for the low molecular weight fraction.

Evaluation of Shelf-aged UHMWPE Components

Samples of γ-ray irradiated (in air), shelf-aged acetabular cups and tibial inserts made of surgical implant grade UHMWPE were evaluated. The results are plotted in Figures 1 through 5. In Figure 1, crystallinity was seen to increase with aging time for irradiated, shelf-aged samples, ranging from 53% to 73%. This increase in crystallinity is believed to be caused by chain scission followed by loose chain or short chain crystallization. In contrast, unirradiated samples showed little change in crystallinity upon shelf-aging. Increasing the crystallinity can cause an increase in density and modulus which may lead to an undesirable increase in the contact stress at the articulating surface. Tensile yield strength also increased with aging time (Figure 2), due to the corresponding increase in crystallinity shown in Figure 1. Both ultimate tensile strength and elongation at break decreased with aging time (Figures 3 and 4), indicating a partial loss of ductility upon irradiation and shelf-aging. Oxidation index increased with aging time in almost a linear relationship, ranging from 0.02 to 0.18 (Figure 5), indicating the occurrence of progressive oxidation at a nearly constant oxidation rate which is described by the rate equations:

$$r = k\,[O_2]\,[p.] \tag{1}$$
$$k = A\,e^{-\Delta H/T} \tag{2}$$

where r is the oxidation reaction rate and $[O_2]$ and $[p.]$ are the oxidant and free radical concentrations, respectively. Since the oxidation reaction during shelf-aging is relatively slow (years), oxygen diffusion into the polymer matrix is not the limiting step and the oxygen concentration should be close to that in air (i.e. 20.8%) throughout the sample during shelf-aging. The ambient temperature during storage can also be assumed to be constant. Therefore, the linear relationship of oxidation-induced property changes with aging time implies that the free radical concentration is fairly constant over time. Literature reports (2) indicate that the slow decomposition of oxidized products, such as hydroperoxides (-OOH groups) can generate new free radicals. An equilibrium concentration of free radicals can be reached when the generation and the recombination of free radicals are in balance. A conclusion that can be drawn from the above results is that oxidation occurred in a slow but steady rate during the shelf-aging after sterilization.

Development of Accelerated Aging

Our first attempt to develop an accelerated aging procedure was made by immersing irradiated UHMWPE samples in a 3% hydrogen peroxide solution at room temperature for 15 days. Subsequent evaluations using DSC, FTIR, and tensile tests showed no significant change in physical, chemical, or mechanical properties. In view of equations (1) and (2) shown above, it was realized that temperature is the key for the acceleration of the oxidation reaction. When temperature increases, the reaction rate constant k

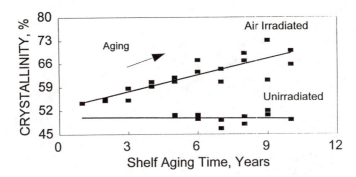
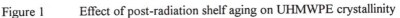

Figure 1   Effect of post-radiation shelf aging on UHMWPE crystallinity

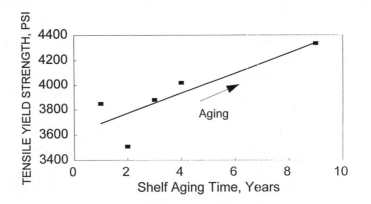

Figure 2   Effect of post-radiation shelf aging on tensile yield strength of
                UHMWPE

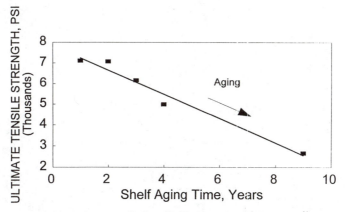

Figure 3   Effect of post-radiation shelf aging on ultimate tensile strength of
                UHMWPE

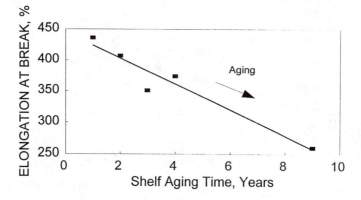

Figure 4   Effect of post-radiation shelf aging on elongation at break of
                  UHMWPE

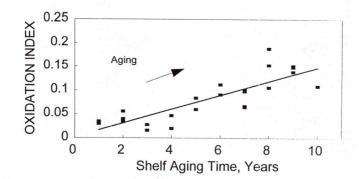

Figure 5   Effect of post-radiation shelf aging on oxidation index of
                  UHMWPE

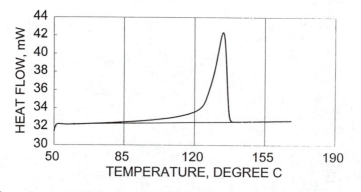

Figure 6   DSC heating curve for a unirradiated surgical implant grade
                  UHMWPE; The onset of melting occurred at ca. 80°C.

increases exponentially. The effective oxygen concentration will increase due to the relaxation of amorphous regions in the polymer which makes the polymer matrix more accessible to oxidants. Diffusivity of oxidants increases as well. The only downside of employing an elevated temperature is that it can also accelerate recombination of free radicals and cause crosslinking under the condition that there is a shortage of oxygen in the polymer matrix. Another concern of using a high temperature for accelerating oxidative aging is that it can potentially cause morphology and property changes which would not occur at room temperature or body temperature. UHMWPE differs from HDPE and LDPE in that it possesses a relatively large number of tie molecules in the amorphous region. These entangled tie molecules are responsible for UHMWPE's excellent toughness, impact, and wear properties and also make the penetration of oxygen more difficult. After irradiation and at the beginning of shelf-aging, many tie molecules still remain intact. As a result, material property changes (such as tensile, impact, and wear properties) are hard to detect. To overcome these difficulties, we have developed an accelerated thermal diffusion oxidative aging (ATDOA) method which comprises the following steps:

i Conduct a DSC heating run and determine the onset point of melting for the polymer. The maximum temperature used for ATDOA must be below the onset temperature to avoid any thermally-induced morphological changes.

ii Develop an optimum heating schedule to raise the sample temperature from room temperature to the aging temperature in air. This step provides adequate oxygen diffusion and minimizes the undesirable recombination of free radicals during the initial stage of ATDOA.

iii Continue ATDOA by holding the sample at the aging temperature for several time periods.

iv Evaluate material properties of ATDOA samples as a function of aging time.

v Evaluate material properties of shelf-aged samples as a function of aging time.

vi Obtain a correlation between ATDOA aging time and shelf-aging time for specific material properties on the basis that ATDOA and shelf-aging will produce similar property changes.

vii Evaluate material properties of clinically retrieved samples as a function of *in vivo* implantation time.

viii Obtain a correlation between ATDOA aging time and implantation time for specific material properties on the basis that ATDOA aging and *in vivo* implantation will produce similar property changes.

Steps i through vi are required to establish the ATDOA method for shelf-aging, while Steps i through viii allow a rapid simulation of long-term sterilization effects for both shelf-aging and *in vivo* performance. In this report, UHMWPE implants were exposed to the ATDOA process. Studies on other biomaterials are to be reported later.

Material Property Changes in ATDOA UHMWPE Samples

Figure 6 shows the DSC heating curve of an UHMWPE sample. The onset of melting occurs at ca. 80°C which was determined to be the maximum aging temperature for ATDOA. Several heating rates have been studied in order to optimize the heating procedure to bring the sample temperature from room temperature to 80°C. Results from a heating rate of 0.6°C/min are reported here which gave satisfactory aging effects. After the sample temperature of 80°C was reached, ATDOA continued for 11 and 23 days, respectively. Two samples included in the study were an UHMWPE sample irradiated in air and a new experimental UHMWPE (γ-ray irradiated in nitrogen followed by a free radical elimination step (*1*)) with improved oxidation resistance. Figure 7 shows the crystallinity change as a function of accelerated aging time. For the sample irradiated in air, crystallinity varied from 57% to 67% upon the accelerated aging. Comparing Figure 7 with Figure 1 (shelf-aged) for samples irradiated in air, it was found that the 11 day (62% crystallinity) and 23 day (67% crystallinity) ATDOA are equivalent to ca. 4 to 6 and 7 to 9 years of shelf-aging, respectively. In contrast, little crystallinity change was found for the new experimental UHMWPE upon aging (Figure 7). Figure 8 shows the ultimate tensile strength change as a function of accelerated aging time. Referring to Figure 3 for the shelf-aging correlation, the 11 day (5200 psi) and 23 day (3200 psi) ATDOA processes corresponds to ca. 5 and 9 years of shelf-aging, respectively. Again, the new experimental UHMWPE shows little change. Figure 9 shows the oxidation index as a function of accelerated aging time. Referring to Figure 5 for shelf-aging, the 11 day (0.06 oxidation index) and 23 day (0.11 oxidation index) ATDOA correspond to ca. 4 to 6 and 7 to 9 years of shelf-aging, respectively. The oxidation level in the new experimental UHMWPE is almost undetectable (below 0.01) and remains low after ATDOA. Figure 10 shows the low molecular weight fraction (below 100,000, determined by high temperature GPC for the soluble portion) as a function of accelerated aging. The increase in the low molecular weight fraction upon aging for the sample irradiated in air, ranging from 27.5% to 47.4%, is consistent with oxidation-induced chain scission. For the new experimental UHMWPE, the low molecular weight fraction remains low (below 18.5%) upon ATDOA.

Material Property Changes in Clinically Retrieved UHMWPE

Oxidative aging of UHMWPE implants *in vivo* may occur at a different rate compared to shelf-aging, due to a higher temperature (37°C vs. room temperature), a lower oxidant concentration (O_2, H_2O_2, enzymes, etc., all at low concentrations), fatigue loading, and wear. When an UHMWPE retrieval is received for evaluation, both shelf-aging (prior to implantation and post-surgery) and *in vivo* aging (actual implantation time) must be considered. The information required includes device manufacturing date, date of surgery, date of retrieval, and date of evaluation. Patient-related parameters, such as body weight and activity level may also play a secondary role in oxidative aging of UHMWPE. In the literature, Eyerer et.al. (3) showed retrieved UHMWPE acetabular cups to possesses an increased density and oxidation level with increasing implantation time. We are currently conducting a systematic study along these lines.

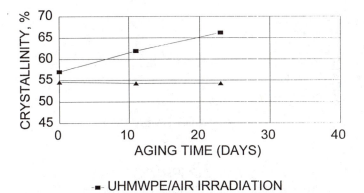

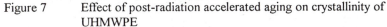

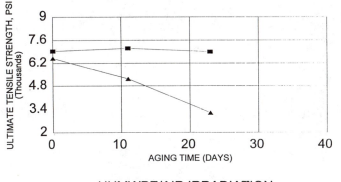

Figure 7   Effect of post-radiation accelerated aging on crystallinity of UHMWPE

Figure 8   Effect of post-radiation accelerated aging on ultimate tensile strength of UHMWPE

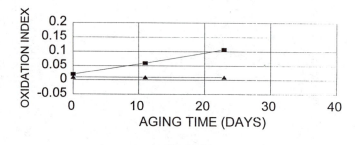

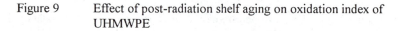

Figure 9   Effect of post-radiation shelf aging on oxidation index of
                UHMWPE

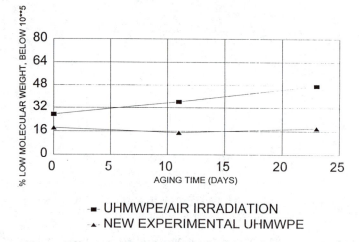

Figure 10   Effect of post-radiation shelf aging on low molecular weight
                fraction of UHMWPE

Conclusion

A general scheme for developing an accelerated aging method for biomaterials is proposed. Using UHMWPE, an accelerated thermal diffusion oxidative aging (ATDOA) method has been developed which allows a rapid evaluation of long-term irradiation effects on the material properties of UHMWPE.

List of Symbols

r	oxidation reaction rate
k	oxidation reaction rate constant
$[O_2]$	oxygen concentration
[p.]	free radical concentration
A	constant
ΔH	activation energy of oxidation reaction
T	absolute temperature

Acknowledgement

The authors would like to thank Mr. G. Schmidig for assistance in preparing all the figures. Thanks are also given to Ms. P. Szabo for preparing the manuscript.

Literature Cited

1. U.S. Patent 5,414,049 (D.C. Sun and C. Stark, "Non-oxidizing Polymeric Medical Implants")
2. Carlsson, D. J., Chmela, S. , and Lacoste, J. in *Radiation Effects On Polymers;* Editors, Clough, R.L. and Shalaby, S.W. , ACS Symposium Series 475, American Chemical Society Books Department, Washington D.C., 1991, pp. 432.
3. Eyerer, P, Kurth, M., McKellop, H.A., and Mittlmeier, T., *J. Biomedical Materials Research*, 1987, vol 21, pp. 275-291.

RECEIVED July 28, 1995

Chapter 27

Performance of Polymeric Materials as Shielding for Cosmic Radiation

M. Y. Kim[1], S. A. Thibeault[2], J. W. Wilson[2], R. L. Kiefer[1], and R. A. Orwoll[1]

[1]Program in Applied Science and Department of Chemistry, College of William and Mary, Williamsburg, VA 23187–8795
[2]Environmental Interactions Branch, Materials Division, Langley Research Center, National Aeronautics and Space Administration, Hampton, VA 23681–0001

NASA LaRC's high-charge, high-energy nuclei transport codes are implemented to calculate the fluence of projectile fragments behind polymeric materials from energetic ion beams and to estimate their performance as shielding from galactic cosmic ray (GCR) for biological systems on long-term space missions. Energetic heavy ions fragment and lose energy upon interaction with shielding materials specified elemental composition, density, and thickness. The shield effectiveness is examined by using conventional quality factors to calculate the dose equivalents and also by using the probability of the neoplastic transformation ratio of shielded C3H10T1/2 mouse cells. The attenuation of biological effects within the shield and body tissues depends on the shielding nuclear properties. The results show that hydrogenous materials are good candidates for high-performance shields. The quantitative results vary with the biological model used.

Humans in a lunar habitat or on a manned mission to Mars will require more protection from GCR than has been used heretofore on shorter missions. The cosmic radiation arriving beyond Earth's magnetic fields is composed of ~98% nuclei, stripped of all their orbital electrons, and ~2% electrons and positrons (*1*). The galactic cosmic rays consist mostly of protons and alpha particles, but have a small but significant component of heavier particles with kinetic energies up to 10^{10} GeV (*2*). The specific ionization of these high-charge, high-energy (HZE) nuclei is unusually high and they may pose a significant health hazard, especially acting as a carcinogen. Galactic heavy ions will probably be the ultimate limiting factor in space operations because their relative dose contributions are comparable to those of light particles and because gross rearrangements and mutations and deletions in DNA from HZE particles are expected (*3*). The purpose of this work is to predict the effectiveness of various polymeric materials as shields from GCR.

As HZE in the GCR interact with a shield, they fragment and deposit energy at rates that depend on the nature and energy of the incident particles, the nature of the shield material, and the depth into the shield. The relationships are complex, so that, in some instances, the "shield" can cause an increase in both the number of particles and the dose due to the production of secondary particle radiation. For example, the dose from HZE particles absorbed by a human behind a 1.3-cm aluminum shield, the traditional structural material for spacecraft, exceeds by 10% the dose absorbed in free space (*4*). This is due to a greater rate of energy transfer at the back side of the shield (a) by the projectile or its fragments after they have been slowed by their passage through the shield, (b) by secondary energetic nuclei and fragments generated in the shield, and (c), to a lesser extent, by secondary particles knocked out of the target material.

GCR Transport

Cosmic ray nuclei are the only direct and measurable sample of matter from outside the solar system. Although GCRs probably include every natural element, not all are important for space radiation protection purposes. The abundances for species heavier than iron (atomic number, $Z > 26$) are typically 2 to 4 orders of magnitude smaller than that for iron (*5*). Figure 1 illustrates the measured spectra at 1 astronomical unit (AU) for hydrogen, helium, and heavy ions up to nickel (atomic number 28) at the 1977 solar minimum modulation from the relatively quiet solar cycle 21 (1975-1986) (*6*).

The propagation and interactions of high-energy ions up to atomic number 28 (Ni) in various target materials were simulated (*4*) using the transport code HZETRN (*3*). The code applies the straight-ahead approximation with velocity conserving fragmentation interactions for high-charge, high-energy nuclei colliding with shield materials. These interactions depend on the shield material, thickness, and the projectile-target interaction parameters, such as nuclear fragmentation cross sections. This code accounts for the fragmentation of the incident HZE ions and nucleons, but neglects the secondary fragments derived from heavy atoms of the shield material. Materials in the target shield are characterized for the computation by their bulk density and elemental composition. Their properties as a shield depend on the atomic and nuclear cross sections.

The primary mechanism for loss of energy by HZE particles is by means of Coulombic interactions with electrons in the target. Thus, high linear-energy transfer (LET) for HZE particles is more easily achieved with materials having large numbers of loosely bound electrons per unit mass. Additional energy is lost through Coulombic interactions and collisions with target nuclei. Although nuclear reactions are far less numerous, their effects are magnified because of the large momentum transferred to the nuclear particles and the impacted nucleus itself. Many of the secondary particles of nuclear reactions are sufficiently energetic to promote similar nuclear reactions and thus cause a buildup of secondary radiation, which may pose an

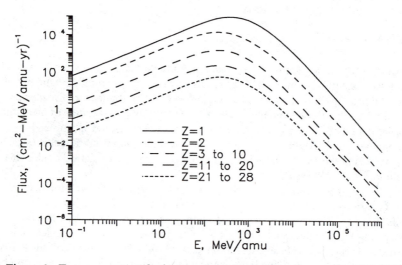

Figure 1. Energy spectra of primary galactic cosmic ray ions at the 1977 solar minimum (data from ref. 6).

increased hazard. Because primary nuclei undergo nuclear fragmentation, less ionizing secondaries produced by fragmentation of HZE may pose a reduced hazard.

With cosmic rays propagating through interstellar space, it is customary, and more useful physically, to express distances by the total mass of all atoms encountered, and to do so in units of grams per square centimeter (g/cm^2). The thickness of an absorber can be converted to a linear thickness by dividing by the density of the matter.

Modeling of Monoenergetic, Single-Ion Beams

The high-energy heavy-ion radiation components are usually attenuated to lower linear energy transfer (LET) as a result of nuclear interactions between projectile and target nuclei. These processes become more significant as the particles penetrate further into the medium. Although heavy nuclei are present in small amounts in GCR, their effects are important because LET is proportional to the square of the ion charge. Radiation within a spacecraft structure, which interacts with onboard personnel or equipment, depends on the shield composition because of differences in atomic cross sections, nuclear attenuation, and the distribution of fragmentation products. Since hydrogen presents the greatest cross section per unit mass from theoretical calculations (*4*), hydrogen-containing materials, such as polymers, are candidates for shielding materials. Additionally, hydrogen is particularly effective in undergoing elastic collisions with the secondary neutrons generated within the shield thereby reducing the neutrons' energy and making them susceptible for absorption by other hydrogen atoms or other elements.

Energetic primary particles suffer nuclear reactions before stopping in a shield medium. The secondary radiation resulting from these reactions yields a broad distributions of energies among the lighter particles. The most energetic secondaries are confined to a narrow cone about the initial direction and are close to the initial beam axis over at least the first mean-free path (*3*). This cone narrows with increasing primary energy. The flux of each secondary radiation with a broad energy distribution was integrated numerically to compute the total ion fluence. This was then compared for different materials.

Calculations were performed for an irradiation with 33.88 GeV ^{56}Fe ions of six shield concepts constructed with the polymeric materials listed in Table I. The projectile was chosen for analysis because relativistic ^{56}Fe nuclei are among the dominant HZE particles in GCR of radiobiological significance for manned spaceflight. The beam energy matched experimental data taken at the Lawrence Berkeley Laboratory. Pilot experiments to validate theoretical results have been performed but data reduction is not yet complete (Thibeault, S. A., private communication). Figures 2a and 2b show the fluence of ^{56}Fe and its fragments from the back face of the shield for two thicknesses, 5 and 18 g/cm^2. For projectile fragments below atomic number 12, there is negligible distinction among the polymers selected. For Mg and above (i.e., for $Z \geq 12$), polyethylene (PE), with its

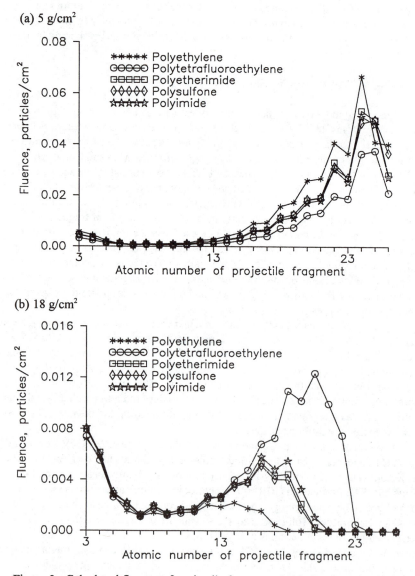

Figure 2. Calculated fluence of projectile fragments behind polymeric shields
with 33.88 GeV ^{56}Fe ions.

Table I. Empirical Formulas and Densities of Six Polymers Studied

Epoxy	$C_{37}H_{42}N_4O_6S$	1.32 g/cm^3
Polyetherimide	$C_{37}H_{24}N_2O_6$	1.27 g/cm^3
Polyethylene	CH_2	0.92 g/cm^3
Polyimide	$C_{22}H_{10}N_2O_5$	1.42 g/cm^3
Polysulfone	$C_{27}H_{22}O_4S$	1.24 g/cm^3
Polytetrafluoroethylene	CF_2	2.17 g/cm^3

high hydrogen density, is the most effective absorber for the thick shield, but the least effective for the thin. Polytetrafluoroethylene (PTFE), which contains heavier fluorine atoms with no hydrogen, lies at the other extreme. PE is the most effective shield material among several polymers at a thickness greater than 18 g/cm^2 for a 33.88 GeV ^{56}Fe beam.

Because lunar soil is a potential construction material for a habitat for long-term lunar missions, its suitability as a shield from HZE particles was studied. A representative sampling of lunar regolith was reported (7) to have a density of 1.5 g/cm^3 and to contain almost exclusively only five elements: O (61.5 mol-%), Si (19.3), Al (7.5), Fe (6.1), and Mg (5.5). As shown in Figures 3a and 3b, the addition of an epoxy (as a possible binder for the regolith) enhances the regolith's shielding capabilities. The degree of shielding can be very sensitive to the thickness of the material. For example, calculations show that increasing the thickness from 16 g/cm^2 (Figure 3a) to 18 g/cm^2 (Figure 3b) would yield significant improvements.

The effects of introducing boron into shielding materials were also studied as a way of capturing neutrons. As reported elsewhere (8), high neutron fluxes are possible inside a spacecraft owing to neutron formation in the nuclear fragmentation processes from the GCR impacting on exterior walls. (Generally, the density of neutrons in free space is negligible owing to their 11-min half life.) These neutral species cannot dissipate their kinetic energy through Coulombic interactions but must do so with elastic collisions with atomic nuclei. As noted above, hydrogen is the most effective nucleus for reducing the energy of neutrons to the thermal region. The boron isotope ^{10}B, which constitutes 19.6 percent of the naturally occurring element, has a large neutron-capture cross section for thermal neutrons and has been tested (8-10) for neutron shielding. Data from reference 8 is plotted here as Figure 4 to illustrate some benefits achievable for neutron capture by boron-loaded polymers.

Calculations were carried out for several polymeric shields containing amorphous, submicron boron powder (having a density of 2.59 g/cm^3 for the naturally occurring distribution of boron isotopes) dispersed uniformly throughout the polymers. The inclusion of boron slightly diminishes the material's capacity to absorb secondary HZE particles, as shown in Figure 5 for a polyetherimide. As the fraction of boron is increased from 5 to 20 wt %, both the density of the material and the initial range of incident particles increase because boron has a higher atomic

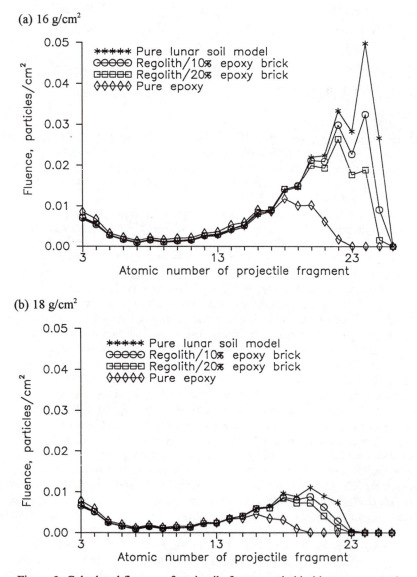

Figure 3. Calculated fluence of projectile fragments behind lunar construction materials with 33.88 GeV ^{56}Fe ions.

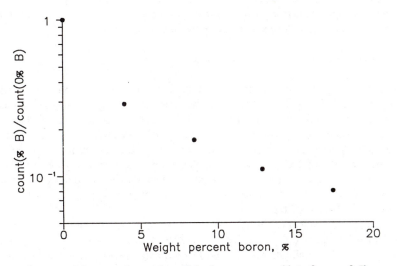

Figure 4. Neutron absorption of the boron-epoxy (data from ref. 7).

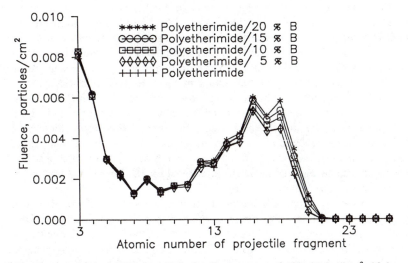

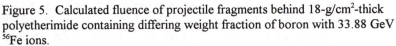

Figure 5. Calculated fluence of projectile fragments behind 18-g/cm²-thick polyetherimide containing differing weight fraction of boron with 33.88 GeV ^{56}Fe ions.

number than hydrogen. It should be noted that most of the contribution to fragmentation comes from a broad range of charges above $Z = 3$ (for Li). The code, LBLBEAM (3), for laboratory ion beams does not include light fragments of $Z < 3$ in any realistic way because a greater knowledge of nuclear fragmentation processes and a corresponding theory are required for these fragments.

Modeling of Galactic Cosmic Radiation

Interaction data for atomic ionization and nuclear reactions were combined in the Boltzmann equation with the 1977 solar minimum cosmic ray spectrum (6) to assess the transmitted environment through various shields for evaluation of biological effects. The shield effectiveness is intimately related to the nature of the nuclear cross sections through the change in the microscopic fluctuations in biological response. Shield effectiveness was examined in terms of two biological models. The first model is the conventional risk assessment method using the quality factor as a function of LET. The second model is a track-structure repair kinetic model (11) for the mouse cell C3H10T1/2.

The dose equivalent $H(x)$, which is obtained by multiplying the absorbed dose at each LET by a corresponding quality factor (12), is a measure of the response of living tissue. The quality factor was used to estimate the dose equivalent because all cells do not absorb energy equally from each LET component. Materials with atoms of low atomic number (e.g., PE) attenuate a very broad range of LET components (4) even though there is a gain in many low LET components. However, the effects from these are due primarily to indirect damage in cellular DNA brought about by OH radicals and are of negligible significance (13). Materials with atoms of higher atomic number (e. g., PTFE) attenuate only the highest LET components (4) at the expense of producing a broad range of LET components for which biological response may be enhanced relative to free space exposures. These results occur for shielding depths of 2 to 10 g/cm^2 of aluminum which are typical for the space program (14). The relative attenuation of dose equivalent $H(x)/H(0)$ with depth is shown in Figures 6a and 6b. It was found that, among the materials studied, PE provides the most effective shielding at all thicknesses. PE is more effective than PTFE even for very thin films because of its greater efficiency in attenuating the heavier ions that are the most destructive to living tissue. The calculations show lunar regolith to be a less effective shield material for HZE particles than the hydrogen-containing polymers studied.

The second model of the response of living cells to the effects of GCR is represented here in terms of occurrences of neoplastic cell transformations $T(x)$ resulting from a one-year exposure behind a shield of thickness x relative to occurrences $T(0)$ in free space. Unlike conventional dosimetric analysis wherein radiation quality is represented by LET-dependent quality factors, the repair kinetics model is driven by track-structure-dependent injury coefficients from experimental data with various ions in the mouse cell C3H10T1/2 (11). The variation in the calculated cell transformation ratio $T(x)/T(0)$, shown in Figures 7a and 7b, shows that the dependence on material is qualitatively similar to that found for $H(x)/H(0)$ as

(a) polymeric shields

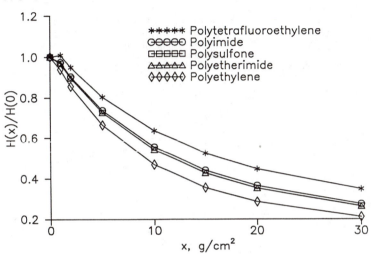

(b) lunar construction materials

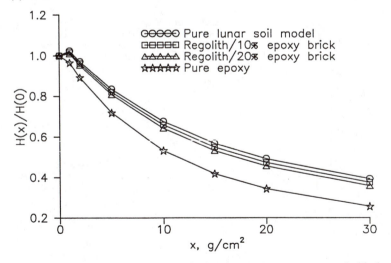

Figure 6. Attenuation of dose equivalent in a one-year exposure behind various shield materials as a function of shield thickness.

(a) polymeric shields

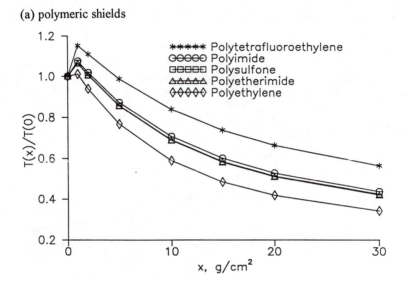

(b) lunar construction materials

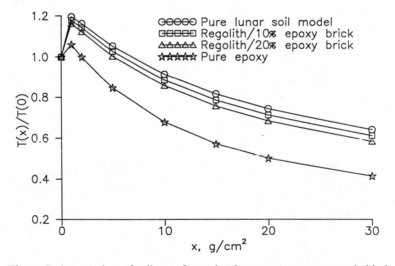

Figure 7. Attenuation of cell transformation in a one-year exposure behind various shield materials as a function of shield thickness.

shown in Figures 6a and 6b although the cell transformation model predicts a noticeable increase in risk for thin shields (1-5 g/cm^2). However, there are important quantitative differences in the protective properties of shield materials dependent on the biological model used. Clearly, many shield materials provide only modest reductions in neoplastic transformation ratios (Figures 7a and 7b); whereas, they show a much greater reduction in dose equivalent (Figures 6a and 6b) for the same shield thickness. Shield optimization must await an improved understanding of biological response as well as a more accurate nuclear database.

Conclusion

A theoretical study was initiated to investigate the interaction and alteration of space radiations by various structural materials in order to select the materials that will provide the best shielding (i.e., significantly reduce the exposure risk from the HZE particles of GCR).

The effects of hydrogen-bearing compounds as potential space structural components were examined by comparing the total ion fluence after passing through the shield. For energetic ion beams, a polyethylene target with its high hydrogen density is the most effective absorber for thick shields, while a polytetrafluoroethylene target with the heavier fluorine atoms appears to be more effective for thin shields with respect to the production of secondary radiation. Adding an epoxy to lunar regolith to bind it into a composite enhances its shielding properties from HZE particles. The inclusion of boron in a polymeric material only slightly diminishes the capacity of the material to absorb HZE particles. Lunar regolith is a less effective shield material for HZE particles than the hydrogen-containing polymers studied. Therefore, a material with a high percentage of lighter atoms such as hydrogen would be effective for thick shields while a material composed of heavier atoms might yet prove to be more effective in thin shields for energetic ion beams with respect to the number of secondary particles without considering their radiation quality.

Radiation biological risks depend on the microscopic fluctuations of energy absorption events in specific tissues (*14*). The number of particles and the energy deposited behind most materials increase for thin shields due to a buildup of secondary radiation, which increases the hazard. Biological effects are reduced efficiently not only by selecting different materials but also by adjusting the thickness of the material. Polyethylene is an efficient shield material at all thicknesses for GCR exposure in spite of the large number of heavy projectile fragments produced. There are important quantitative differences in the predicted biological effects between the two different biological models. Uncertainties in the nuclear database exist for the calculation of the radiation field modified by different polymeric materials. The greatest uncertainty in biological response is expected from high LET components (*15*). The appropriate shield material optimization against GCR must await an improved understanding of biological response as well as the development of an adequate nuclear cross section database.

Literature Cited

1. Simpson, J. A. In *Composition and Origin of Cosmic Rays*; Shapiro, M. M., Ed.;
NATO ASI Series C, V. 107; D. Reidel Publ. Co.: Dordrecht, Holland, **1983**; pp
1-24.
2. Choppin, G. R.; Rydberg, J. *Nuclear Chemistry - Theory and Application*;
Pergamon: New York, NY, **1990**: p 78.
3. Wilson, J. W.; Townsend, L. W.; Schimmerling, W.; Khandelwal, G. S.; Khan, F.;
Nealy, J. E.; Cucinotta, F. A.; Simonsen, L. C.; Shinn J. L.; Norbury, J. W.
Transport Methods and Interactions for Space Radiations; NASA RP-1257;
Washington, DC, **1991**
4. Kim, M. Y.; Wilson, J. W.; Thibeault, S. A.; Nealy, J. E.; Badavi, F. F.; Kiefer, R.
L. *Performance Study of Galactic Cosmic Ray Shield Materials*; NASA TP-3473;
Washington, DC, **1994**
5. Adams J. H., Jr.; Silberberg, R.; Tsao, C. H. NRL Memo. Rep. 4506-Pt. I, U.S.
Navy; **1981**
6. Badhwar, G. D.; Cucinotta, F. A.; O'Neill, P. M. *Radia. Res.* **1993**, Vol. 134, pp
9-15.
7. *Conceptual Design of a Lunar Colony*; Dalton, C.; Hohmann, E., Eds.; NASA
CR-129164; **1972**.
8. Thibeault, S. A.; Long, Jr., E. R.; Glasgow, M. B.; Orwoll, R. A.; Kiefer, R. L.
Polym. Prep. **1994**, Vol. 35(2), pp 954-955.
9. Kraus, W.B.; Glasgow, M. B.; Kim, M. Y.; Olmeijer, D. L.; Kiefer, R. L.; Orwoll,
R. A.; Thibeault, S. A. *Polym. Prep.* **1993**, Vol. 34(1), pp 592-593.
10. Stephens, J. M.; Glasgow, M. B.; Kiefer, R. L.; Orwoll, R. A.; Long, S. A. T.
Polym. Prep. **1992,** Vol. 33(1), pp 1152-1153.
11. Wilson, J. W.; Cucinotta, F. A.; Shinn, J. L. In *Cell Kinetics and Track
Structure*; Swenberg, C. E.; Horneck, G.; Stassinopoulos, E. G. Ed.; Biological
Effects and Physics of Solar and Galactic Cosmic Radiation, Part B; Plenum Press:
New York, NY, **1993**; pp 295-338.
12. International Commission on Radiological Protection, *1990 Recommendations
of the International Commission on Radiological Protection*; ICRP Publ. 60;
Pergamon: New York, NY, **1991**; p 81.
13. Billen, D. *Radia. Res.* **1990,** *124,* pp 242-245.
14. Wilson, J. W.; Kim, M. Y.; Schimmerling, W.; Badavi, F. F.; Thibeault, S. A.;
Cucinotta, F. A.; Shinn, J. L.; Kiefer, R. L. *Health Phys.* **1995**, *68(1)*, pp 50-58.
15. Schimmerling, W.; Wilson, J. W.; Nealy, J. E.; Thibeault, S. A.; Cucinotta, F.
A.; Shinn, J. L.; Kim, M. Y.; Kiefer, R. L. *Adv. Space Res.*; 1995 COSPAR;
Pergamon: Great Britain, **1996**, Vol. 17(2), pp (2)31-(2)36.

RECEIVED June 20, 1995

Application of Radiation in Lithography

Chapter 28

Advanced Materials and Forms:
Photosensitive Metathesis Polymers

A. Mühlebach and U. Schaedeli

Materials Research, Ciba-Geigy Ltd., CH–1723 Marly 1, Switzerland

The synthesis and ring opening metathesis homo- and copolymerization (ROMP) of the exo-7-oxa-norbornene- and exo-norbornene-carboximid esters, **1-3** and **1'-3'** respectively, have been investigated. Molecular weights were controlled with 2-butene-1,4-diol as chain transfer agent. These homo- and copolymers were used to formulate very sensitive positive tone high resolution resists. They are the first positive working photoresists based on a metathesis polymer backbone.

Cycloolefins with strained rings can be polymerized by the so called ROMP-reaction (= ring-opening-metathesis-polymerization) yielding linear chains with one double bond per repeat unit: Scheme 1. Whereas classical catalysts like $WCl_6/AlEt_2Cl$ (*1*) have a very limited tolerance towards functional groups such as ketones, esters, amides and imides, and the recently discovered Schrock-type catalysts (Mo- and W-carbenes) (*2*) have a somewhat broader tolerance, certain ruthenium(II)-complexes allow the polymerization of almost any functionalized monomers even in aqueous or ethanolic solution.

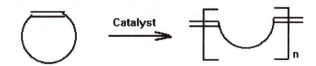

Scheme 1: Principle of the ring opening metathesis polymerization (ROMP).

Therefore, polymers bearing almost any functional groups useful for specific applications can be designed. There is a fast increasing demand for speciality polymers for applications in many high technology fields. For example, homo- and copolymers containing specific acid cleavable side groups are used in today's chemically amplified positive tone photoresists (*3*). The excellent physical properties like high T_g-values and good thermostabilities of ROMP polymers derived from exo-7-oxabicyclo[2.2.1]hept-5-ene-2,3-dicarboximides (*4*) attracted our attention. Therefore we decided to synthesize specific exo-(7-oxa-)norbornene-carboximids with photochemically or thermally cleavable

0097–6156/96/0620–0364$12.00/0
© 1996 American Chemical Society

protecting groups. [Ru(H$_2$O)$_6$]tos$_2$ (tos=p-toluenesulfonate) (*5*) was used as ROMP catalyst because it turned out to be one of the most active catalysts known so far for aqueous ROMP (*6*), although photochemically activated Ru(II)-sandwich or Ru(II)-acetonitrile complexes allow also a photoinduced ring opening metathesis polymerization (PROMP) as we discovered recently (*7*).

An easy and straightforward synthesis of the hitherto unknown monomers exo-(7-oxa-)norbornene-carboximid esters **1-3'** (Scheme 2) will be reported in this paper. Homo- as well as copolymerizations with the other (oxa-)norbornene type monomers **4-8** using [Ru(H$_2$O)$_6$]tos$_2$ as catalyst will be described. In the last chapter, these homo- and copolymers with controlled molecular weights will be used, inter alia, to formulate sensitive positive tone high resolution microresists. To the best of our knowledge, metathesis polymers have not yet been described as backbones for such high performance positive resists.

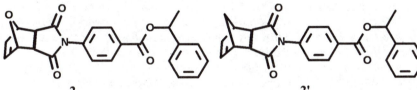

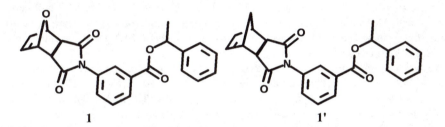

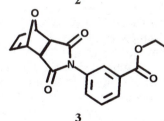 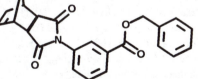

 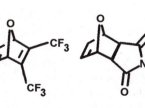 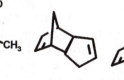

Scheme 2: Monomers used in this study.

Experimental

Materials. Laboratory grade reagents and solvents were used as received (Fluka, Aldrich, Merck). cis-5-Norbornene-exo-2,3-dicarboxylic anhydride was synthesized by thermal isomerization (1h, 200°C) of the endo derivative (Merck) and subsequent recrystallization from toluene; m.p.: 109°C. [Ru(H$_2$O)$_6$]tos$_2$ was synthesized according to the literature (5).

Physical measurements. ^{1}H- and ^{13}C-NMR spectra were recorded in CDCl$_3$ on a Bruker AC 300 instrument; chemical shifts (δ) are given in ppm relative to internal TMS. IR-spectra were measured on a Nicolet 20 SX with 2 cm^{-1} resolution. Differential scanning calorimetric measurements (DSC) and thermogravimetric analyses (TGA) were performed on a Mettler TA 3000 system, heating rate: 10°C/min. Melting points were measured either by DSC or on a Reichert Thermovar microscope with hot stage. Gel permeation chromatographic measurements (GPC) were performed in THF on a Waters 600E apparatus from Millipore Corp. (Milford, MA, USA) with Maxima 820 software. The M$_n$- and M$_w$-values (g/mol) refer to polystyrene-standards, which were used to calibrate the system. The scanning electron microscopy analysis (SEM) was executed on a Cambridge S120 stereoscan instrument.

Monomers. The monomers *exo-3-(3,5-dioxo-10-oxa-4-aza-tricyclo[5.2.1.0^{2,6}]dec-8-en-4-yl)-benzoic acid 1-phenyl-ethyl ester (1)*, *exo-4-(3,5-dioxo-10-oxa-4-aza-tricyclo[5.2.1.0^{2,6}]dec-8-en-4-yl)-benzoic acid 1-phenyl-ethyl ester (2)*, *exo-3-(3,5-dioxo-10-oxa-4-aza-tricyclo[5.2.1.0^{2,6}]dec-8-en-4-yl)-benzoic acid benzyl ester (3)* and *exo, exo-7-oxa-bicyclo[2.2.1]hept-5-ene-2,3-dicarboxylic acid dibenzyl ester (4)* were synthesized as described before (8). *2,3-Bistrifluoromethyl-7-oxa-bicyclo[2.2.1]hepta-2,5-diene (5)* (9) *4-methyl-10-oxa-4-aza-tricyclo[5.2.1.0^{2,6}]dec-8-ene-3,5-dione (6)* (4) and *exo-dicyclopentadiene (7)* (10) were synthesized according to literature procedures, *bicyclo[2.2.1]hept-2-ene (2-norbornene) (8)* was used as received from Fluka.

exo-3-(3,5-Dioxo-4-aza-tricyclo[5.2.1.0^{2,6}]dec-8-en-4-yl)-benzoic acid 1-phenyl-ethyl ester (1'): 2.8 g (0.017 mol) cis-5-norbornene-exo-2,3-dicarboxylic anhydride were dissolved in 50 ml DMSO. 2.34 g (0.017 mol) 3-Aminobenzoic acid were added and the homogeneous red solution was stirred for 24h under nitrogen. 10.39 g (0.068 mol) 1,8-Diazabicyclo-[5,4,0]undec-7-ene (DBU) was than added dropwise followed by 12.63 g (0.068 mol) 1-phenylethyl bromide (PEB). Stirring was continued for another 24 h and the solution poured into 100 ml water. Filtration, washings with water and cold EtOH and drying at 50°C in vacuo gave the product in 89% yield (6.61 g). It was further purified by recrystallization from toluene. M.p.: 148°C; Elemental analysis: calc. (found): C: 74.40 (73.69), H: 5.46 (5.62), N: 3.62 (3.65); ^{1}H-NMR: δ 1.50, 1.63 (2xd, J=9.9Hz, 2H: CH$_2$(10), 1.68 (d, J=6.6Hz, 3H: CH$_3$), 2.88 (s, 2H, CH(1), CH(7)), 3.42 (s, 2H, CH(2), CH(6)), 6.13 (q, J=6.6 Hz, 1H, CH (phenylethyl ester)), 6.36 (s, 2H, CH(8), CH(9)), 7.30-7.58 (m, 7H), 7.97 (s, 1H), 8.11 (d, J=7.6Hz, 1H, totally 9 aromat.H).

exo-4-(3,5-Dioxo-4-aza-tricyclo[5.2.1.0^{2,6}]dec-8-en-4-yl)-benzoic acid 1-phenyl-ethyl ester (2'): Synthesis similar to *1'*, using 4-aminobenzoic acid instead of 3-aminobenzoic acid. The pure product was obtained by recrystallization from EtOH in 74% yield. M.p.: 132°C; Elemental analysis: calc. (found): C: 74.40 (73.54), H: 5.46 (5.33), N: 3.62 (3.45); ^{1}H-NMR: δ 1.47, 1.61 (2xd, J=9.9Hz, 2H: CH$_2$(10), 1.68 (d, J=6.6Hz, 3H: CH$_3$), 2.88 (s, 2H, CH(1), CH(7)), 3.42 (s, 2H, CH(2), CH(6)), 6.13 (q, J=6.6 Hz, 1H, CH (phenylethyl ester)), 6.36 (s, 2H, CH(8), CH(9)), 7.30-7.44 (m, 7H), 8.17 (d, J=7.6Hz, 2H), totally 9 aromat.H); IR (KBr): 1775 cm^{-1}, 1708 cm^{-1} (C=O imide and ester).

exo-3-(3,5-Dioxo-4-aza-tricyclo[5.2.1.0^{2,6}]dec-8-en-4-yl)-benzoic acid benzyl ester (3'):
Synthesis similar to *1'*, using benzylbromide instead of 1-phenylethyl bromide. The
product was recrystallized from toluene. Yield: 77%; m.p.: 180°C; Elemental analysis:
calc. (found): C: 73.98 (73.27), H: 5.13 (5.28), N: 3.75 (3.81); ^{1}H-NMR: δ 1.50, 1.64
(2xd, J=9.9Hz, 2H: CH$_2$(10), 2.88 (s, 2H, CH(1), CH(7)), 3.42 (s, 2H, CH(2), CH(6)),
5.37 (s, 2H, CH$_2$ (benzyl)), 6.36 (s, 2H, CH(8), CH(9)), 7.32-7.57 (m, 7H), 7.98 (s, 1H),
8.11 (d, J=7.6Hz, 1H), totally 9 aromat.H);

Homo- and Copolymers. In a typical example, 1-2 g of the monomer mixture and 100
mg 2-butene-1,4-diol were dissolved in ca. 10 ml Ar saturated dioxane/EtOH (1:1) and 15
mg [Ru(H$_2$O)$_6$]tos$_2$ (tos=p-toluenesulfonate) were added. The solution was kept under
Ar at 60°C for 48 h. The polymer was precipitated in 100 ml MeOH. Washing with water
and MeOH and drying 24 h at 50°C in vacuo gave the pure product. Further details are
given elsewhere (*6, 8*).

Lithography. Resist films were obtained by spin casting of the resist formulation on 3
inch silicon wafers (lithography) or 1 inch quarz disks (UV measurements). Resist film
thickness was measured using a Zeiss Axiospeed FT instrument. Visible absorption spectra
were recorded on a Varian Cary 1E UV/visible spectrometer. Exposures were performed
on an Oriel contact printing tool through a 254 nm narrow band pass filter.

Results and Discussion

Monomer Synthesis. The synthesis of polyimides with pendent 1-phenylethyl ester
groups, using a direct one-pot polyaddition/imidization/esterification reaction of an aromatic tetracarboxylic acid dianhydride with 3,5-diamino-benzoic acid and a DBU/PEB
mixture in DMSO was recently described (*11*). Surprisingly, this reaction sequence
worked also with *aliphatic* anhydrides as starting materials, e.g. exo-3,6-epoxy-Δ^4-
tetrahydro-phthalic anhydride (the Diels-Alder adduct of furane and maleic anhydride) (*12*)
or the exo-bicyclo[2.2.1]hept-5-ene-2,3-dicarboxylic anhydride and therefore gave an easy
and straightforward access to the desired monomers **1-3'**: Scheme 3.

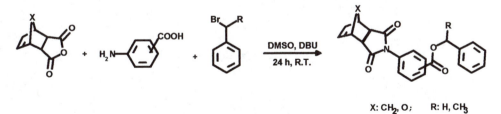

X: CH$_2$, O;   R: H, CH$_3$

Scheme 3: Monomer synthesis.

Homo- and Copolymer Synthesis. *Homo- and copolymerizations* were performed with
1-2% [Ru(H$_2$O)$_6$]tos$_2$ in monomer saturated EtOH/dioxane solution (the monomers **1-3'**
are not sufficiently soluble in pure EtOH): Scheme 4. Fig. 1 shows the ^{1}H-NMR of poly-1
(b) compared with the monomer 1 (a).

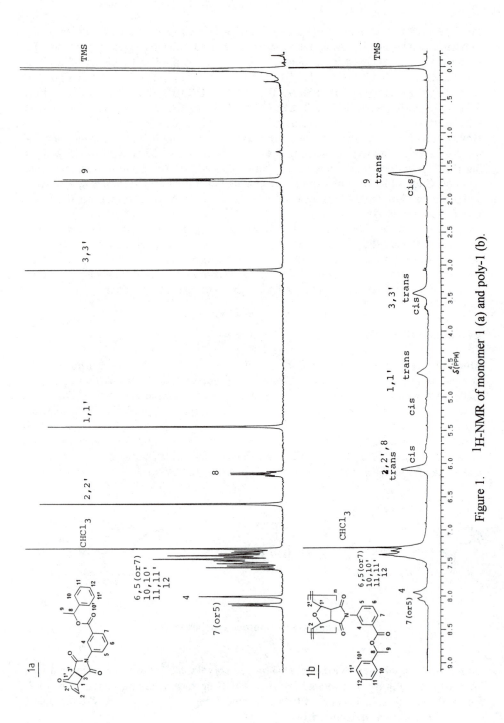

Figure 1.   ¹H-NMR of monomer 1 (a) and poly-1 (b).

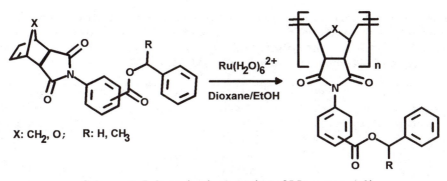

X: CH$_2$, O;   R: H, CH$_3$

Scheme 4: Polymerization reaction of Monomers 1-3'.

Control of molecular weights. The control of molecular weights in polymers is crucial for most applications. For example, in microresist technology a 1μm thick film is generated on a silicon wafer by solvent casting. Polymers with high molecular weights - usually observed in ROMP with Ru-catalysts (6) - show very high solution viscosities which make their application difficult if the spincoating technology is used: films tend to be much thinner than the specified thickness. Furthermore, studies on resist polymers with different molecular weights show clearly, that the microresist properties have their optimum in a distinct molecular weight range, usually M_n=10-30,000 g/mol.

We used the ring-opening-metathesis-polymerization of exo,exo-5,6-bis(methoxycarbo-nyl)-7-oxabicyclo[2.2.1]hept-2-ene as a test system for different chain transfer agents (CTA), having investigated its polymerization in great detail (6): Scheme 5. The following olefins were tested as CTA's for their influence on molecular weights, polydispersities and yields in the polymerization reaction: 2-Butene-1,4-diol, cyclohexane, endo-dicyclopenta-diene, 2,3-dimethyl-2-butene, 3-butene-1-ol (allyl alcohol) and 1-hexene.

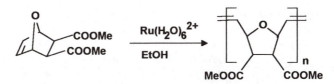

Scheme 5: ROMP of exo, exo-5,6-bis(methoxycarbonyl)-7-oxabicyclo[2.2.1]hept-2-ene.

Only 3-butene-1-ol and especially 2-butene-1,4-diol showed good results. The molecular weight was considerably lowered without significant changes in the polymerization yield and the polymer structure (% trans double bonds). However, the molecular weight distribution tends to become broader: Tab. 1. These results are in good agreement with literature data on a similar monomer/CTA-system (13).

Table 1: Control of molecular weights in oxa-norbornene derivatives: Influence of 2-butene-1,4-diol on the ROMP of exo,exo-5,6-bis(methoxycarbonyl)-7-oxa-bicyclo[2.2.1]hept-2-ene with [Ru(H$_2$O)$_6$]tos$_2$. Experimental conditions: 2 mmol monomer and 0.008 mmol catalyst in 25 ml EtOH, 24h at 60°C under Ar.

Weight % 2-butene-1,4-diol (rel. to monomer)	Polym. Yield (%)	M$_n$ (g/mol)	M$_w$ (g/mol)	M$_w$/M$_n$	Mol% trans double bonds
0	90-98	236000	340000	1.4	ca. 50
0.1	91	112000	306000	2.7	48
2.0	97	13800	63400	4.6	47
20.0	83	7700	44400	5.7	47

10% 2-Butene-1,4-diol was added as CTA in the homo- and copolymerization of the ester-imides **1-3'** which led to M$_n$=20-50000 g/mol and molecular weight distributions (M$_w$/M$_n$) of 1.5 to 2.7, see Figure 2 and Tables 2 and 3. Without this regulator, the molecular weights tend to be about one order of magnitude higher, as can be seen in the homopolymerization of **1** or in the copolymerization of **1** and **8** (Tab. 2 and 3). On the other hand, polymerization yields dropped in the presence of 2-butene-1,4-diol, especially for the 2-norbornene derivatives without the oxygen bridge (**1'**, **3'**). This could be due to some kind of blocking of the catalyst precursor by the CTA, as was suggested in a previous paper (*6*). This blocking would be considerably influenced by the monomer/catalyst/solvent system. Polymerization yields also depend strongly on the monomer concentration (*6*) and therefore the low yield of poly-**2** is stemming from the lower monomer concentration in the reaction mixture (because of the low solubility of **2** in EtOH/dioxan 50/50). Table 2 lists yields, molecular weights and glass transition temperatures (T$_g$) of the homopolymers and Tab. 3 those of the copolymers of **1** with **4-8**.

Table 2: Homopolymerizations

Polymer	Yield (%)	M$_n$ (g/mol)	M$_w$ (g/mol)	M$_w$/M$_n$	T$_g$ (°C)
poly-**1**[a]	95	100000	240000	2.4	115
poly-**1**	52	24000	36000	1.5	112
poly-**2**	33	10600	18600	1.8	110
poly-**3**	57	7900	15400	1.9	140
poly-**1'**	45	20000	42000	2.1	130
poly-**3'**	49	9700	23000	2.4	129

[a] without 2-butene-1,4-diol.

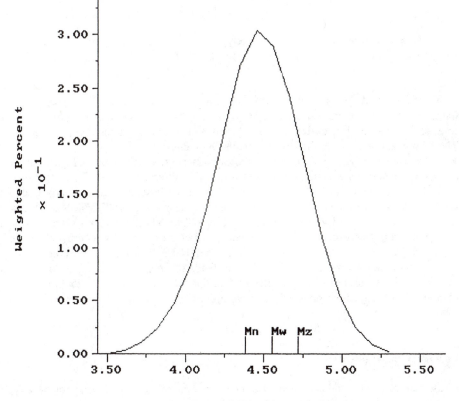

Figure 2.   GPC of poly-1 (M_n=24'000, M_w=36'000, M_w/M_n=1.5).

Table 3: Copolymerizations of 1 with different comonomers

Como-nomer	Yield (%)	Mol % of 1 in feed	Mol % of 1 in copolymer	M_n (g/mol)	M_w (g/mol)	M_w/M_n	T_g (°C)
4	52	65	52	22000	41000	1.9	60
5	27	54	51	19000	52000	2.7	114
6	76	48	44	36000	64000	1.8	135
7	19[a]	41	38	55000	160000	1.9	108
8	31	20	14	60000	160000	2.7	70
8[b]	66	62	61	230000	520000	2.3	117

[a] plus variable amounts of insoluble polymer.
[b] copolymerization was done *without* 2-butene-1,4-diol.

Although the yields lie far below 100%, the copolymer composition corresponds well to the feed ratio of the monomers. Therefore we assume that the copolymerization of **1** leads to a mainly random distribution of the different monomer units along the polymer chain. The following observations further support this hypothesis:

- Molecular weight distributions of the copolymers are monomodal and quite narrow.
- Only one T_g-value in the range of 60 - 135°C is observed.
- The presence of AB sequences (beside AA- and BB-sequences) is clearly visible in the NMR spectra of the copolymers, see e.g. Fig. 3a displaying the ^{1}H-NMR spectrum of the copolymer between **1** and **8** (second last entry in Tab. 3). This spectrum is not a simple superposition of the respective homopolymer spectra (as expected for a completely blocky copolymer) as revealed by a comparison with the ^{1}H-NMR spectra of pure poly-**1** (Fig. 1b) and pure poly(norbornene) (Fig. 3b). Instead, the peaks belonging to the olefinic protons of the **1**-unit are slightly shifted and additional peaks, such us a multiplett at 5.6 ppm and broad signals at 4.9 and 4.5 ppm, are visible, indicating the formation of AB-sequences.

Lithographic Applications

In the semiconductor industry there has been an ever increasing demand for new type, photostructurable materials having improved physical properties, as compared to the established systems. Especially for printed circuit board applications, where improved electrical and water resorbing properties would be desired, as well as for microlithography, where the potential of polymeric backbones different from novolaks / poly(4-hydroxy styrene)'s would be of interest to explore.

The newly introduced ring-opening-metathesis-polymers represent a new class of potentially photostructurable materials. These polymers cover a broad range of different physical properties: pure hydrocarbon polymers as well as polymers having a wide variety of functional side groups have been successfully synthesized to date. In other words, the design of new materials, tailormade for well defined specific applications, seems feasible. There are several ways of introducing photosensitivity to formulations based on metathesis polymers:

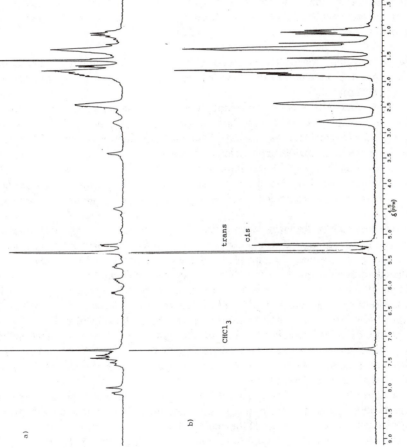
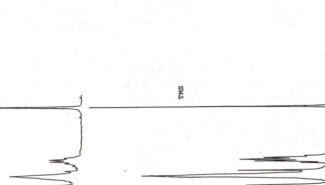

Figure 3.   (a) ^{1}H-NMR spectrum of the copolymer between 1 and 8 (second last entry in Tab. 3); (b) ^{1}H-NMR spectrum of poly(norbornene).

• A first approach is based on the attachment of acid labile side groups to the polymer backbone. These are cleaved in the presence of photolytically generated Broensted acid, thereby leading to a change in polarity between exposed and unexposed zones. This approach can be used for the design of positive tone, chemically amplified photoresist systems.

• Another possibility is the use of photoactive metathesis catalysts (7), where the active species is generated exclusively in the illuminated zones. As a result, the built-up in molecular weight is observed in these zones, whereas the nonirradiated zones consist of low molecular weight oligomers or monomers. Thus, negative tone resist systems are obtained.

• Crosslinking of linear metathesis polymers with, for example, bisazides is well known. An experimental resist system based on norbornene homo- and copolymers was found to yield negative tone images of the mask (14).

All different imaging concepts mentioned above are based on the light induced differentiation of the dissolution properties between exposed and non-exposed zones of the photoactive formulation.

The usefulness of metathesis polymers in lithography will be demonstrated on one specific example. The homopolymer of 1 was choosen as the candidate material, since it bears labile side groups. Styrene can be cleaved of thermally or in the presence of Broensted acid, thereby generating imidobenzoic acid functionalities: Figure 4B. Whereas poly-1 is insoluble in aqueous base, rapid dissolution occurs with the deblocked material. A resist formulation was prepared by dissolving 97 parts poly-1, along with 3 parts triphenyl sulfonium triflate (a substance which generates triflic acid upon irradiation), in 1800 parts dioxane.

Films of one micrometer thickness were prepared by spin casting the resist formulation onto 3 inch silicon wafers and subsequent solvent removal by heating for 1 minute at 120°C on a hot plate. the imaging principle is outlined in Figure 4: in the first step the films are irradiated through a quartz mask with light of 254 nm wavelength. Triflic acid is photogenerated in the irradiated zones of the resist film (Fig. 4A). The wafer is then baked on the hotplate for one minute at 110°C. The presence of acid lowers the decomposition temperature from 180°C to 100°C: Figure 5. Therefore, deblocking of poly-1 (Fig. 4B) occurs exclusively in the irradiated zones of the resist film. Since the acid induced deblocking reaction is of catalytic nature, chemical amplification (one proton giving rise to many chemical events) is achieved. Resist materials based on this type of deblocking chemistry typically exhibit high photosensitivity. The wafer is finally developed in a 1N aqueous solution of sodium hydroxide. Whereas the unirradiated zones of the film withstand the developer attack, the irradiated parts are rapidly dissolved and washed away, leaving behind a positive tone image of the mask patterns.

The contrast plot is depicted in Figure 6, revealing a contrast γ of 4.5. E_0, the minimum dose for complete clearing, is found to be 55 mJ/cm^2, indicating that the lithographically useful dose for this resist system lies around 80 mJ/cm^2. SEM micrographs (Figure 7) of a patterned film demonstrate that submicrometer structures are resolved. The UV absorbance spectra of the nonexposed film, as well as films after irradiation and postbake, are plotted in Figure 8. Interestingly, no significant change in optical density is detected after irradiation. However, the absorption increases after a subsequent baking step. The optical density of the 1.1 micrometer thick resist film, measured at 254 nm, is found to be 0.84, indicating that only a small fraction of the incoming light reaches the bottom layers of the film. This might be the cause for the formation of sloped, non-vertical resist profiles.

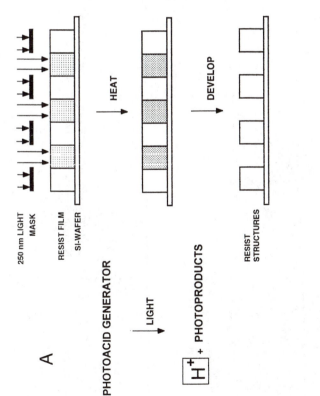

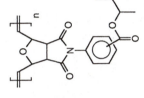

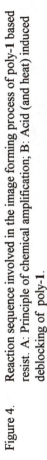

Figure 4.   Reaction sequence involved in the image forming process of poly-1 based resist. A: Principle of chemical amplification; B: Acid (and heat) induced deblocking of poly-1.

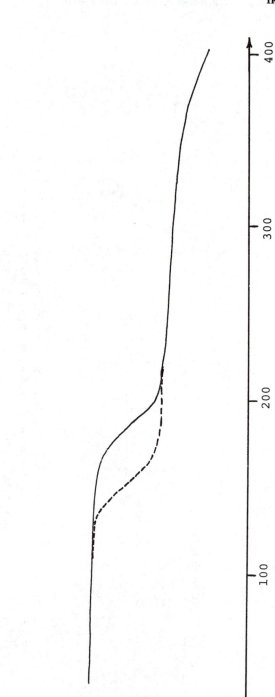

Figure 5.   TGA trace of poly-**1** without (——) and with 2 mol-% CF$_3$SO$_3$H (---).
              Heating rate: 10°C/min. (for both curves).

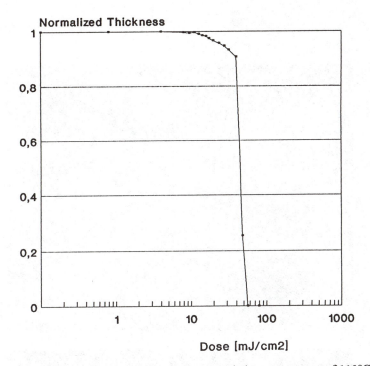

Figure 6.   Contrast curve recorded at a postbake temperature of 110°C.

It is interesting to note that no residues, which might result from particles with reduced dissolution rates, are detected in the developed parts of the wafer, see Figure 7. Obviously, styrene, formed in the irradiated zones of the resist film, is leaving the film before its polymerization to unpolar, nonsoluble homopolymers, induced thermally or by photolytically generated triflic acid, can effectively take place. Evidence comes from the fact that a dose dependent shrinkage of the resist film is observed, corresponding well to the evaporation of styrene from the film: Figure 9.

Conclusion

A simple and straightforward synthesis of ring-opening-metathesis homo- and copolymers with dicarboximide-ester functionalities and controlled molecular weights is presented. Although the use of these tailor-made polymers for certain applications (e.g. in microresist technology) are promising due to their attractive properties, potential problems arising from residual catalyst in the polymer (6) need to be addressed further.

Acknowledgments

The authors would like to thank Mrs. D. Guerry and Mr. E. Tinguely for their substantial contribution.

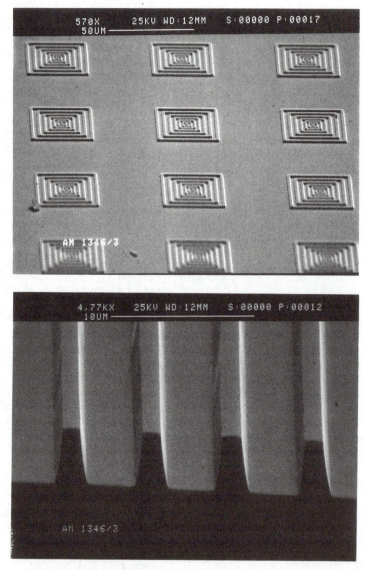

Figure 7.   a) and b): SEM micrographs of micropatterns from resist based on poly-**1** (97%) and triphenyl sulfonium triflate (3%). Softbake 120°C/60s, dose 80 mJ/cm², postbake 110°C/60s, develop 60s 1N NaOH.

ABSORBANCE

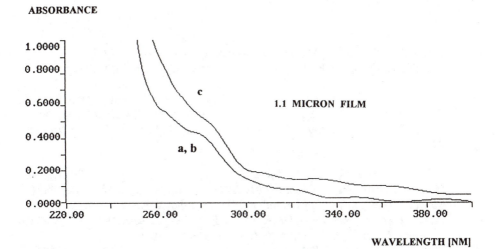

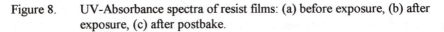

Figure 8.   UV-Absorbance spectra of resist films: (a) before exposure, (b) after exposure, (c) after postbake.

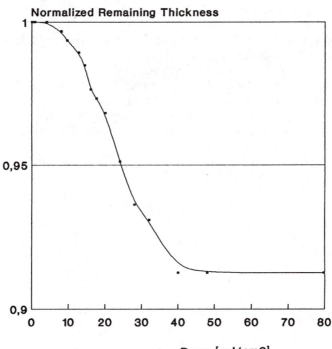

Figure 9.   Film thickness loss of poly-1 based resist film after exposure and postbake at 120°C for 60 seconds.

References

(*1*)   Natta, G.; Dall'Asta, G.; Mazzanti, G. *Angew. Chemie* **1964**, *76*, 765.
(*2*)   Schrock, R.R.; DePue, R.; Feldman, J.; Schaverien, C.J.; Dewan, J.C.; Liu, A.H. *J. Am. Chem. Soc.* **1988**, *110*, 1423.
(*3*)   (a) Reiser, A. in "Photoreactive Polymers, The Science and Technology of Resists", Wiley Interscience, New York, 1991. (b) Ito, H.; Willson, C.G. *Polym. Eng. Sci.* **1983**, *23*, 1012. (c) Fréchet, J.M.J.; Eichler, E.; Ito, H.; Willson, C.G. *Polymer* **1983**, *24*, 995.

(*4*)   (a) Hillmyer, M.A.; Lepetit, C.; McGrath, D.V.; Grubbs, R.H. *Polym. Prepr. (Am. Chem. Soc., Div. Polym. Chem.)* **1991**, *32(1)*, 162. (b) Hillmyer, M.A.; Lepetit, C.; McGrath, D.V.; Novak, B.M.; Grubbs, R.H. *Macromolecules* **1992**, *25*, 3345.
(*5*)   (a) Bernhard, P.; Bürgi, H.B; Hauser, J.; Ludi, A. *Inorg. Chem.* **1982**, *21*, 3936. (b) Bernhard, P.; Biner, M.; Ludi, A. *Polyhedron* **1990**, *9*, 1095.
(*6*)   Mühlebach, A.; Bernhard, P.; Bühler, N.; Karlen, T.; Ludi, A. *J. Mol. Catal.* **1994**, *90*, 143.
(*7*)   (a) Mühlebach, A.; Bernhard, P.; Hafner, A.; Karlen, T.; Ludi, A. *Europ. Pat. Appl.*, *Prio. 10.9.93.* (b) Karlen, T.; Ludi, A.; Mühlebach, A.; Bernhard, P.; Pharisa, C. *J. Polym. Sci., Polym. Chem. Ed.* **1995**, *33*, 1665.

(*8*)   Mühlebach, A.; Schaedeli, U.; Bernhard, *P. Polym. Prepr.* **1994**, *35*, 963.
(*9*)   Prinzbach, H.; Bingmann, H.; Markert, J.; Fischer, G.; Knothe, L.; Eberbach, W.; Brokatzky-Geiger, J. *Chem. Ber.* **1986**, *119*, 589.
(*10*)   Fritz, H.E. et al., *US Patent 3,235,614,* **1966**.
(*11*)   Iizawa, T.; Seno, E. *Polym. Journal* **1992**, *24*, 1169.
(*12*)   Woodward, R.B.; Baer, H. *J. Am. Chem. Soc.* **1948**, *70*, 1161.
(*13*)   France, M.B.; Grubbs, R.H.; McGrath, D.V.; Paciello, R.A. *Macromolecules*, **1993**, *26*, 4742.
(*14*)   Benedikt, G.M. (to B.F. Goodrich, Co.) *Europ. Pat. Appl. 140319, Prio. 24.10.83.*

RECEIVED August 11, 1995

Chapter 29

Polymeric Imaging Material Based on an Acid-Catalyzed Main-Chain Cleavage Reaction

M. Hiro[1] and J. M. J. Fréchet

Baker Laboratory, Department of Chemistry, Cornell University, Ithaca, NY 14853–1301

A novel polymeric imaging material containing polyacetal and photoacid generator has been designed. The polyacetal having acetals in its main chain was prepared by condensation of a monomer containing both a carbonyl and a diol group and also it was reverted to the monomer through an main chain cleavege reaction catalyzed with photogenerated acid.

Photogenerated acid in the design of resist materials

The design of polymers that can be modified under the influence of radiation is of prime importance in the fields of imaging and resist chemistry. In recent years the use of photogenerated catalysts (1) such as radicals, acids, or bases has been used extensively to develop new chemically amplified resist and imaging materials that show extremely high sensitivities to UV or other forms of irradiation. Perhaps the best known example of chemically amplified resist operating on the basis of photogenerated acid is the so-called t-BOC resist (2, 3) developed more than a decade ago. This material consists of a blend of poly(4-t-butyloxy-carbonyloxystyrene) and a thermally stable photoacid generator such as triphenylsulfonium hexafluoro-antimonate (4). Upon irradiation, acid is released within the polymer creating a latent image that can be developed in a thermal process that releases poly(4-hydroxystyrene), isobutylene and carbon dioxide. Image development by differential dissolution can provide either a positive or a negative tone image depending on the choice of the development solvent. Numerous

[1]Current address: Ibaraki Research Laboratory, Hitachi Chemical Company Ltd., 4–13–1, Higashi-cho, Hitachi, Ibaraki 317, Japan

0097–6156/96/0620–0381$12.00/0

resists based on this concept of radiation-induced modification of the side-chain of a polymer have been developed *(5, 6)* and used in practical applications.

In a related approach, photogenerated acids may be used to cleave the main-chain of a polymer leading to the disappearance of the solid materials with formation of volatile by products.   For example we have described polycarbonates *(7)*, polyesters *(8)*, and polyethers *(9)* that can be cleaved to volatile by-products under the action of acid photogenerated *in situ*.   All these systems involve the simple protonation of a reactive site on the polymer main-chain by the photogenerated acid.   This initial step is followed by chain cleavage in a process that affords stabilized tertiary, allylic, or benzylic carbocationic intermediates. Subsequent elimination from these intermediate species results in the release of a proton that is able to perpetuate the reaction leading to multiple cleavage of the polymer main-chain with formation of volatile products.   Because the photogenerated acid can participate in numerous successive reactions, the process is said to be chemically amplified.

Design of photocleavable polyacetals

Acetal and ketals are readily obtained by treatment of carbonyl compounds with alcohols under dehydrating conditions in the presence of trace of acid. The reaction is reversible and may be used for the design of polymers that may be capable of photoreversion to monomer under appropriate conditions.   This report concerns the design of new monomers that can be used in the acid-catalyzed formation of photocleavable polyacetals.

Polyacetals and polyketals may be prepared by different routes involving either the condensation of a monomer containing two carbonyl groups with a second monomer containing 1,2- or 1,3-diol groups, or condensation of a single monomer containing both a diol and an appropriately placed carbonyl group. We selected to test the second alternative as it provides for the ideal stoichiometric balance of the two complementary functionalities needed for the step-growth polymerization. Scheme 1 shows two general classes of monomers that can form six or five-membered acetal or ketal rings respectively as part of the polymer main-chain.

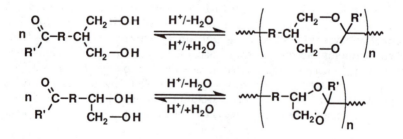

Scheme 1

Preparation of polyacetal

Although several monomers were prepared in the context of this study, only one, 3-(2', 3'-dihydroxypropoxy)benzaldehyde, is reported here.   The monomer prepared from readily available 3-hydroxybenzaldehyde.   Polymerization carried out in the presence of a trace of acid with removal of water using a Dean-Stark apparatus affords the desired polymer with five-membered acetal rings in its main-chain (Scheme 2). The polymer that is obtained is characterized by ^{1}H-NMR and FT-IR as well as elemental analysis.   Gel permeation chromatography (calibration with polystyrene standards) suggests a relative molecular weight $M_W = 24{,}000$ with a broad polydispersity.

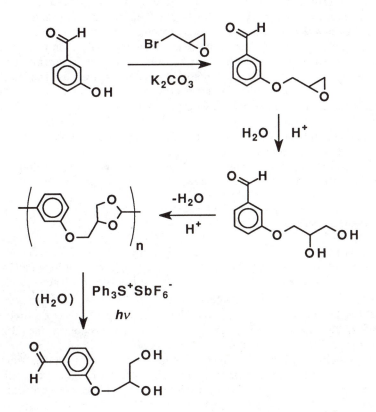

Scheme 2

Photocleavege of the polyacetal

The lability of the polyacetal was the first ascertained by performing reactions with the polyacetal dissolved in a solvent containing aqueous acid.   These model reactions confirmed that the polymer is degraded rapidly in the presence of aqueous acid with liberation of the monomer, 3-(2', 3'-dihydroxypropoxy) benzaldehyde, as well as small amounts of oligomers.

The photocleavage reaction was then carried out in the solid state using films of the polyacetal containing a small percentage of a photoactive triarylsulfonium salt acid generator. Following irradiation at 254 nm using a dose of 10 mJ/cm^2, the film was heated to 110°C and the chemical change caused by the photoexposure was monitored by FT-IR spectrometry. Figure 1 shows the change that occurs when the unexposed polyacetal film [Figure 1(a)] is exposed and heated [Figure 1(b)]. Obvious changes such as the appearance of a strong hydroxyl band near 3400 cm^{-1} as well as a carbonyl band near 1719 cm^{-1} confirm that cleavage has taken place. In an attempt to ascertain the extent of main-chain cleavage, the exposed polymer was dissolved in THF and its GPC trace [Figure 2(b)] was compared to that of the starting unexposed polymer [Figure 2(a)]. The GPC data documents the cleavage of the main-chain as lower molecular weight products are clearly formed. However, cleavage is only partial and does not proceed all the way to the monomer under the condition of this reaction (without water presence). In a second series of experiments, the film containing the photoacid generator was exposed and heated as above but the post-exposure heating step was carried out in the presence of water droplets deposited at the surface of the polymer film. The cleavage reaction monitored by IR [Figure 1(c)] as well as GPC [Figure 2(c)] proceeds essentially to completion with release of the monomer 3-(2', 3'-dihydroxypropoxy) benzaldehyde confirming that water is required to effect the complete photoinitiated reversion of the polyacetal to monomer. These experiments demonstrate the validity of our design for polyacetal that can be cleaved by an indirect photochemical process involving a photogenerated catalyst.

Experimental section

Photoinitiated cleavage of a film of the polyacetal   A 20 wt% cyclo-hexanone solution of the polyacetal containing the triphenylsulfonium hexafluoroantimonate (5 wt% with respect to polyacetal) was filtered through a 0.45 μm Teflon filter and was spincoated onto silicon wafers to afford 1 μm thick film. After heating to 90°C to remove the solvent, the film was exposed to 10 mJ/cm^2 at 254 nm light, then heated to 110°C. After cooling, the film was dissolved into THF and the product was analyzed by FT-IR and GPC. Control experiments without irradiation showed that the polyacetal was not depolymerized in the absence of irradiation. In some experiments, the heating step following exposure was carried out after the surface of the film had been covered with water droplets.

Acknowledgments

Thanks are due to Hitachi Chemical Co.Ltd for financial support of this project, as well as a leave of absence for M. Hiro's studies at Cornell University. This project made use of the facilities of the Cornell National Nanofablication facility, as well as the MRL central facilities (polymer characterization) funded by the National Science Foundation.

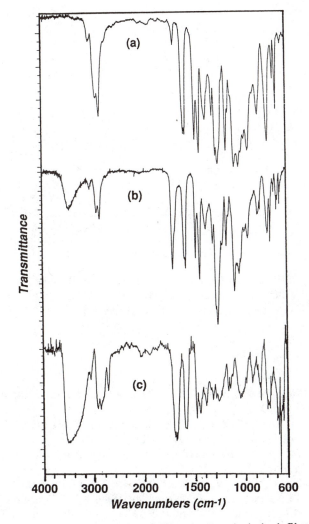

Figure 1. IR spectra of polyacetal film (curve a), baked film without additional water (curve b) and of baked film with additional water (curve c).

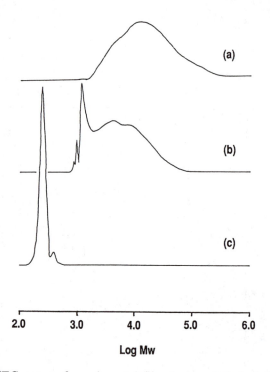

Figure 2. SEC curves for polyacetal film (curve a), baked film without additional water after irradiation at 254 nm (curve b) and of baked film with additional water after irradiation at 254 nm (curve c).

References

1. Fréchet, J.M.J. *Pure Appl. Chem.*, **1992**, *64*, 1239
2. Fréchet, J.M.J., Ito H., Wilson, C.G. *Proceedings Microcircuit Engineering 1982, Grenoble*, **1982**, 260
3 Wilson, C.G., Ito H., Fréchet, J.M.J., Houlihan, F. *Proceedings IUPAC 28th Macromol. Symp., Amherst, Mass.* **1982**, 448
4. Wilson, C.G., Ito H., Fréchet, J.M.J. *Proceedings Microcircuit Engineering 1982, Grenoble*, **1982**, 261
5. MacDonald, S.A., Wilson, C.G., Fréchet, J.M.J. *Acc. Chem. Res.*, **1994,** in press
6. Reichmanis, E., Thompson, L.F. *Chem. Rev.* **1989**, *89*, 1273
7. Fréchet, J.M.J., Bouchard, F., Eichler, E., Houlihan, F.M., Iizawa, T., Kryczka, B., Wilson, C.G. *Polym. J.*, **1987**, *19*,31
8. Fréchet, J.M.J., Kryczka, B., Matuszczak, S., Reck, B., Stanciulescu, M., Wilson, C.G. *Photopolym. Sci. & Technol.* **1990**, *3*, 235
9. Fréchet, J.M.J., Stanciulescu, M., Iizawa T., Wilson, C.G. *Polym. Mat. Sci. Eng.*, **1989**, *60*, 170

RECEIVED June 20, 1995

Chapter 30

X-ray Lithography with Environmentally Stable Chemical Amplification Positive Resist

Hiroshi Ito[1], Greg Breyta[1], Don Hofer[2], Andrew Pomerene[2], Karen Petrillo[2], and David Seeger[2]

[1]IBM Research Division, Almaden Research Center, 650 Harry Road, San Jose, CA 95120–6099
[2]IBM Research Division, T. J. Watson Research Center, Yorktown Heights, NY 10598

Synchrotron x-ray lithography of a new environmentally stable chemically amplified positive resist (ESCAP) is described. The resist consists of a thermally and hydrolytically stable resin and acid generator and employs high temperature bake processes to achieve resistance to airborne contamination. The resist is environmentally stable due to reduction of the free volume by annealing and affords true single layer x-ray lithography, eliminating a need for a protective overcoat. Its design concept and x-ray lithographic performance are discussed.

As the minimum feature size of semiconductor devices shrinks to below 0.5 μm, short wavelength lithographic technologies such as deep UV (<300 nm), electron beam, and x-ray exposure systems play a critical role in manufacture of integrated circuits. In these new lithographic technologies, chemical amplification resist systems (*1*) are likely to be the key player while the conventional diazoquinone/novolac resists fail to perform adequately due to their low sensitivity (and a strong unbleachable absorption in the case of deep UV lithography). Although the main body of the chemical amplification resist activities is directed toward supporting KrF excimer laser (248 nm) lithography (*2*), the high sensitivity and high resolution capabilities of this class of resists have prompted their use in e-beam and x-ray imaging technologies (*3*). In fact, the high resolution capability has been convincingly demonstrated by e-beam nanolithography using the tBOC resist, which produced <30 nm images (*4*).

The chemical amplification concept is based on the use of acids (generated by irradiation) as catalytic species to carry out a cascade of chemical transformations in resist films. Among many acid-catalyzed imaging mechanisms reported so far, the side chain deprotection chemistry to change the polarity and the solubility, as embodied in the IBM tBOC resist, has attracted the most attention in efforts to design aqueous base developable positive resists for replacement of the diazoquinone/novolac resists. Because of the catalytic nature of the imaging mechanisms, however, chemical amplification resists are extremely susceptible to contamination by a trace amount, on

0097–6156/96/0620–0387$12.00/0
© 1996 American Chemical Society

the order of 10 ppb, of airborne basic substances such as amines and *N*-methylpyrrolidone (NMP), which results in a line width change and/or formation of a skin or "T-top" profile upon standing, especially before postexposure bake (PEB) (*5*). The PEB delay problem or the latent image instability has been alleviated by purifying the enclosing atmosphere using activated carbon filtration (*5*), applying a protective overcoat (*6,7*), or incorporating stabilizing additives in resist formulation (*8,9*). However, there still exists a serious need for developing a chemical amplification resist that is resistant to environmental contamination.

The systematic studies on the propensity of thin polymer films to absorb NMP by using a ^{14}C labeling technique have indicated that NMP uptake is primarily governed by glass transition temperatures (T_g) (*10*); lower T_g polymer films absorb NMP at a much slower rate due to better annealing and reduced free volume. Our annealing concept for environmental stabilization is schematically presented in Scheme I.

Good Annealing→Reduced Free Volume→
Reduced Uptake of Airborne Contaminants→Environmental Stabilization

Scheme I. Annealing concept for environmental stabilization

We have proved the validity of our annealing concept for environmental stabilization by comparing chemical amplification resists prepared with lower T_g *meta*-isomers with more common higher T_g *para*-isomers (*11,12*). However, lower T_g resist systems suffer from pattern degradation during high temperature fabrication processes.

In this paper are reported an environmentally stable chemical amplification positive resist (ESCAP) and its x-ray lithographic performance. The ESCAP design, based on our annealing concept, requires high temperature postapply bake (PAB) for its maximum environmental stabilization, which in turn demands a thermally and hydrolytically stable resin and photochemical acid generator (PAG) (*13*).

Experimental

Materials. 4-Acetoxystyrene (ACOST) and *t*-butyl acrylate (TBA) employed in the copolymer synthesis were commercially obtained. The PAG used in the resist formulation was *N*-camphorsulfonyloxynaphthalimide (CSN), which had been synthesized by treating camphorsulfonyl chloride with *N*-hydroxynaphthalimide in the presence of pyridine. The casting solvent for the resist formulation was propylene glycol methyl ether acetate (PMA). The developer employed in the positive imaging was a 0.21 N tetramethylammonium hydroxide (TMAH) aqueous solution.

Copolymer Synthesis. We employed a two-step procedure for the preparation of our resist copolymer (Scheme II). ACOST and TBA were copolymerized using benzoyl peroxide (BPO) in toluene at 60 °C to ca. 80 % conversion to afford homogeneously random copolymers. After purification by precipitation and drying, the copolymers were base-hydrolyzed selectively with ammonium hydroxide in methanol to prepare the desired 4-hydroxystyrene (HOST) copolymers. The hydrolysis mixture was neutralized

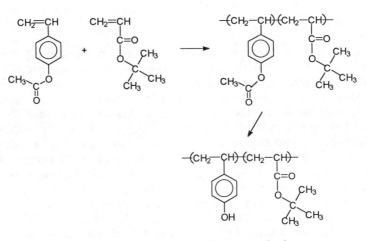

Scheme II. Two-step copolymer synthesis

with acetic acid and poly(HOST-co-TBA) was purified by precipitation in water, isolated by filtration, and dried in a vacuum oven at 50 °C.

Measurements. NMR spectra were obtained on an IBM Instruments NR-250/AF spectrometer in $CDCl_3$ or in acetone-d_6. The inverse gated decoupling technique was applied to polymer samples to minimize the nuclear Overhauser effect for better quantitative analysis in conjunction with the use of $Cr(acac)_3$ as the relaxing agent. Molecular weight determination was made by gel permeation chromatography (GPC) using a Waters Model 150-C chromatograph equipped with 4 ultrastyragel columns at 40 °C in tetrahydrofuran. Thermal analyses were performed on a Perkin Elmer TGS-2 at a heating rate of 5 °C/min for thermogravimetric analysis (TGA) and on a Du Pont 910 at 10 °C/min for scanning calorimetry (DSC) under N_2. IR spectra were measured on an IBM IR/32 FT spectrometer using thin (ca. 1 μm thick) polymer films spin-cast on 1-mm thick NaCl substrates. UV spectra were recorded on a Hewlett-Packard Model 8450A spectrometer using thin films cast on quartz plates. Film thickness was measured on a Tencor alpha-step 200. Accurate thickness and refractive index measurements were carried out using the wave-guide technique.

Lithographic Evaluation. Silicon wafers were vacuum-vapor-primed with hexamethyldisilazane, coated with the ESCAP resist film to ca. 1 μm thickness, and then baked at 150-160 °C for 1-2 min. Exposures were performed using a Suss XRS-200/3 x-ray step-and-scanner. X-ray flux was generated from the Oxford Helios compact synchrotron operating at 687 MeV with 200 mA of stored beam current and a critical wavelength of 9.0 Å. More detailed description of the x-ray exposure system can be found elsewhere (*14*). The exposed wafers were subjected to PEB at 150-140 °C for 1-2 min and developed with a 0.21 N TMAH aqueous solution.

Design Concept

A typical chemical amplification positive resist design is schematically illustrated in Scheme III. The system consists of a base-soluble functionality, a protected dissolution inhibiting group, and a PAG. The base soluble and protected functionalities can be chemically bonded in the form of copolymers or mechanically blended. The unexposed film is not soluble in aqueous base due to the dissolution inhibiting effect of the protected functionality, which is converted to a base soluble form by reaction with an acid (generated by irradiation) and subsequent PEB, providing positive images. Heating such a resist film above its T_g for annealing cannot be performed in many cases as the acid-labile protecting group cannot survive such a high temperature bake in the presence of the *acidic* base-solubilizing phenol or carboxylic acid functionality which raises the T_g through a hydrogen-bonding interaction. Figure 1 illustrates the reduced thermal stability of the tBOC (*t*-butoxycarbonyl) group, which is the most commonly employed protecting group in the positive chemical amplification resists, in the presence of the phenolic functionality. PolyHOST partially protected with the tBOC group undergoes thermal deprotection at ca. 130 °C while the fully protected polymer is stable to ca. 190 °C and the T_g of the copolymers is higher than the deprotection temperature due to the high T_g (170-180 °) of polyHOST (*15*).

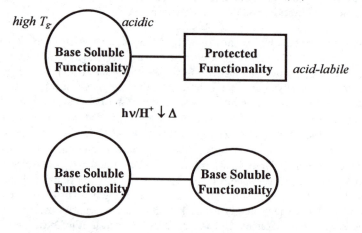

Scheme III. Design of chemically amplified positive resist.

Our ESCAP resist consists of a base soluble poly(TBA-co-HOST) and a sulfonic acid generator. The copolymer (M_n=~17,000 and M_w=~28,000 by GPC) has a T_g of ca. 150 °C and exhibits the onset of thermal deprotection at ca. 180 °C (Figure 2). Thus, the copolymer film can be baked to even 180 °C without decomposing the protecting group, allowing good annealing. The CSN PAG employed in this formulation as also very stable thermally with its decomposition commencing at ca. 250 °C, as the TGA curve in Figure 3 indicates in comparison with one of the most stable PAG triphenylsulfonium trifluoromethanesulfonate. Its thermal stability as studied by

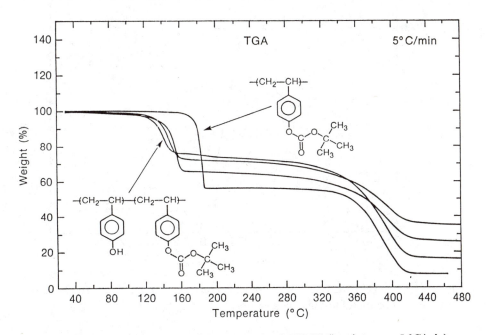

Figure 1 TGA curves of partially protected polyHOST (heating rate: 5 °C/min).

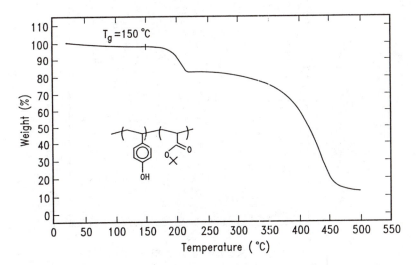

Figure 2 TGA curve of poly(TBA-co-HOST) (heating rate: 5 °C/min).

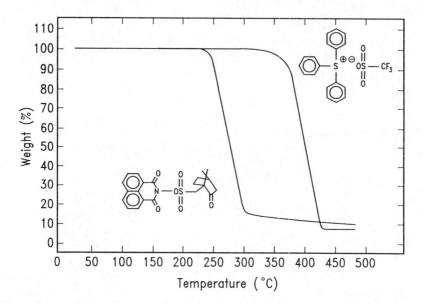

Figure 3   TGA curves of photochemical acid generators (CSN and triphenylsulfonium trifluoromethanesulfonate) (heating rate: 5 °C/min).

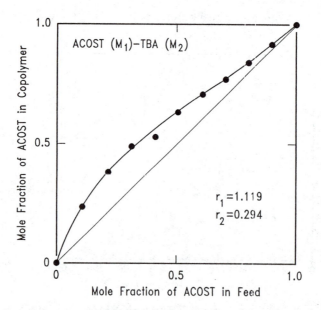

Figure 4   Copolymer composition vs. feed ratio in radical copolymerization of ACOST and TBA in toluene at 60 °C.

TGA is only slightly reduced in the presence of a phenolic group, indicating its high hydrolytic stability.

Results and Discussion

Copolymer Synthesis. The two-step resin synthesis as depicted in Scheme II is highly robust. The composition curve of the radical copolymerization of ACOST (M_1) and TBA (M_2) in toluene is presented as Figure 4. All the copolymerizations were terminated at ca. 10 % conversion to allow use of the differential copolymerization equations. The copolymer compositions were determined by inverse-gated ^{13}C NMR spectroscopy using Cr(acac)$_3$ as the relaxing agent and the monomer reactivity ratios were determined by the Kelen-Tüdös linear method (*16*) as $r_1=1.12$ and $r_2=0.29$. Thus, ACOST is significantly more reactive than TBA in this copolymerization. However, according to our computer simulation of the copolymerization from 0 to 100 % conversion, which was carried out using the reactivity ratio values, the copolymer compositions, sequence lengths, and microstructures remain relatively unchanged as long as the conversion is kept below 90 % (Figure 5). Thus, the reproducibility of the copolymerization is pleasingly high.

The second base hydrolysis step is also very robust and extremely selective. In Figure 6 are compared inverse-gated ^{13}C NMR spectra in Cr(acac)$_3$-containing acetone-d_6 of poly(ACOST-co-TBA) (a) and the base hydrolysis product (HOST-TBA copolymer) (b). While the acetate carbonyl resonance at 167.4 ppm (Figure 6a) has been completely removed by the base hydrolysis, the absence of a resonance due to the carboxylic acid carbonyl group at 177.5 ppm and no reduction of the intensity of the *t*-butyl quarternary carbon resonance at 77.8 ppm (Figure 6b) unequivocally indicate the excellent selectivity of the base hydrolysis step. Prolonged hydrolysis in refluxing methanol does not destroy the *t*-butyl ester.

NMP Uptake in ESCAP Resin Film. Because of the excellent thermal and hydrolytic stabilities of the resin and PAG, the ESCAP resist film can be baked above its T_g, which achieves good annealing and results in reduced free volume. In consequence, NMP uptake as measured by the ^{14}C labeling technique is very low in the ESCAP resin film (*10-12*). The ESCAP resin film absorbs NMP at a much slower rate (7.8 ng/min) from airstream containing 10 ppb NMP even when baked below its T_g than a conventional chemical amplification resist (18 ng/min), which is presumably due to the incorporation of the acrylate structure (solubility parameter consideration) (*10*). However, a dramatic reduction of NMP uptake down to 1.8 ng/min is achieved by baking the ESCAP resin film at 170 °C (above its T_g of ~150 °C), which should be compared with 1.1 ng/min for a bare silicon wafer (*10-12*). Thus, it is clear that the high temperature bake minimizes the absorption of NMP in the ESCAP resin film, which we believe is due to the reduced free volume. The spin-cast ESCAP resin film is completely isotropic even when baked at 100 °C ($n_{||}=1.5388\pm0.0001$ and $n_{\perp}=1.5395\pm0.0001$) and contains 7.9 wt% residual PMA. Heating the resin film to 170-180 °C reduces the PMA residue to 0.51 wt% and results in 9.4 % thickness reduction, which is accompanied by an increase in the refractive indices by 0.6-0.4 % ($n_{||}=1.5481\pm0.0003$ and $n_{\perp}=1.5485\pm0.0001$). As reported previously (*10-12*), a small change in the

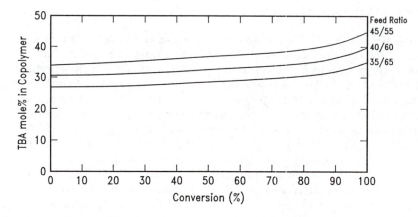

Figure 5 Effect of conversion on ACOST-TBA copolymer composition for feed ratio of ACOST/TBA=60/40, 65/35, and 70/30.

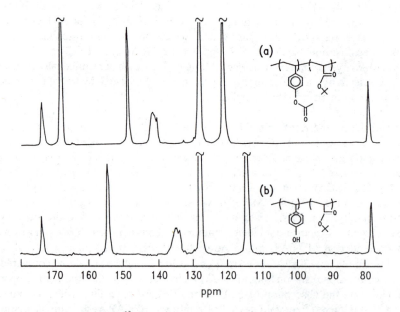

Figure 6 Inverse-gated ^{13}C NMR spectra in acetone-d$_6$ with Cr(acac)$_3$ of (a) poly(ACOST-co-TBA) and (b) poly(HOST-co-TBA) prepared by base hydrolysis of (a).

refractive index (density) can result in an extremely large change in the diffusivity of small molecules in polymer films. The diffusivity is an exponential function of the free volume.

Contamination Resistance of ESCAP.   The much reduced contaminant absorption of the ESCAP resist film due to annealing can result in dramatically improved lithographic PEB delay stability. While a conventional chemical amplification resist typically exhibits a skin or T-top formation after standing less than 30 min before PEB is carried out, the ESCAP resist can print fine images even when baked ~18 hr after exposure. Figure 7 presents scanning electron micrographs of 0.35 μm line/space patterns obtained on a Micrasan II deep UV exposure tool without a PEB delay and after overnight standing before PEB (0.255 N TMAH developer). Two images are essentially indistinguishable, indicating the excellent environmental stability of the ESCAP resist.

X-Ray Lithography of ESCAP. While the previous acid-catalyzed system required a protective overcoat to seal out airborne basic substances (*17*), the ESCAP resist can provide a true single layer x-ray imaging, enhancing the advantage of x-ray lithography, which requires no antireflection coating (ARC) and is therefore simpler than the deep UV technology (Scheme IV).

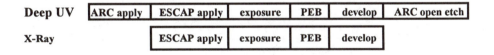

Scheme IV. Comparison of deep UV and x-ray lithographic process sequences

Although the ESCAP resist contains only a small amount of PAG which generates a rather weak camphorsulfonic acid, its sensitivity to synchrotron x-ray radiation is quite high. The high thermal stability of the resist and the high temperature PAB allow PEB to be carried out at elevated temperatures, which contributes to its high radiation sensitivity in conjunction with conversion of the ester structure to a highly soluble carboxylic acid functionality. When PAB and PEB were carried out at 160 and 140 °C, respectively, clean development to the substrate was observed at 80-90 mJ/cm^2 (●, Figure 8). Under the optimum bake conditions (PAB: 160 °C and PEB: 150 °C, O, Figure 8), the dose is cut in half with little change in contrast and an increase in exposure latitude from 24 to 31 % is achieved for 300 nm images. Although the consistently highest exposure latitude of 27-32 % (for 300 nm features) tends to be obtained with the dose range of 35-45 mJ/cm^2, adjusting the process conditions to achieve 15-25 mJ/cm^2 sensitivity provides usable exposure latitude of >20 %. Thus, an almost 3x boost in sensitivity can be achieved if needed and the ESCAP resist has an application towards point source x-ray systems at 14 Å with an estimated sensitivity of 3-6 mJ/cm^2.

Figure 9 shows PEB temperature latitude plots for 200 and 300 nm isolated lines. The slopes are -15.8 and -14.5 nm/°C for 200 and 300 nm features, which is about 2x improvement over the previous chemical amplification resist.

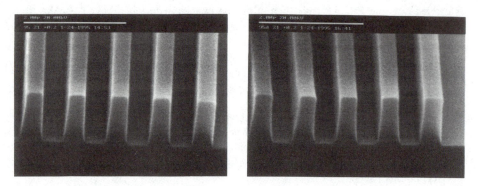

Figure 7   Scanning electron micrographs of ESCAP resist images (0.35 μm line/space) obtained with no (left) and overnight PEB dealy (right).

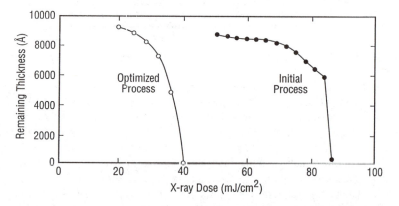

Figure 8   Synchrotron x-ray sensitivity curves of the ESCAP resist under two different bake conditions.

A scanning electron micrograph of 150 nm line/350 nm space images printed at 46 mJ/cm2 in the ESCAP resist on a nitride substrate is presented as Figure 10. Due to the high T_g and the high deprotection temperature of the resin, the ESCAP image is devoid of thermal flow at 150 °C for at least 15 min, which is a very important advantage in device fabrication in comparison with other chemical amplification positive resists which suffer from image degradation due to thermal deprotection at <130 °C.

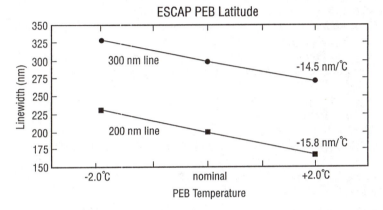

Figure 9   Effect of PEB temperature variation on line width of ESCAP images (200 and 300 nm).

Figure 10   Scanning electron micrograph of 150 nm line/350 nm space positive images printed in ESCAP on a nitride substrate at 46 mJ/cm^2 of synchrotron x-ray exposure.

Acknowledgment

The authors thank H. Truong for her GPC measurments and thermal analyses and R. Johnson for his NMR support. The Yorktown portion of this work was sponsored by ARPA and administered by NAVAIR contract No. N00019-91-C-0270.

References

1. Ito, H; Willson, C. G. in *Polymers in Electronics*; Davidson, T., Ed.; Symposium Series 242; American Chemical Society: Washington, D. C., 1984, p. 11.
2. Ito, H. in *Radiation Curing in Polymer Science and Technology*; Fouassier, J. P.; Rabek, J. F., Eds.; Elsevier: London, 1993, Vol, 4, Chapter 11.
3. Ito, H. in *Radiation Effects on Polymers*; Clough, R. L.; Shalaby, S. W., Eds.; Symposium Series 475; American Chemical Society: Washington, D. C., 1991, p. 326.
4. Umbach, C. P.; Broers, A. N.; Willson, C. G.; Koch, R.; Laibowitz, R. B. *J. Vac. Sci. Technol.* **1988**, *B6*, 319.
5. MacDonald, S. A.; Clecak, N. J.; Wendt, H. R.; Willson, C. G.; Snyder, C. D.; Knors, C. J.; Deyoe, N. B.; Maltabes, J. G.; Morrow, J. R.; McGuire, A. E.; Holmes, S.J. *Proc. SPIE* **1991**, *1466*, 2.
6. Nalamasu, O.; Cheng, M.; Timko, A. G.; Pol, V.; Reichmanis, E.; Thompson, L. F. *J. Photopolym. Sci. Technol.* **1991**, *4*, 299.
7. Kumada, T.; Tanaka, Y.; Ueyama, A.; Kubota, S.;Koezuka, H.; Hanawa, T.; Morimoto, H. *Proc. SPIE* **1993**, *1925*, 31.
8. Röschert, H.; Przybilla, K.-J.; Spiess, W.; Wengenroth, H.; Pawlowski, G. *Proc. SPIE* **1992**, *1672*, 33.
9. Funhoff, D. J. H.; Binder, H.; Schwalm, R. *Proc. SPIE* **1992**, *1672*, 46.
10. Hinsberg, W.; MacDonald, S. A.; Clecak, N. J.; Snyder, C.; Ito, H. *Proc. SPIE* **1993**, *1925*, 43.
11. Ito, H.; England, W. P.; Clecak, N. J.; Breyta, G.; Lee, H.; Yoon, D. Y.; Sooriyakumaran, R.; Hinsberg, W. D. *Proc. SPIE* **1993**, *1925*, 65.
12. Ito, H.; England, W. P.; Sooriyakumaran, R.; Clecak, N. J.; Breyta, G.; Hinsberg, W. D.; Lee, H.; Yoon, D. Y. *J. Photopolym. Sci. Technol.* **1993**, *6*, 547.
13. Ito, H.; Breyta, G.; Hofer, D.; Sooriyakumaran, R.; Petrillo, K.; Seeger, D. *J. Photopolym. Sci. Technol.* **1994**, *7*, 433.
14. Pomerene, A.; Petrillo, K.; Seeger, D.; Ito, H.; Breyta, G.; Hofer, D. *Proc. SPIE* **1994** *2194*, 162.
15. Ito, H. *J. Polym. Sci., Polym. Chem. Ed.* **1986**, *24*, 2971.
16. Kelen, T.; Tüdös, F. *J. Macromol. Sci., Chem.* **1975**, *A9(1)*, 1.
17. Pomerene, A. T. S.; Seeger, D.; Blauner, P. *Proc. SPIE* **1993**, *1024*, 2.

RECEIVED June 20, 1995

Chapter 31

Silylating Reagents with High Silicon Contents for Dry-Developed Positive-Tone Resists for Extreme-UV (13.5 nm) and Deep-UV (248 nm) Microlithography

David Wheeler[1], Eric Scharrer[1], Glenn Kubiak[1], Richard Hutton[2], Susan Stein[2], Ray Cirelli[2], Frank Baiocchi[2], May Cheng[2], Craig Boyce[2], and Gary Taylor[2]

[1]Sandia National Laboratories, Albuquerque, NM 87185-0368
[2]AT&T Bell Laboratories, Murray Hill, NJ 07974-0636

Recent results in the use of disilanes as silylating reagents for near-surface imaging with deep-UV (248 nm) and EUV (13.5 nm) lithography are reported. A relatively thin imaging layer of a photo-cross-linking resist is spun over a thicker layer of hard-baked resist that functions as a planarizing layer and antireflective coating. Photoinduced acid generation and subsequent heating crosslinks and renders exposed areas impermeable to an aminodisilane that reacts with the unexposed regions. Subsequent silylation and reactive ion etching afford a positive-tone image. The use of disilanes introduces a higher concentration of silicon into the polymer than is possible with silicon reagents that incorporate only one silicon atom per reactive site. The higher silicon content in the silylated polymer increases etching selectivity between exposed and unexposed regions and thereby increases the contrast. Additional improvements that help to minimize flow during silylation are also discussed, including the addition of bifunctional disilanes. We have resolved high aspect ratio, very high quality 0.20 μm line and space patterns at 248 nm with a stepper having a numerical aperture (NA)= 0.53, and have resolved ≤ 0.15 μm line and spaces at 13.5 nm.

Design rules for microlithography are moving into the sub 0.35 μm regime. This generates more stringent requirements for the entire lithographic process. Deep UV (248 nm) lithography has demonstrated the ability to print sub 0.35 μm features. However, deep UV lithography at 248 nm will probably be unable to meet the design rules that are expected to be needed at the turn of the century. To meet these needs, exposure tools will probably employ shorter wavelength radiations. Extreme ultraviolet lithography, EUVL, at 13.5 nm has demonstrated the capability to print very fine features, and it is predicted that the trend to shorter wavelengths will eventually employ 13.5 nm radiation about the year 2010. Unfortunately, radiation at

0097–6156/96/0620–0399$12.00/0
© 1996 American Chemical Society

13.5 nm, formerly called soft x-ray lithography, is strongly attenuated by almost all materials. Since very thin films are not adequate for the manufacture of devices, some sort of surface imaging scheme will be needed for EUVL resists. Even 193 nm lithography may require surface imaging techniques due to the high absorption of most photoresists at this wavelength. While there are a few photoresist materials for 193 nm lithography that are transparent enough to allow imaging in a single layer, there are no such materials for EUV lithography. This is due to the atomic nature of the absorption for EUV as opposed to the molecular absorption for deep UV wavelengths. Surface imaging provides a method for overcoming the problems of optical density, but introduces additional processing variables. However, surface imaging relaxes many of the other processing issues and requirements such as those associated with topography and reflection.

Surface imaging schemes can be broken down into two basic categories based on the thickness of material used for imaging. Near surface imaging schemes are typified by processes such as DESIRE[2] and SAHR[3], which utilize the top few thousand angstroms of resist. At-the-surface-imaging schemes employ only the uppermost 100Å and are typified by such processes as the recently described photo-definable electroless nickel deposition process.[4]　We recently described a near surface imaging scheme that employed disilanes and a bilayer resist scheme to improve silicon contrast (Figure 1).[5]

The rational for choosing to examine disilanes can be seen in Table 1. A summary of commonly used mono- and difunctional silylating agents is shown in Table 1. The swelling was calculated from the change in number of atoms in the repeat unit of the polymer realative to the number of atoms in the silylated polymer. While admittedly simplistic, the calculations matched experimental results fairly closely. Experimentally measured volume increases were smaller than calculated. This is probably due to the fact that photoresists are not pure polymers and contain species which are not involved in silylation. Difunctional silylating agents were assumed to react completely for the calculations. From the table it is clear that despite the higher boiling points, disilanes hold the promise of large weight percent silicon incorporation without enormous swelling. It is further evident that pure difunctional silylating agents, while causing little swelling, add little silicon to the resist. Increasing the silicon content increases the etch selectivity and processing latitudes.

Our silylation scheme is similar to the SAHR process. Briefly, it employs photoinduced acid generation in the top imaging layer and subsequent post exposure baking to crosslink the acid-containing regions and render these areas impermeable and unreactive to N,N-dimethylaminopentamethyldisilane, (DMAPMDS). This disilane, despite a boiling point of 155 °C and large steric size, is as reactive under our silylation conditions as hexamethyldisilazane. A brief chlorine/argon descum followed by reactive ion etching transfers the surface pattern into the bulk of the resist. By employing etching conditions that effect both sidewall deposition and etching at the same time, nearly vertical side walls can be achieved with minimum undercut.[6] The descum step is needed since there is some silylation, approximately 200 Å, of the crosslinked areas. The descum step removes all residue and improves the reproducibility and linearity of the process. We believe that the surface silylation in the crosslinked areas occurs because of acid loss during post exposure baking by volatilization from the surface of the resist.

Imaging Layer
Processing Layer

Substrate with Topography

EXPOSURE

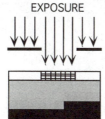

Acid catalyzed
cross-linking during
post exposure bake

SILYLATION

Volume increase
during silylation

PLASMA DESCUM

Thin Silylation of
Crosslinked Region
Removed

PLASMA ETCHING

Linewidth controlled
by bilayer

Figure 1: Schematic of bilayer silylation process using disilanes.

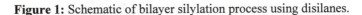

Table 1: Commonly used silylation agents and properties. Weight percent silicon was calculated by assuming complete silylation of a cresol novolac. Volume Increase was calculated as atom percent increase in a cresol novolac. Reagents marked with an * are difunctional. Etching selectivity is silylated resist vs. hard baked photoresist etched under the following conditions: source power, 2500 W; chuck power, 75 W; pressure, 2.5 mTorr; flow, 100 sccm O_2; temperature, 25 °C.

Silylating Agent	Binding Group	b.p. (°C)	Weight % Silicon	Volume Increase (%)	Etching Selectivity
Me_3SiNMe_2	$-SiMe_3$	86	14.6	71	18
$Me_2HSiNMe_2$	$-SiHMe_2$	67	15.7	47	27
$Me_5Si_2NMe_2$	$-Si_2Me_5$	155	22.4	124	45
$Me_2Si_2H_3NMe_2$	$-Si_2Me_2H_3$	125	26.9	71	~60
$Me_2Si_2H_3NEt_2$	$-Si_2Me_2H_3$	147	26.9	71	~60
——————————— Difunctional Silylating Agents: ———————————					
*$(Me_2N)_2SiMe_2$	$-SiMe_2-$	128	9.5	21	~10
*$(Me_2N)_2SiMeH$	$-SiMeH-$	112	9.9	12	~10
*$Me_4Si_2(NMe_2)_2$	$-Me_4Si_2-$	192	15.8	47	~15

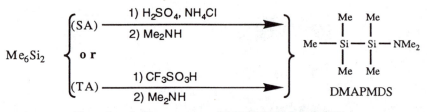

Equation 1: Two routes to DMAPMDS, SA and TA.

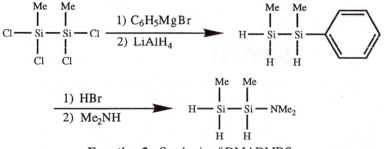

Equation 2: Synthesis of DMADMDS.

Synthesis of Disilanes:

Most of our work has been with N,N-dimethylaminopentamethyldisilane (DMAPMDS). Two syntheses of this material are shown in Equation 1. A single methyl group of hexamethyldisilane, Me_6Si_2, can be replaced with a chlorine by reacting the disilane with excess sulfuric acid followed by ammonium chloride (65% yield).[7] This we termed the sulfuric acid (SA) route. Alternatively, Me_6Si_2 reacts with an equivalent of triflic acid to give pentamethyldisilyl triflate (80%). We term this the triflic acid (TA) route. A second equivalent of triflic acid gives 1,1,2,2-tetramethyldisilane ditriflate in a somewhat lower yield.[8] Aminolysis with excess dimethylamine or methylamine gives the aminodisilane (>80%). The reaction of trifilic acid with hexamethyldisilane can be carried out neat. The reaction can therefore be carried out in a single pot. While aminolysis of the ditriflate gives 1,2-bis(N,N-dimethylamino)tetramethyldisilane, we have found that it is easier to synthesize it from the dichloride. The dichloride is available in over 95% yield from the reaction of two equivalents of the aluminum trichloride and acetyl chloride with hexamethyldisilane.   This   is   an   adaptation   of   the   procedure   for tetrachlorodimethyldisilane.[9]

In the course of our work, we also desired a silylating agent that would produce less swelling than DMAPMDS. We decided to synthesize the "smaller" N,N-dimethylamino-1,2-dimethyldisilane (DMADMDS). This material is moisture-sensitive but not pyrophoric. While many of the synthetic routes examined did provide the product, they often were not amenable to scale up or provide for adequate purification. Equation 2 shows a synthesis of DMADMDS. An aromatic ring is used to protect one of the halogens of 1,2-dimethyltetrachlorodisilane[10] during reduction by lithium aluminum hydride. Addition of phenyl magnesium bromide to 1,2-dimethyltetrachlorodisilane provides phenyl-1,2-dimethyltrichlorodisilane which is reduced without isolation which provides phenyldimethyldisilane in almost 85% isolated yield. The phenyl group of phenyldimethyldisilane can be removed by many routes including anhydrous HBr at -78 °C.[11] However, careful Schlenk or high vacuum techniques must be employed. Additionally, rigorous control of the reaction time, starting material purity, and other reaction conditions is needed in order to avoid pyrophoric by-products. Aminolysis provided DMADMDS in 54% overall yield from phenyldimethyldisilane.

Results:

The low boiling point of DMADMDS (110 °C) allows use of lower temperatures during silylation. DMADMDS also causes less swelling of the polymer network (65% theoretical volume increase for a pure cresol novolac) upon silylation than does DMAPMDS (100%). The calculated swelling associated with DMADMDS is identical to that of a trimethysilyl group. DMADMDS yields a silylated polymer with a higher weight percent of silicon (26.8%, calculated for complete silylation of a pure cresol novolac) as compared to DMAPMDS, (22.3%). The increased weight percent silicon provides greater etch selectivity and increased processing latitudes during reactive ion etching, RIE. However, imaging results with this material have not been as good as expected. DMADMDS diffuses into the crosslinked areas of the photoresist during silylation. This lowers the silicon contrast and consequently the imaging quality.

DMAPMDS diffuses through a layer of photoresist XP-8844 in much the same manner as do other silylated agents.[12] Rutherford backscattering (RBS) data (Figure 3) indicate that the diffusion of DMAPMDS is non-Fickian. Sharp delineation between the unsilylated and silylated regions is always observed. This indicates that under the conditions employed for the silylation step, DMAPMDS reacts very rapidly with the resin relative to its diffusion rate. RBS data indicate that the planarizing layer used is not silylated. This prevents silicon from diffusing under exposed regions. Diffusion of silicon under the crosslinked areas can cause loss of line width control.[13] Figure 2 is a standard contrast curve from early studies in our process development. It indicates that the contrast is very high and, as expected, is related to the degree of overetch. More silicon incorporation increases overetch latitude and the contrast as well.

An important observation was that the DMAPMDS produced via the two different routes (SA and TA) gave very different results. Silylations using material prepared via triflic acid were more rapid than silylations using material produced via the sulfuric acid method. Additionally, the lithographic patterns achieved with the triflic acid material were much poorer than those produced with the sulfuric acid material. We suspected that the reason for the differences might be due to an impurity in the SA produced material.

Synthesis of chloropentamethyldisilane (b.p. 134 °C) from sulfuric acid and ammonium chloride as described in the literature required two distillations in order to separate the product, chloropentamethyldisilane, from the unreacted hexamethyldisilane and 1,2-dichlorotetramethyldisilane (b.p. 148 °C) which forms during a competing, second protonolysis as shown in Equation 3. The protonolysis of the first methyl group is claimed to be 10 times faster than the protonolysis of the second methyl group. We found the rates differed by a factor of 5. The similarity of the boiling points of chloropentamethyldisilane and 1,2-dichlorotetramethyldisilane complicates the separation and suggested that SA produced DMAPMDS might be contaminated with a small amounts $\sim 1\%$ of 1,2-bis(N,N-dimethylamino)tetramethyldisilane. We verified its presence and quantified the amount of 1,2-bis(N,N-dimethylamino)tetramethyldisilane by GC and NMR. In order to unambiguously determine that small amounts of difunctional disilane and crosslinking were the source of the differences, we added 1-2 percent of bis(dimethylamino)dimethylsilane, another bifunctional crosslinking silane, to pure TA produced DMAPMDS. This addition restored resolution similar to that of the SA produced DMAPMDS. Finally, we synthesized 1,2-bis(N,N-dimethylamino)tetramethyldisilane and added it to the TA prepared DMAPMDS, again restoring resolution. We believe that small percentages of crosslinker in the DMAPMDS decrease flow of the silylated regions by lightly crosslinking the silylated regions and raising the T_g of the silylated resist. This increase in T_g, modulus, and viscosity due to crosslinking helps ensure that the volume increase is anisotropic. This is shown clearly in the SEM's of cross-sectioned wafers that were silylated with different batches of DMAPMDS, Figure 4.

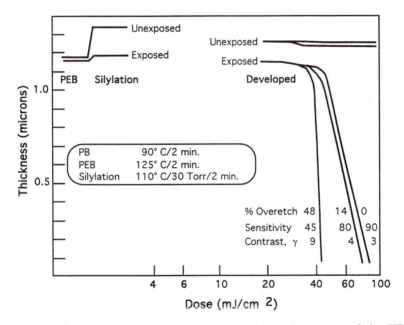

Figure 2: Effect of overetch on the sensitivity and contrast of the XP-8844/SPR 1811 bilayer resist system. Silylation conditions were 110 °C for 2 min. at 30 Torr DMAPMDS. Development used Helicon etching with a source power of 2500 Watts, chuck power of 75W, 25 °C, 2.5 mTorr, 100 sccm oxygen. The initial film thickness was 1.1 μm. Post exposure bake (PEB) was 125 °C for 2 minutes. Prebake (PB) was 2 minutes at 90 °C. The percentage of overetch is indicated along with the sensitivity (mJ/cm²) and contrast (g).

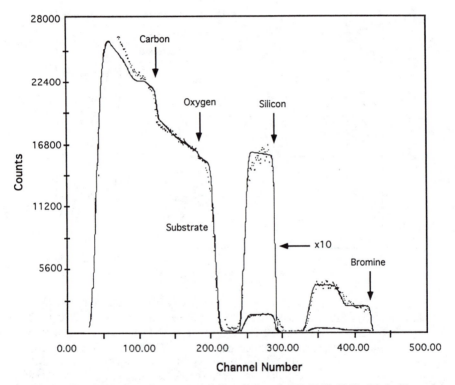

Figure 3: RBS data from a partially silylated (DMAPMDS) thick film of Shipley XP-8844, (XP-8844 is the imaging layer shown in Figure 1).

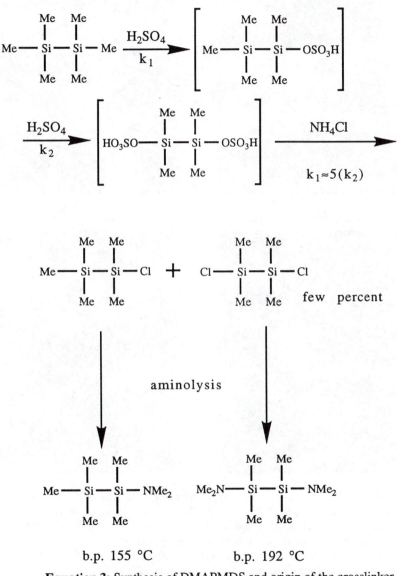

Equation 3: Synthesis of DMAPMDS and origin of the crosslinker 1,2-bis(dimethylamino)tetramethyldisilane.

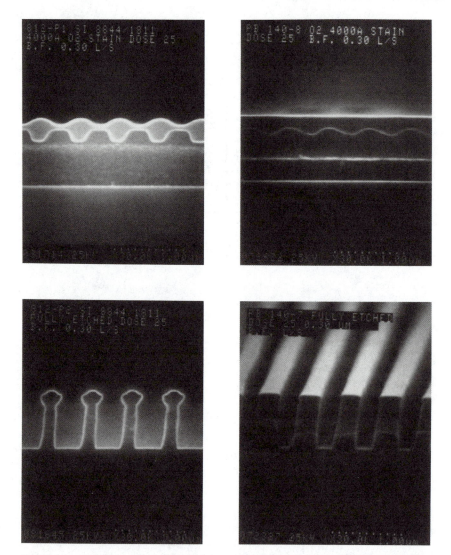

Figure 4: The effect of crosslinker on silylation with DMAPMDS on 0.30 μm lines and spaces, cleaved and stained. A) Top left: silylation with crosslinker present. B) Top right: silylated without crosslinker present, flow evident. C) Lower left, patterns resolved after descuming and oxygen plasma etching of silylated patterns containing crosslinker. The undercut results since there is less erosion of the silicon masking layer and the effect of neutrals is observed. D) Lower right, patterns resolved after descuming and oxygen plasma etching of patterns silylated without crosslinker. The masking layer erodes at rate comparable to the undercut rate giving straight sidewalls but CD control is harder.

(Reproduced with permission from SPIE. Copyright 1995.)

Discussion

Flow in silylated resists was identified as a problem very early in the development of silylation imaging schemes. The reduction of T_g upon silylation and the associated volume increase make flow very likely. A number of different techniques to deal with this flow problem have been developed. They range from techniques to prevent flow during silylation to techniques to remove the effects of flowed resist afterwards. These techniques include plasma descuming[14] and partial solution development of the wafer prior to silylation.[15] Presilylation aqueous development can be used to create a channel in which the silylated resist expands.[16] While all of these schemes contribute to alleviating the effects of silylated resist flow, we feel that the presence of the crosslinker in the disilane silylating agent during silylation achieves the same results, but with a simplicity that the multistep methods can not match.

The significance of the crosslinker is that it dramatically modifies the lithographic behavior even in low concentrations. We believe that this is the first example of using small amounts of vapor phase crosslinking agents to control flow during silylation, but this is certainly not the first example of the use of bifunctional reagents. The use of bifunctional silylation reagents such as hexamethylcyclotrisilazane has been reported previously.[17] Typically, these materials are used in solution silylation schemes as they are not very volatile.[18] A comparison of solution and vapor phase silylation has been made.[19] Bifunctional reagents consume up to two phenolic OH's for every reagent molecule, resulting in less silicon incorporation, lower etching selectivity, lower silicon contrast and narrower processing windows. Previous use of crosslinking silylating agents has employed them either as the sole component in silylation or in a single step of a two step silylation prior to silylation with monofunctional silylation reagents. By employing only a small amount of crosslinker, fewer phenolic sites bind to the same silylating agent. This results in more silicon in the photoresist and increased etching selectivity and silicon contrast. Use of pure monosilicon difunctional crosslinker would result in decreased silicon in the silylated polymer since two phenolic sites are used per each silicon agent.

The CARL process addresses this problem by using bifunctional oligomers containing multiple silicon atoms.[20] The oligomers are bifunctional, having a reactive amine group at each end of the chain, and deliver multiple silicon atoms to each binding site in the polymer. However, the oligomers are so large that a great deal of additional organic material is delivered and much greater swelling occurs. This results in the swelling and line feature expansion that is the distinctive feature of the CARL process.

A two step silylation process to control flow that also employs crosslinking has been discussed.[21] In this process, the imaging layer is first presilylated by bis(dimethylamino)dimethylsilane in the gas phase to form a crosslinked "skin" on the surface of the resist that inhibits flow during the second silylation step which employs a monofunctional reagent. Our single step crosslinking silylation process is similar, but offers the advantages of simplicity (1 step) and increased silicon content due to the significantly lower level of crosslinker and use of disilane silylation reagents.

In order to test the utility of the silylation techniques that we have developed, we have extensively examined cross-sectioned and plasma-stained samples of wafers to learn more about the effect of the crosslinker concentrations on imaging results. Figure 4 shows the effect of the crosslinker in DMAPMDS on 0.30 μm equal lines and spaces after silylation and oxygen plasma development in a helicon etcher. With pure DMAPMDS made by the TA method, the surface of the silylated top layer was planarized by flow, Figure 4B. With SA prepared DMAPMDS the swelling was still evident but was much more anisotropic, Figure 4A. The effect of this difference in silicon contrast on the patterns developed after descum and oxygen plasma etching is dramatic. The sidewalls have less undercut for material silylated with the TA produced DMAPMDS, Figure 4D. This anisotropy results from erosion of the thin silylated masking layer during etching. It occurs because there is less silicon present at the edges of features. As the edges of the masking layer are slowly etched away, any undercut is rendered invisible because the silicon mask lateral erosion rate is comparable to the undercut rate. Although this affords attractive micrographs, it causes poor CD control during overetch. Additionally, erosion of the masking layer will cause feature size and feature geometry nonlinearities. The micrographs in Figure 4C clearly show undercut due to etching by neutrals which are responsible for isotropic etching. Features such as these, Figure 4C, would be unacceptable for device manufacture.

In order to improve the imaging and decrease undercut, which is clearly present in Figure 4C, we examined etching conditions that afforded sidewall deposition during etching. Again the effect of the crosslinker is dramatic. In Figure 5, 0.25 μm line and space patterns silylated with and without crosslinker were etched with SO_2 and O_2 under our improved etching conditions. Under these new conditions with a 50% overetch, the patterns silylated with TA produced DMAPMDS do not develop at all, while those silylated with DMAPMDS containing crosslinker develop and have straight sidewalls. Integration of all of these advance (disilane silylation agents containing a few percent of crosslinking disilane, and etching conditions that afford sidewall deposition) provides a surface imaging scheme that is very reproducible and capable of very high quality features. Figure 7 shows three micrographs of line and space patterns developed using this integrated process. The three micrographs in Figure 7 (0.30, 0.25 and 0.20 μm line and space patterns) were all achieved with identical doses, 38 mJ/cm^2.

To address the resolution limits of this dry-developed imaging system, we examined wafers exposed at 13.5 nm (EUV). Figure 6 shows 0.20 μm and 0.15 μm lines and spaces obtained by EUV exposures. The EUV images are inferior to the DUV 248 nm images (Figure 7 contains 0.20 μm line and space patterns) even though the EUV images should have resolved features as small as 80 nm. Experiments with 60 nm thickness films of ZEP-320 have given even poorer results than those obtained with our dry developed scheme.[22] This suggests that the resolution is not resist limited, but rather is aerial image limited. At this point we can only say that our silylated resist is capable of a least 150 nm resolution.

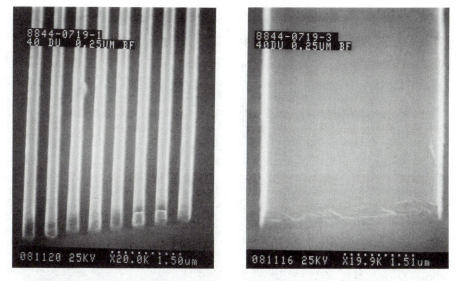

Figure 5: The effect of the crosslinker on 0.25 μm lines and spaces images silylated with DMAPMDS using SO_2/O_2 etching conditions that afford sidewall deposition during etching. The patterns at right were silylated without crosslinker and did not develop. The images at left were achieved with the crosslinker present and have straight sidewalls. A 50% overetch was used.

(Reproduced with permission from SPIE. Copyright 1995.)

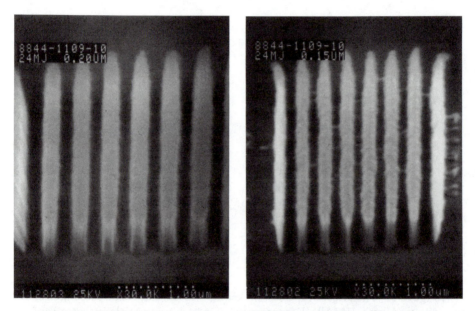

Figure 6: EUVL (13.5 nm) exposures of 0.20 μm and 0.15 μm line and space patterns developed using DMAPMDS containing a crosslinker and using the improved etching conditions with a dose of 24 mJ/cm^2.
(Reproduced with permission from SPIE. Copyright 1995.)

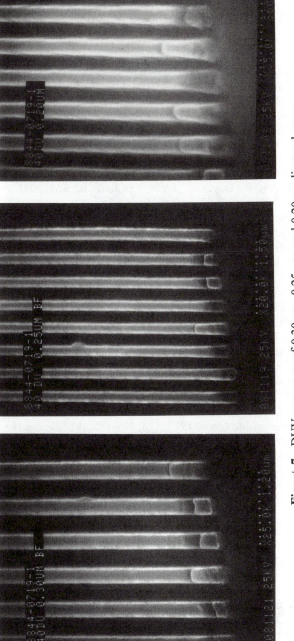

Figure 7: DUV exposures of 0.30 μm, 0.25 μm, and 0.20 μm line and space patterns developed using DMAPMDS with 5% weight percent of 1,2-bis(N,N-dimethylamino)tetramethyldisilane present and developed with SO_2/O_2 etching conditions that afford sidewall deposition during etching. A 50% overetch was employed. All features were printed with the same dose, 38 mJ/cm^2.

(Reproduced with permission from SPIE. Copyright 1995.)

Literature Cited

1   This work was performed at Sandia National Labs supported by the U.S. Department of Energy under contract DE-AC04-94AL85000.

2   (a) Coopmans, F.; Roland, B. *Proc. SPIE*, **1986**, *631*, 34. (b) Garza, C. M.; Misium, G. R.; Doering, R. R.; Roland, B.; Lombaerts, R. *Proc. SPIE*, **1989**, *1086*, 229.

3   (a) Thackery, J. W.; Bohland, J. F.; Pavelchek, E. K.; Orsula, G. W.; McCullough, A. W.; Jones, S. K.; Bobbio, S. M. *Proc. SPIE*, **1989**, *1185*, 2-11. (b) Pavelchek, E. K.; Bohland, J. F.; Thackery, J. W.; Orsula, G. W.; Jones, S. K.; Dudley, B. W.; Bobbio, S. M.; Freeman, P. W. *J. Vac .Sci. Technol.* **1990**, *B8*, 1497-1501.

4   Calvert, J. M.; Koloski, T. S.; Dressick, W. J.; Dulcey, C. S.; Peckerar, M. C.; Cerrina, F.; Taylor, J. W.; Suh, D.; Wood II, O.R.; MacDowell, A. A.; D'Souza, R; *Optical Engineering*, **1993**, *32(10)*, 2437-2445.

5   Wheeler, D. R.; Hutton, R.; Boyce, C.; Stein, S.; Cirelli, R.; Taylor, G. *Proc. SPIE*, **1995**, *2438,* 762.

6   Taylor, G. N.; Hutton, R. S.; Stein, S. M.; Boyce, C. H.; Wood, O. R.; LaFontaine, B.; MacDowell, A. A.; Wheeler, D. R.; Kubiak, G. D.; Ray-Chaudhuri, A. K.; Berger, K.; Tichenor, D. *Proc. SPIE*, **1995**, *2437*, 308.

7   (a) Dubowchik, G. M.; Gottschall, D. W.; Grossman, M. J.; Norton, R. L.; Yoder, C. H. *J. Am. Chem. Soc.*, **1982**, *104*, 4211-14. (b) Engelhardt, G.; Radeglia, R.; Kelling, H.; Stendel, R. *J. Organomet. Chem.* **1981**, *212(1)*, 51-8. (c) Hengge, E.; Pletka, H. D.; Hoefler, F. *Monatsh. Chem.*, **1970**, *101(2)*, 325-36. (d)Kumada, M.; Yamaguchi, M.; Yamamoto, Y.; Nakajima, J.; Shhna, K. *J. Org. Chem.* **1956**, *24*, 1264-1268.

8   (a) Matjaszewski, K.; Chen, Y. L. *J. Organometallic Chem.* **1988**, *340*, 7-12. (b) Bassindale, A. R.; Stout, T. *J. Organometallic Chem.* **1984**, *271*, C1-C3.

9   Sakurai, H.; Watanabe, T.; Kumada, M. *J. Organometallic Chem..* **1967**, *7*, P15-P16.

10   (a) Watanabe, H.; Kobayashi, M. Koike, Y.; Nagashima, S.; Matsumoto, H.; Nagai, Y. *J. Organometallic Chem.* **1977**, *128*, 173-175. (b) Sakurai, H.; Watanabe, T.; Kumada, M. *J. Organometallic Chem.* **1967**, *7*, P15-P16.

11   (a) Schmidbauer, H.; Zech, J.; Rankin, D. W. H.; Robertson, H. E.; *Chem. Ber.* **1991**, *124*, 1953-1956. (b) Hagcr, R. Stcigclmann, O.; Müller, G. Schmidbauer, H. *Chem. Ber.* **1989**, *122*, 2115-2119.

12   (a) Hartney, M. A.; Rothschild, M.; Kunz, R. R.; Ehrlich, D. J.; Shaver, D. C. *J. Vac. Sci. Technol.* **1990**, *B 8(6)*, 1476-1480. (b) Baik, K.; Van den hove, L.; Goethals, A. M.; Op de Beeck, M; Roland, B. *J. Vac. Sci. Technol.* **1990**, *B 8(6)*, 1481-1487.

13   Baik, K-H.; Ronse, K.; Van den Hove, L.; Roland, B. *Proc. SPIE*, **1993**, *1925*, 302.

14   Goethals, A. M.;Baik, K.H.; Van den hove, L.; Tedesco, S. *Proceedings SPIE,* **1991**, *1466*, 604.

15   Pavelchek, E. K.; Calabrese, G. S.; Bohland, J. ; Dudley, B. W. ; Jones, S. K. ; Freeman, P. W. *Optical Eng.* **1993**, *32(10),* 2376.

16   W. Han, W.; Lee, J.; Park, J. ; Park, C.; Kang, H.; Koh, Y.; Lee, M. *Proceedings SPIE* , **1993**,*1925*, 291.

17   Shaw, J. M.; Hatzakis, M.; Babich, E. D.; Paraszczak, J. R.; Witman, D. F.; Stewart, K. J. *J. Vac. Sci. & Technol. B* , **1989**, *7*, 1709.

18   Baik, K-H.; Ronse, K.; Van den Hove, L.; Roland, B. *Proc. SPIE*, **1993**, *1925*, 302.

19   Hartney, M. A.; Kunz, R. R.; Eriksen, L. M.; LaTulipe, D. C. *Optical Eng.* **1993**, *32(10),* 2382.

20   Sebald, M.; Bertold, J.; Beyer, M.; Leuschner, R.; Nölscher, Ch.; Scheler, U.; Sezi, R.; Ahne, H.; Birkle, S. *Proceedings SPIE,* **1991**, *1466*, 227.

21   Pavelchek, E. K.; Calabrese, G. S.; Bohland, J. ; Dudley, B. W. ; Jones, S. K. ; Freeman, P. W. *Optical Eng.* **1993**, *32(10),* 2376.

22   Taylor, G.N.; Hutton, R. S.; Stein, S. M.; Boyce, C. H.; LaFontaine, B.; MacDowell, A. A.; Wheeler, D. R. *Proceedings SPIE,* **1995**, *2437*, 308.

RECEIVED December 1, 1995

INDEXES

Author Index

Affiliation Index

Subject Index

Production: Meg Marshall
Indexing: Deborah H. Steiner
Acquisition: Anne Wilson
Cover design: Michelle Telschow

Printed and bound by Maple Press, York, PA

Bestsellers from ACS Books

The ACS Style Guide: A Manual for Authors and Editors
Edited by Janet S. Dodd
264 pp; clothbound ISBN 0–8412–0917–0; paperback ISBN 0–8412–0943–X

Understanding Chemical Patents: A Guide for the Inventor
By John T. Maynard and Howard M. Peters
184 pp; clothbound ISBN 0–8412–1997–4; paperback ISBN 0–8412–1998–2

Chemical Activities (student and teacher editions)
By Christie L. Borgford and Lee R. Summerlin
330 pp; spiralbound ISBN 0–8412–1417–4; teacher ed. ISBN 0–8412–1416–6

Chemical Demonstrations: A Sourcebook for Teachers,
Volumes 1 and 2, Second Edition
Volume 1 by Lee R. Summerlin and James L. Ealy, Jr.;
Vol. 1, 198 pp; spiralbound ISBN 0–8412–1481–6;
Volume 2 by Lee R. Summerlin, Christie L. Borgford, and Julie B. Ealy
Vol. 2, 234 pp; spiralbound ISBN 0–8412–1535–9

Chemistry and Crime: From Sherlock Holmes to Today's Courtroom
Edited by Samuel M. Gerber
135 pp; clothbound ISBN 0–8412–0784–4; paperback ISBN 0–8412–0785–2

Writing the Laboratory Notebook
By Howard M. Kanare
145 pp; clothbound ISBN 0–8412–0906–5; paperback ISBN 0–8412–0933–2

Developing a Chemical Hygiene Plan
By Jay A. Young, Warren K. Kingsley, and George H. Wahl, Jr.
paperback ISBN 0–8412–1876–5

Introduction to Microwave Sample Preparation: Theory and Practice
Edited by H. M. Kingston and Lois B. Jassie
263 pp; clothbound ISBN 0–8412–1450–6

Principles of Environmental Sampling
Edited by Lawrence H. Keith
ACS Professional Reference Book; 458 pp;
clothbound ISBN 0–8412–1173–6; paperback ISBN 0–8412–1437–9

Biotechnology and Materials Science: Chemistry for the Future
Edited by Mary L. Good (Jacqueline K. Barton, Associate Editor)
135 pp; clothbound ISBN 0–8412–1472–7; paperback ISBN 0–8412–1473–5

For further information and a free catalog of ACS books, contact:
American Chemical Society
Customer Service & Sales
1155 16th Street, NW, Washington, DC 20036
Telephone 800–227–5558